# Integrated Sustainable Design of Buildings

# Integrated Sustainable Design of Buildings

*Paul Appleby*

earthscan
from Routledge

First published by Earthscan in the UK and USA in 2011

For a full list of publications please contact:
Earthscan
2 Park Square, Milton Park, Abingdon, Oxfordshire OX14 4RN
711 Third Avenue, New York, NY 10017

First issued in paperback 2015

*Earthscan is an imprint of the Taylor & Francis Group, an informa business*

**Notices**
Practitioners and researchers must always rely on their own experience and knowledge in evaluating and using any information, methods, compounds, or experiments described herein. In using such information or methods they should be mindful of their own safety and the safety of others, including parties for whom they have a professional responsibility.

Product or corporate names may be trademarks or registered trademarks,and are used only for identification and explanation without intent to infringe.

ISBN 13: 978-1-138-97284-1 (pbk)
ISBN 13: 978-1-84971-117-3 (hbk)

Typeset by Domex e-Data, India
Cover design by Yvonne Booth

A catalogue record for this book is available from the British Library

Library of Congress Cataloging-in-Publication Data
Appleby, Paul.
 Integrated sustainable design of buildings / Paul Appleby.
   p. cm.
 Includes bibliographical references and index.
 ISBN 978-1-84971-117-3 (hardback)
1. Sustainable architecture. 2. Sustainable design. 3. Architectural design. I. Title.
 NA2542.36A67 2010
 720'.47--dc22
                        2010027872

To Liz, with a big thank you

# Contents

# List of Figures, Tables and Boxes

## Figures

## Tables

# Boxes

# Acknowledgements

I would like to thank the following individuals without whom this endeavour would not have been possible:

Ivan Rodriguez and the Building Sustainability team at URS Corporation Ltd and Jeremy Castle of Treasury Holdings for the case history on Battersea Power Station in 2.1 Sustainable Communities.

Paul Freathy of RWDI Anemos for help with the section dealing with wind environment in 2.3 Massing and Microclimate.

Elena Di Biase of URS Corporation Ltd for help with 2.4 Social Sustainability.

Juliette Seddon of URS Corporation Ltd for the contents of an EIA for 1.4 Environmental Impact Assessment.

Cressida Curtis of Quintain plc for help with the case history on Wembley City in 3.11 Waste Management and Recycling.

The Office of Government Commerce for the case history on Lion House in 4.1 Tendering Process.

Along with all those too numerous to mention for supplying images for use in this book.

# Abbreviations and Acronyms

| | |
|---|---|
| 10YFP | 10 year framework of programmes |
| ABGR | Australian Building Greenhouse Rating |
| ACA | Association of Consultant Architects |
| ACPO | Association of Chief Police Officers |
| ADF | average daylight factor |
| AECB | Association of Energy Conscious Builders |
| AFC | alkaline fuel cell |
| AGC | Associated General Contractors of America |
| AIA | American Institute of Architects |
| ALO | Architectural Liaison Officer |
| AMC | Appalachian Mountain Club |
| ANSI | American National Standards Institute |
| APSH | annual probable sunlight hours |
| AQO | air quality objectives |
| AQUA | *Alta Qualidade Ambientale* (Brazil) |
| ASHRAE | American Society of Heating, Refrigeration and of the Airconditioning Engineers |
| ASTM | American Society for Testing and Materials |
| ATES | aquifer thermal energy storage |
| AVM | anti-vibration mounting |
| BAPCT | best available pollution control technology |
| BASIX | Building Sustainability Index (Australia) |
| BCA | Building and Construction Authority (Singapore) |
| BCO | British Council of Offices |
| BEES | Building for Environmental and Economic Sustainability (US) |
| BEMA | Building Emergency Management Assessment |
| BEMS | Building Energy Management System |
| BER | Building Emission Rating |
| BERR | Department for Business, Enterprise and Regulatory Reform |
| BFF | Best Foot Forward |
| BIPV | building integrated photovoltaic systems |
| BIQ™ | Building Intelligence Quotient |
| BMS | Building Management System |
| BOMA | Building Owners and Managers Association |
| BREEAM | Building Research Establishment Environmental Assessment Method (UK) |
| BRERA | Brownfields Revitalization and Environmental Restoration Act (US) |
| BSI | British Standards Institute |
| C&D | construction and demolition |
| CAA | Civil Aviation Authority (UK) |
| CABA | Continental Automated Buildings Association |
| CABE | Commission for Architecture and the Built Environment |
| CaRB | Carbon Reduction in Buildings Programme |
| CASBEE | Comprehensive Assessment System for Built Environmental Efficiency (Japan) |
| CCC | Committee on Climate Change |
| CCHP | combined cooling heat and power (or trigeneration) |
| CCS | Considerate Constructors' Scheme |
| CDC | Centers for Disease Control and Prevention (US) |

| | |
|---|---|
| CDM | Construction Design and Management (Regulations) |
| CEEQUAL | Civil Engineering Environmental Quality Assessment and Award Scheme |
| CEPAS | Comprehensive Environmental Performance Assessment Scheme (Hong Kong) |
| CEQA | California Environmental Quality Act |
| CEQR | City Environmental Quality Review (New York City) |
| CERCLA | Comprehensive Environmental Response, Compensation and Litigation Act (US) |
| CFC | Chlorofluorocarbons |
| CFD | computational fluid dynamics (modelling) |
| CHP | combined heat and power (or cogeneration) |
| CIBSE | Chartered Institution of Building Services Engineers |
| CIE | Commission Internationale l'Eclarage |
| CIEB | Continual Improvement Assessment for Existing Buildings |
| CIRIA | Construction Industry Research and Information Association |
| CITES | Convention on International Trade in Endangered Species of Wild Flora and Fauna (UN) |
| CLG | Department for Communities and Local Government |
| CMO | commissioning management organization |
| CMS | conservation of migratory species |
| CNI | Critical National Infrastructure |
| $CO_2$ | carbon dioxide |
| CONQUAS | Construction Quality Assessment System (Singapore) |
| CoP | coefficient of performance |
| COSU | Cabinet Office Strategy Unit |
| CPDA | Crime Prevention Design Adviser |
| CPNI | Centre for Protection of National Infrastructure |
| CRI | Carpet and Rug Institute (US) |
| CRSD | circadian rhythm sleep disorders |
| CSCS | Construction Skills Certification Scheme |
| CSD | Commission on Sustainable Development |
| CSH | Code for Sustainable Homes |
| CSS | Context Sensitive Solutions |
| DBIS | Department of Business Innovation and Skills |
| DECC | Department of Energy and Climate Change |
| DEFRA | Department for Environment, Food and Rural Affairs |
| DER | dwelling emission rates |
| DF | daylight factor |
| DfT | Department for Transport |
| DGNB | *Deutsche Gesellschaft fur Nachhaltiges Bauen* (German Sustainable Building Council) |
| DHW | domestic hot water |
| DMFC | direct methanol fuel cell |
| DNO | District Network Operator |
| DPD | Development Plan Documents |
| DR | draught rating |
| DSE | display screen equipment |
| DSM | dynamic simulation modelling |
| DSY | Design Summer Year |
| DTI | Department for Trade and Industry |
| EA | Environment Agency (England and Wales) |
| EAT | Economic Appraisal Tool |

| | |
|---|---|
| EC | European Community |
| ECE | external combustion engine |
| EDSL | Environmental Design Solutions Limited |
| EEA | European Environment Agency |
| EH | English Heritage |
| EMAS | eco-management and audit scheme |
| EMEP | European Monitoring and Evaluation Programme |
| EMS | Environmental Management System |
| END | Environmental Noise Directive (EC) |
| EPA | Environmental Protection Agency (US) |
| EPBD | Energy Performance of Buildings Directive (EU) |
| EPC | energy performance certificate |
| ES | Environmental Statement |
| ESCo | Energy Services Company |
| ESL | Eco-Synthesis Limited |
| ETAP | Environmental Technology Action Plan |
| ETP | environmental technology platforms |
| ETS | Emissions Trading System (EU) |
| EWC | European Waste Catalogue |
| FAA | Federal Aviation Authority (US) |
| FCU | fan coil unit |
| FEMA | Federal Emergency Management Agency (US) |
| FFC | Federal Facilities Council (US) |
| FHWA | Federal Highways Administration (US) |
| FICFB | fast internal circulating fluidized bed |
| FIDIC | International Federation of Consulting Engineers |
| FiT | feed-in tariff |
| FLEGT | Forest Law Enforcement Governance and Trade |
| FM | facilities management |
| FOG | fats, oil and grease |
| FSC | Forest Stewardship Council |
| FTE | full-time equivalent |
| GBC | Green Building Council |
| GBCA | Green Building Council of Australia |
| GBCE | Green Building Council España |
| GBI | Green Buildings Initiative (US) |
| GC | Government Contracts |
| GCCP | Government Construction Client Panel |
| GHG | greenhouse gas(es) |
| GIS | Geographic Information System |
| GLA | Greater London Authority |
| GMIS | Green Mark Incentive Scheme (Singapore) |
| GPP | Green Public Procurement |
| GRD | grease recovery devices |
| GRIHA | Green Rating for Integrated Habitat Assessment (India) |
| GSHP | ground source heat pumps |
| GWP | global warming potential |
| HAWT | horizontal axis wind turbine |
| HCDA | Housing and Community Development Act (US) |
| HCFC | hydrochlorofluorocarbons |
| HFC | hydrofluorocarbons |

| | |
|---|---|
| HK-BEAM | Hong Kong Building Environmental Assessment Method |
| HIR | Home Information Pack |
| HKSAR | Hong Kong Special Administrative Region |
| HQE | *Haute Qualite Environnementale* (France) |
| HSC | Health & Safety Commission |
| HUD | Housing and Urban Development (US) |
| HUDU | Healthy Urban Development Unit |
| HVAC | heating, ventilation and air conditioning |
| IAQ | indoor air quality |
| ICAO | International Civil Aviation Organisation |
| ICE | Institution of Civil Engineers |
| ICOLD | International Commission on Large Dams |
| IEA | International Energy Agency |
| IEQ | indoor environmental quality |
| IESNA | Illuminating Engineering Society of North America |
| iiSBE | International Initiative for a Sustainable Building Environment |
| ILE | Institution of Lighting Engineers |
| IMD | Index of Multiple Deprivation |
| IPCC | Intergovernmental Panel on Climate Change |
| ISC | Interagency Security Committee (US) |
| IUCN | World Conservation Union |
| JaGBC | Japan GreenBuild Council |
| JCT | Joint Contracts Tribunal Ltd |
| KPI | Key Performance Indicators |
| LA | Local Authority |
| LCA | Life Cycle Analysis |
| LDF | Local Development Framework |
| LED | light emitting diode |
| LEED | Leadership in Energy & Environmental Design (US) |
| LZC | low or zero carbon |
| MCFC | molten carbonate fuel cell |
| MEA | Millennium Ecosystem Assessment |
| MGBC | Mexico Green Building Council |
| MMC | modern methods of construction |
| MOVES | Motor Vehicle Emission Simulator (US EPA) |
| MUSCo | Multi-Utility Services Company |
| NABERS | National Australian Building Environmental Rating Scheme |
| NAHP | National Affordable Homes Programme |
| NAO | National Audit Office (UK) |
| NAS | National Audubon Society |
| NC | noise criterion |
| NCN | National Cycle Network (Sustrans) |
| NEC | New Engineering Contract |
| NEC | noise exposure categories |
| NEPA | National Environmental Policy Act (US) |
| NEPS | North East Premises Solution (DEFRA) |
| NGO | non-governmental organization |
| NIST | National Institution of Standards and Technology (US) |
| $NO_x$ | nitrous oxides |
| NPV | net present value |
| NR | noise rating |

| | |
|---|---|
| NRDC | National Resources Defense Council (US) |
| NREL | National Renewable Energy Laboratory (US) |
| NSPE | National Society of Professional Engineers |
| NSR | New Source Review (US) |
| O&M | operation and maintenance |
| OA | Opportunity Area |
| ODP | ozone depletion potential |
| ODPM | Office of the Deputy Prime Minister |
| ODS | ozone-depleting substances |
| OECD | Organisation for Economic Co-operation and Development |
| OGC | Office of Government Commerce |
| OHSA | Occupational Health and Safety Agency |
| OPEC | Oil Producing and Exporting Countries |
| ORC | Organic Rankine Cycle |
| PAFC | phosphoric acid fuel cell |
| PAH | polycyclic aromatic hydrocarbons |
| PATH | Partnership for Advancing Technology in Housing (US) |
| PCM | phase change material |
| PEFC | Programme of Endorsement of Forest Certification |
| PEM | proton exchange membrane/polymer electrolyte membrane |
| PF | power factor |
| PIQ | property information questionnaire |
| PMV | predicted mean vote |
| PPD | predicted percentage dissatisfied |
| PPE | personal protective equipment |
| PPP | public/private partnerships |
| PTAL | Public Transport Accessibility Level |
| PV | photovoltaic |
| PWN | private wire network |
| QI | Quality Index |
| RC | Room Criterion |
| REACH | Registration, Evaluation and Authorisation of Chemicals (EU) |
| REPA | Resource and Environmental Profile Analysis (US) |
| RH | relative humidity |
| RIDDOR | Reporting of Injuries, Diseases and Dangerous Occurrences Regulations |
| ROCs | Renewable Obligation Certificates |
| RTFO | Renewable Transport Fuels Obligation |
| RVO | recycled vegetable oil |
| SAC | Special Area of Conservation |
| SAD | seasonal affective disorder |
| SAFED | Safe and Fuel Efficient Driving |
| SAP | Standard Assessment Procedure |
| SBD | Secured by Design (Association of Chief Police Officers) |
| SBEM | Simplified Building Energy Model |
| SBS | sick building syndrome |
| SCALDS | Social Cost of Alternative Land Development Scenarios (US) |
| SCP | sustainable consumption and production |
| SCR | selective catalytic reduction |
| SDC | Sustainable Development Commission |
| SEEDA | South East of England Development Agency |
| SEQRA | State Environmental Quality Review Act (New York State) |
| $SF_6$ | sulphur hexafluoride |

| | |
|---|---|
| SIPS | structural insulated panel system |
| SLL | Society of Light and Lighting |
| SMART | 'simple multi-attribute rating technique' |
| $SO_2$ | sulphur dioxide |
| SOFC | solid oxide fuel cell |
| SOGE | Sustainable Operation of the Government Estate |
| SPA | Special Protection Area |
| SPARTACUS | System for Planning and Research in Towns and Cities for Urban Sustainability (EU) |
| SpEAR | Sustainable Project Appraisal Routine |
| SPG | supplementary planning guidance |
| SPV | Special Purpose Vehicle |
| SQE | suitably qualified ecologist |
| SRF | State Revolving Fund (US) |
| SSSI | Site of Special Scientific Interest |
| SUDS | sustainable urban drainage system |
| SWEA | Severn Wye Energy Agency (UK) |
| SWMP | Site Waste Management Plan |
| T&CP EIA Regs | Town and Country Planning (Environmental Impact Assessment) Regulations 1999 |
| TCPA | Town and Country Planning Act |
| TCSP | Transportation, Community and System Preservation Program (US) |
| TER | target emission rates |
| TERI | The Energy Resources Institute (India) |
| TGBRS | Teri Green Building Rating System (India) |
| TIA | Transport Impact Assessment |
| TQL | Transport for Quality of Life |
| TRY | Test Reference Year |
| UCO | used cooking oil |
| UKERC | UK Energy Research Centre |
| UN | United Nations |
| UNCTD | UN Commission on Trade and Development |
| UNDESA | United Nations Department of Economic and Social Affairs |
| UNECE | United Nations Economic Commission for Europe |
| UNEP | United Nations Environment Programme |
| UNESCO | UN Educational, Scientific and Cultural Organization |
| UNFCCC | UN Framework Convention on Climate Change |
| USE-IT | Urban Sustainability Evaluation and Interpretation Tool |
| USGBC | United States Green Building Council |
| USGSA | US General Services Administration |
| VAV | variable air volume |
| VAWT | vertical axis wind turbine |
| VE | value engineering |
| VIP | vacuum insulation panels |
| VOCs | volatile organic compounds |
| VSC | vertical sky component |
| WC | water closet |
| WDA | Welsh Development Agency |
| WHO | World Health Organization |
| WLC | whole life costing |
| WRAP | Waste & Resources Action Programme |

'Cooperate for the well-being of the body and the survival of the human species.'
*Avicenna* AD *980–1037*

'We can no longer consume the world's resources without regard to effect.
For the world has changed, and we must change with it.'
*President Barack Obama, December 2008*


**'It wasn't my fault'**

I killed someone last night
I drove a car too fast
and killed someone
I didn't mean to
I took all the necessary precautions
I kept to the speed limit
I fastened my seat belt
I looked in my mirror
and stopped at traffic lights

but:

It was my exhaust that
polluted the atmosphere
heated the planet
melted the ice cap
flooded Bangladesh
and drowned a thousand people

I admit it
It was me

It was my lead that poisoned the brains of children
I let the child starve for the price of a gallon of petrol

It wasn't my fault
I couldn't stop
I didn't see it coming

*Paul Appleby (1992)*

# Part 1

# Background

# 1.1

# Introduction and Scope

This is first and foremost a 'how to' book for people involved in the design of buildings, but it is the author's view that buildings cannot be designed sustainably without creating a foundation of sustainability in the planning and masterplanning of development. A development comprises a number of buildings set in a landscaped environment that has to fit within the context of its neighbourhood. This book covers sustainability in the broadest sense of the word, including all aspects of design and planning that impact on the long-term viability of buildings and the developments of which they form a part. Viability in this context refers to the 'triple bottom line' of environmental, social and economic impacts, from the global level down to the health and comfort of the building user.

Dictionary definitions of sustainability usually refer to the root component, the verb 'to sustain', the definition of which depends on context, but for our purposes means 'to be maintained or kept in existence'. As this implies, context is important when defining sustainability, hence it is usual to think in terms of sustainable development, sustainable construction, sustainable buildings, etc.

A classic definition of sustainable development was coined by Gro Harlem Brundtland for the UN World Commission on Environment and Development in her 1987 report *Our Common Future* as:

> development that meets the needs of the present without compromising the ability
> of future generations to meet their own needs. (Brundtland Commission, 1987)

This definition can be used as a test for all aspects of sustainability, whether it be environmental, social or economic, and their global, regional or local impacts.

The best known example of the global environmental impact of a building is long-term global warming, melting of the ice caps and a rise in sea levels, to which buildings contribute by virtue of their energy consumption and associated carbon emissions from burning fossil fuels. Local environmental impacts include flooding caused by excessive run-off of rainwater from impermeable surfaces, whilst negative impacts on occupants might include, for example, summer overheating and thermal discomfort due to poorly designed facades.

One example of a global socio-economic impact relates to the procurement of materials, which can impact on job creation and economic prosperity for the countries from which the products and materials are obtained. Of course this can have a positive impact on economies local to the development if products are procured locally, depending on whether this benefits suppliers only or manufacturers, suppliers of raw materials, etc.

Until quite recently the terms 'sustainable design' and 'green design' of buildings have been virtually interchangeable with low energy building design, but in recent years, and in particular since the advent of the UK's Building Research Establishment Environmental Assessment Method (BREEAM) and the American Leadership in Energy & Environmental Design (LEED) schemes, a broader range of performance criteria have been introduced. The use and importance of assessment methodologies will be discussed in detail in Chapter 1.3.

For much of the 20th century the developed world squandered natural resources on the operation of buildings at an alarming rate. Reasons for this included an increasing demand for comfort in poorly insulated buildings, the invention of air conditioning and the abundance of cheap energy. Before it became possible to artificially control the indoor climate and before the technology and infrastructure became available to generate and distribute energy economically, architecture had evolved so that buildings responded to their external environment and people adapted to the resultant conditions. For example the Anasazi Indians, living in what is now the Four Corners region of the US, built stone villas that featured passive solar design, rainwater harvesting and local materials during the first 1400 years or so after the birth of Christ (see Figure 1.1). It is interesting to note that modern exemplar sustainable buildings minimize or eliminate the need for air conditioning and require occupants to adapt to varying indoor conditions in tune with the prevailing external climatic conditions.

It is self evident, therefore, that the so-called developing world is in general closer to living sustainably than the developed world. This is dramatically illustrated if we consider the per capita carbon footprint for individual countries as an approximate indicator of sustainable lifestyle. In 2007 the average North American had a carbon footprint of 19.1 tonnes $CO_2$ per annum compared with 0.04 for the Democratic Republic of Congo and 0.08 for Ethiopia.[1] Of course this hides the fact that many people in these latter blighted countries are living in extreme poverty with carbon footprints close to zero.

**Figure 1.1** *Stone villas of the Anasazi Indians in North America*

*Source*: Copyright Duncan pilbert/istockphoto.com

Hence the challenge that this book addresses is to answer the question: 'how do we in the developed countries live sustainably whilst maintaining the quality of life and standard of living that we are used to?'. In answering this question there are lessons to be learnt from our ancestors, and we will also examine how modern technologies can be applied economically and on a large scale.

Scale is an important issue in being able to implement sustainable principles effectively. There are numerous examples of individual houses that have attempted to push the boundaries of sustainable design, usually with an emphasis on low energy design. For example, the Sage Residence in Eugene, Oregon and the Kingspan Lighthouse zero carbon house on the Building Research Establishment site at Watford, England (see Figures 1.2 and 1.3 and Chapter 3.2 Operational Energy and Carbon for more detail). However, these have generally been experimental, and hence one-off, architectural designs for private individuals or part of a competition or pilot. To achieve exemplar performance has required a considerable investment in most cases, whereas achieving this performance on a large scale demands both the mass production of high performance building elements and services and a sustainable energy and water infrastructure for entire developments.

This not only applies to new developments but to existing buildings, for there is an urgent need to reduce the carbon emissions from buildings in general, not just new buildings, which represent only a small proportion of carbon emissions at any given time. The problem with new buildings is that they don't always replace an existing one, hence most of the new buildings constructed in the future will add to operational carbon emissions, as well as the embodied carbon within the materials used in their construction.

Although there remains a need to prototype ideas in new designs, the main focus of this book is a strategy for the wholesale introduction of sustainable principles into the development process; from planning and masterplanning

**Figure 1.2** *The Sage LEED Platinum house, Eugene, Oregon, US*

*Source:* Copyright Michael J Dean 2009

**Figure 1.3** *The Kingspan zero carbon house at the BRE site, Watford, England*

*Source:* Peter White, BRE

(Part 2) through the detailed design process (Part 3), implementation through sustainable construction (Part 4) and operation of the completed building through to refurbishment or demolition. Part 1 sets out the legislative framework and drivers for sustainability, along with the tools that are used for assessing likely sustainable performance and financial viability.

## Note

1   These have been taken from the latest International Energy Agency statistics available as a download from www.iea.org/co2highlights.pdf.

## Reference

Brundtland Commission, 1987. *Report of the World Commission on Environment and Development: Our Common Future.* UN documents. Available at www.un-documents. net/wced-ocf.htm (accessed 16 January 2010)

# 1.2

# Policy, Legislation and Planning

## International overview

### Sustainability and the environment

It could be argued that the roots of modern-day sustainability and energy policy lie in theories developed by 18th-century political economist Thomas Malthus, who in 1798 published *An Essay on the Principle of Population* (Malthus, 1798). In this he predicted that population growth would outpace agricultural production in a scenario that for obvious reasons is also known as the Malthus Catastrophe or Nightmare. There are numerous variables that contribute to both population growth and agricultural production, and since the Industrial Revolution, the World Wars, numerous catastrophic events and the steady divergence of the developed and developing world, the way in which Malthus' predictions are evolving has varied considerably globally. It could be argued that his 'catastrophe' has already happened in African countries such as Ethiopia and the Sudan.

Malthus' model was extended by a group of authors commissioned by the Club of Rome in their book The Limits to Growth published in 1972 (Meadows et al, 1972). The Club of Rome was founded in April 1968 by Aurelio Peccei – an Italian industrialist – and Alexander King – a Scottish scientist – as a think tank to consider 'the predicament of mankind'. Donella and Dennis Meadows and colleagues modelled this predicament by extending Malthus' list of variables to include:

- world population;
- industrialization;
- pollution;
- food production;
- resource production and depletion.

This model has been updated as recently as 2004 and numerous scenarios have been developed over the years, resulting in such concepts as 'One Planet Living', (BioRegional and WWF International, 2008) which uses such attention grabbing statements as 'If everyone in the world lived like the average North American we would need 5 planets to live on' to illustrate that some of us are indeed living beyond our means.

The follow-up to the *Limits of Growth, Mankind at the Turning Point*, was first published in 1975 (Mesarovic and Pestel) when the world was still in the middle of an oil crisis and associated economic repercussions and included the following blueprint for sustainable development:

1 A world consciousness must be developed through which every individual realizes his role as a member of the world community... It must become part of the consciousness of every individual that the basic unit of human cooperation and hence survival is moving from the national to the global level.

2 A new ethic in the use of material resources must be developed which will result in a style of life compatible with the oncoming age of scarcity... One should be proud of saving and conserving rather than of spending and discarding.

3 An attitude toward nature must be developed based on harmony rather than conquest. Only in this way can man apply in practice what is already accepted in theory – that is, that man is an integral part of nature.

4 If the human species is to survive, man must develop a sense of identification with future generations and be ready to trade benefits to the next generations for the benefits to himself. If each generation aims at maximum good for itself, Homo sapiens are as good as doomed. (Mesarovic and Pestel, 1974, Epilogue, p147)

The Brundtland Commission echoed this sentiment in slightly less dramatic language in their 1987 report *Our Common Future* (Brundtland Commission, 1987):

Over the course of this century, the relationship between the human world and the planet that sustains it has undergone a profound change. When the century began, neither human numbers nor technology had the power radically to alter planetary systems. As the century closes, not only do vastly increased human numbers and their activities have that power, but major, unintended changes are occurring in the atmosphere, in soils, in water, among plants and animals, and in the relationships among all of these. The rate of change is outstripping the ability of scientific disciplines and our current capabilities to assess and advise.

In parallel with the activities of the Club of Rome, the United Nations organized a Conference on the Human Environment which met in Stockholm in June 1972. This resulted in the *Stockholm Declaration* (United Nations Environment Programme, 1972), which took the form of 26 principles to 'inspire and guide the peoples of the world in the preservation and enhancement of the human environment'. The spirit of Malthus permeates this declaration also, referring as it does to the fact that 'the natural growth of population continuously presents problems for the preservation of the environment, and adequate policies and measures should be adopted, as appropriate, to face these problems.' As with Brundtland some 15 years later there is an emphasis on protecting the planet for future generations, as stated for example in Principle 2: 'The natural resources of the earth, including the air, water, land, flora and fauna and especially representative samples of natural ecosystems, must be safeguarded for the benefit of present and future generations...'. These were translated into an action plan as a series of 109 detailed recommendations.

Probably the most important action to come out of Stockholm was the setting up of the United Nations Environment Programme (UNEP) that was established to become:

> The leading global environmental authority that sets the global environmental agenda, that promotes the coherent implementation of the environmental dimension of sustainable development within the United Nations system and that serves as an authoritative advocate for the global environment. (UNEP, 1972)

Together with the World Conservation Union (IUCN) and WWF, the UNEP produced the World Conservation Strategy in 1980 which in their publication *Caring for the Earth*, published in 1991, first gave currency to the term 'sustainable development', stating that 'humanity ... has no future unless nature and natural resources are conserved' (IUCN, 1991).

It is perhaps a sign of how little progress has been made in heeding this warning that the Millennium Ecosystem Assessment (MEA) commissioned by the UN Secretary General in 2000 took 4 years and input from 1360 experts to conclude that 'human actions are depleting earth's natural capital, putting a strain on the environment that the ability of the planet's ecosystems to sustain future generations can no longer be taken for granted' (MEA, 2005).

UNEP has published a Medium Term Strategy for 2010–2013 (UNEP, undated), which takes as its basis six 'cross-cutting thematic priorities':

1   Climate change;
2   Disasters and conflicts;
3   Ecosystem management;
4   Environmental governance;
5   Harmful substances and hazardous waste;
6   Resource efficiency – sustainable consumption and production.

A series of accomplishments have been set out under each heading by which UNEP's effectiveness will be measured.

As one might expect since their formation in 1945, the United Nations (UN) has been the main driving force globally for environmental protection and sustainability. Apart from the Stockholm Declaration and Brundtland Report, the UN has been responsible for most of the landmark environmental events of the past 60 years, although the focus has gradually changed as more detail becomes available on the issues and pressures that the planet faces. The best known events are the conferences in Montreal, Rio de Janeiro, Kyoto and Johannesburg, but as can be seen from Box 1.1 there have been numerous events sponsored by the UN and others going back as far as 1948.

## Energy and climate change

Although the 1949 conference at Lake Success was primarily about the future utilization of natural resources such as oil and coal rather than its conservation, the protection of flora and fauna was the primary area for concern, particularly

## Box 1.1 Landmark events in the history of sustainability and environmental protection

1798 Malthus Theory published

1948 International Union for the Protection of Nature (currently called World Conservation Union (IUCN)) founded at Fontainebleu, Paris

1949 UN Scientific Conference on the Conservation and Utilization of Resources, Lake Success, New York

1954 First nuclear power plant commissioned, Obninsk, USSR

1961 World Wildlife Fund (WWF) founded, later (1986) this became the World Wide Fund for Nature and subsequently (2000) just WWF

1963 IUCN Red List (conservation status of plant and animal species) first published

1968 UN Economic and Social Council – Swedish Initiative on Human Environment which led to Stockholm Conference

1968 UN Educational, Scientific and Cultural Organization (UNESCO) Biosphere Conference, Paris – Rational Use and Conservation of Resources of the Biosphere

1968 Club of Rome founded

1970 First Earth Day, US

1972 Club of Rome – Limits to Growth published

1972 UN Stockholm Conference and Declaration

1972 UN Environment Programme founded

1973 Oil crisis

1974 UN Convention on International Trade in Endangered Species of Wild Flora and Fauna (CITES) Washington, US

1974 UNEP/UN Commission on Trade and Development (UNCTD) Conference: Models of Resource Utilization: A Strategy for the Environment and Development – Cocoyoc, Mexico. Cocoyoc Declaration

1976 UN Habitat Conference, Vancouver

1977 UN Water Conference, Mar Del Plata, Argentina

1979 Three Mile Island nuclear reactor accident

1979 World Meteorological Organization Greenhouse Effect Conference: World Climate Conference, Geneva

1980 Brandt Commission: Environmental Impact Assessment for Development Proposals

1980 IUCN/UNEP/WWF World Conservation Strategy: Call for sustainable development

1980 US Global 2000 Report to the President

1982 IUCN World Charter for Nature

1986 Chernobyl nuclear reactor disaster

1987 Brundtland Report

1987 Montreal Protocol

1988 Intergovernmental Panel on Climate Change (IPCC) founded

1990 Bergen Conference on Sustainable Development

1991 IUCN/UNEP/WWF: Caring for the Earth – A Strategy for Sustainable Living

1991 Global Environmental Facility: World Bank/UNEP/UN Development
     Programme
1992 Rio Conference: Earth Summit – Rio Declaration and Agenda 21
1992 UN Commission on Sustainable Development founded
1992 UN Framework Convention on Climate Change
1994 UN Population Conference, Cairo
1996 OECD Conference – Shaping the 21st Century
1997 Kyoto Conference – UN Framework Convention on Climate Change (UNFCCC)
2000 Millennium Development Goals
2001 International Forum on National Sustainable Development Strategies,
     Ghana
2002 Johannesburg: World Summit on Sustainable Development – Earth
     Summit
2003 Marrakech Conference: Towards a global framework for action on
     sustainable consumption and production (SCP) – UNEP and UN Department
     of Economic and Social Affairs.
2005 Millennium Ecosystem Assessment report
2006 An Inconvenient Truth – Seminal Al Gore film on climate change
2006 Stern Review: The Economics of Climate Change
2009 UNFCCC Copenhagen

during a parallel United Nations Educational, Scientific and Cultural
Organisation (UNESCO) conference on the 'Protection of Nature', which also
involved the embryonic International Union for the Protection of Nature,
formed in Fontainebleu the year before. The focus of environmentalists
remained flora and fauna and population growth for the next 20 years,
reinforced by the founding of the World Wildlife Fund (WWF) in 1961.

With the first nuclear power plant connected to a grid commissioned in
Obninsk in the USSR in 1954, followed by the first commercial plant in Calder
Hall, Sellafield in the UK in 1956 and the first in the US at Shippingport
Pennsylvania in 1957, the world was persuaded that limitless power would be
available for the foreseeable future. However, development costs and safety
issues slowed its growth, although the oil crisis in the early 1970s led to highly
oil-dependent countries such as France and Japan instigating major reactor
building programmes. The Three Mile Island disaster in 1979 and the
catastrophic failure at Chernobyl in 1986, along with a number of smaller
accidents and concerns about long-term storage resulted in a loss of faith in the
technology, from which it has not fully recovered even today.

The first conference to seriously discuss the greenhouse effect was organized
by the World Meteorological Organization in Geneva in 1979, although the
phenomenon was first identified by a Swedish scientist Svante Arrhenius in
1896. He predicted that a doubling of carbon dioxide concentration in the
atmosphere would result in an average air temperature rise on earth of 5 to 6K.
Although he also predicted that this would take around 3000 years. Most
recent predictions indicate that this could happen during the 21st century, but
with a temperature rise closer to 4K.

In the 100 years or so since Arrhenius there has been a gradual increase in the understanding of atmospheric physics, initially by scientists studying the ice ages and using necessarily simple models based on limited data on long-term changes in atmospheric $CO_2$ concentrations, climate changes and the influence of oceans. By the early 1960s, however, scientists in Sweden and Russia were warning of drastic global warming from human activity within the next century.

The Intergovernmental Panel on Climate Change (IPCC) wasn't formed until 1988 however, followed in 1992 by the UN Framework Convention on Climate Change (UNFCCC) and the Earth Summit in Rio which included a whole series of actions under Agenda 21 by which governments were to drive home improvements in energy efficiency.

It wasn't until the Kyoto protocol to the UNFCCC was agreed by most countries in 1997 that targets were set for reducing carbon emissions. However, the US has remained outside of the Kyoto protocol and, until recently, equivocal on the relationship between human activity and climate change. The Earth Summit in Johannesburg in 2002 saw Russia amongst others ratifying the protocol, and by February 2009 183 countries had ratified Kyoto, with the important exception of the US.

The Stern Review, which reported on 'The Economics of Climate Change' in October 2006 had an even greater impact than Al Gore's film 'An Inconvenient Truth' of the same year. Together they highlighted the importance of human impact on climate, whilst Stern, in the dry and logical language of an economist, concluded in simple terms that 'the benefits of strong, early action considerably outweigh the costs' (Stern, 2006). He goes on to assert in stark terms that:

> Our actions over the coming few decades could create risks of major disruption to economic and social activity, later in this century and in the next, on a scale similar to those associated with the great wars and the economic depression of the first half of the 20th century. And it will be difficult or impossible to reverse these changes. Tackling climate change is the pro-growth strategy for the longer term, and it can be done in a way that does not cap the aspirations for growth of rich or poor countries. The earlier effective action is taken, the less costly it will be. (Stern 2006)

At the time of writing the crucial meeting of the UNFCCC in Copenhagen in December 2009 is deemed to have achieved little, despite the ever increasing, although still disputed, evidence that the worst case scenarios of temperature and sea level rises will occur unless dramatic cuts are made to carbon dioxide emissions by all countries, regardless of development plans.

## Pollution and air quality

Pollution takes many forms, impacting on local air quality, atmospheric constituents, water quality, ground conditions and ecology. Global warming is in part due to the emission of pollution from human activities – not only from emissions of carbon dioxide but other greenhouse gases such as methane, nitrous oxide ($NO_x$), sulfur hexafluoride ($SF_6$) and hydrofluorocarbons (HFCs). In terms of total global emissions $CO_2$ is by far the dominant pollutant associated directly with human activities, however, methane emissions associated with farming and the decay of vegetation are significant. The potential for methane to create global warming (GWP) is around 25 times that of $CO_2$. The greatest concern is over the estimated 900 gigatonnes of methane and $CO_2$ that is trapped in the melting permafrost and ice being released as temperatures rise. Global warming legislation currently focuses on reducing carbon emissions. There are no limitations globally on emissions of methane, $NO_x$ or $SF_6$, whilst HFCs are currently being used as a substitute for ozone-depleting refrigerants and insulation blowing agents, such as the chlorofluorocarbons (CFCs and HCFCs), that are being phased out under the Montreal Protocol.

The link between certain free radical gases and depletion of ozone and high frequency ultra violet radiation was discovered in the laboratory in 1973. This was followed by studies which predicted a thinning of the ozone layer of around 7 per cent in 60 years, which resulted in the US banning the ozone-depleting halon gas as a propellant in aerosols in 1978. However, by the time the British Antarctic Survey reported a 10 per cent reduction in ozone concentrations in the stratosphere over their Halley station in 1985, serious attention was being paid to this phenomenon, resulting in the Vienna Convention for the Protection of the Ozone Layer in the same year, from which the treaty known as the Montreal Protocol was published in 1987. This treaty has been used globally to phase out all ozone-depleting substances, including chlorofluorocarbons (CFCs), hydrochlorofluorocarbons (HCFCs), halons, etc., having a major impact on the manufacture of refrigerants, foam insulation, degreasing agents and fire extinguishing gases. Most ozone-depleting substances (ODS) have already been phased out, apart from HCFCs which are due for complete elimination by 1 January 2030, with 65 per cent having been eliminated by 1 January 2010.

Ambient air quality has been a concern since the Middle Ages, particularly as conurbations grew alongside polluting industries, exacerbated in winter by coal and wood burning and, once the internal combustion engine began to replace horses, pollution from vehicle exhausts. Londoners started burning coal in the 13th century and there is record of coal burning being prohibited in London as early as 1273 because of it being prejudicial to health. From 1845 in England a whole series of Acts was introduced in an attempt to reduce smog. Despite this a smog incident is recorded in 1892 resulting in the death of around a thousand Londoners. Incidents such as this occurred throughout the industrialized world until the second half of the 20th century and, even with the introduction of smoke free zones in London in 1946 4000 Londoners died due to the notorious 'killer fog' of 1952. England introduced a Clean Air Act in 1956, followed in 1963 by the first Federal Clean Air Act in the US.

Some parts of the developing world still face these problems and the United Nations in para 6.39 of Agenda 21 to the 1992 Rio Declaration sought to provide a framework for tackling pollution to air, water and land, noise, ionizing and non-ionizing radiation. Air pollution remains a concern even in the developed world, where most countries have followed the UK and US and introduced legislation to reduce emissions and monitor impacts.

The World Health Organization (WHO) has taken a leading role in developing standards for ambient air quality and strategies for implementation globally. They first published air quality guidelines in 1987: intended for use in Europe these have now been adapted for use worldwide and are being introduced through a series of initiatives such as the Asian Clean Air Initiative. The EU has integrated them into the 2008 Directive on Ambient Air Quality, or the Clean Air for Europe (CAFE) Directive.

WHO have also published Indoor Air Quality Guidelines on Dampness and Mould (2008) and published guidance for specific pollutants at the end of 2009.

## Waste and water

The way in which a society manages its waste is a strong indicator of how developed it is and, like many other sustainability indicators, the developing world is facing the same problems that the developed world faced during its evolution. All major conurbations have developed strategies for managing solid, liquid, municipal and industrial wastes.

The earliest evidence of indoor plumbing comes from the Minoan culture on Crete, where remains of flush toilets, baths and underground sewers dating from between 1700 and 1500 BC have been excavated. The Romans developed plumbing into a fine art and spread the technology across their empire, including Britain after it was conquered in AD 43.

As far back as 500 BC the Greeks brought in a law that required waste to be dumped no less than one mile from a city boundary, whilst in 1388 the English Parliament brought in an Act that no waste be dumped in public waterways. However, up until the late 19th century the River Thames was effectively an open sewer, resulting in widespread cholera and culminating in 1858 with the 'Great Stink'. Legislation came into force in 1848 making it an offence for new houses to discharge sewage into an open ditch. However, this didn't deal with all of the existing open sewers. So between 1859 and 1865 Joseph Bazalgette was tasked to build a network of sewers under London, including pumping stations at Chelsea Embankment, Greenwich, Erith marshes and Abbey Mills in the Lea valley, discharging untreated sewage into the Thames at the Beckton and Crossness pumping stations on the north and south banks respectively. The aim of all of this work was to divert the sewage clear of London and out to sea. Paris on the other hand had closed sewers built under Rue Montmartre in 1370, but an extensive system wasn't started until 1812. However, it wasn't sufficiently extensive by 1832 to avoid a massive cholera epidemic that killed 20,000 people resulting from the advent of flushing toilets causing sewage to be discharged into the remaining open street sewers.

The modern approach to sewage disposal and treatment came out of the British Royal Commission into Sewage Disposal that reported in 1915. This developed the concept of primary and secondary treatment and set standards for effluent discharged to natural watercourses that gained widespread acceptance globally.

However, even in the UK, up until the second half of the 20th century most sewage was either being discharged into rivers or streams, or directly into the sea, or soaking into the land and watercourses via cesspools. As the pollution of rivers became a problem sewage pipes were extended further downstream. However, as concerns grew about the quality of water for bathing, the effect of water quality on fish stocks and the demand for drinking water increased, greater attention was paid to the cleaning up of sewage. In London this resulted in Beckton becoming the largest water treatment works in Europe by 1974. Work started on the Crossness sewage works in 1956. Other cities in the developed world had already built water treatment plants, with many North American cities having plants built during the 1930s and 1940s.

It was a prefect in Paris who introduced the first proper municipal waste management system in 1884 – one Eugene Poubelle. Indeed, the French word for waste bin remains *la poubelle*.

Recycling of waste is not a modern invention. For example there is evidence, from a large barrel of broken glass found at the bottom of the Adriatic in the wreck of the Iulia dating from the 2nd century AD, that the Romans recycled broken glass. Also, in an embryonic United States of America, waste paper and rags used for the manufacture of paper in Pittsburgh as far back as 1690. Whilst New York commissioned the first facility for sorting and recycling industrial waste in 1898.

Apart from the accidental burning of waste in municipal dumps the first purpose-built incinerator was constructed in Nottingham in 1874, whilst a similar plant was built in New York in 1885.

The United Nations position on waste and sewage is set out in Chapter 21 of Agenda 21 under the headings of minimizing waste, re-use and recycling, environmentally sound waste disposal and treatment and extending waste service coverage (within the developing world). Progress in meeting the objectives set out in Agenda 21 is being tracked by the Division of Technology, Industry and Economics of UNEP as part of its project reported in its 'International Source Book on Environmental Sound Technologies (ESTs) for Municipal Solid Waste Management (MSWM)' (UNEP, Division of Technology, Industry and Economics unknown date).

## Land use and biodiversity

The interrelationship between humans and the rest of the natural world has been at the heart of environmentalism since its embryonic years following the Second World War, with the formation of the International Union for the Protection of Nature in 1948 and the United Nations' first conference on conservation in Lake Success the following year. Momentum gathered with the birth of the WWF in 1961 and the publication of Rachel Carson's *Silent Spring* in 1962. This book is frequently credited with providing the catalyst for the modern environmental movement, as well as the banning of DDT, which was described by Carson as having wide scale impacts on wildlife, leading to the titular silent spring of the book.

An interest in nature other than a source of food and fuel is not new: the writings of Aristotle's student Theophastrus from the 4th century BC reveal an understanding of the interrelationship between animals and between animals and their environment. However, in more recent times Alexander Humbolt is recognized as the father of ecology in his publication *Ideas for a Plant Geography* dating from 1805, which looked at the relationship between species

> ## Box 1.2 United Nations bodies whose remit includes issues relating to sustainability and the environment
>
> ### UNEP – United Nations Environment Programme
>
> - Division of Technology, Industry and Economics (DTIE)
> - Division of Environmental Policy Implementation (DEPI)
> - Division of Environmental Law and Conventions (DELC)
> - Division of Global Environmental Facility Coordination (DGEF)
> - Division of Early Warning and Assessment (DEWA)
>
> ### UNDP – United Nations Development Programme
>
> - Global Environmental Facility Implementation
> - Millennium Development Goals
>
> ### UNDESA – United Nations Department of Economic and Social Affairs
>
> - Division of Sustainable Development (DSD)
> - Commission on Sustainable Development (CSD)
> - Division of Social Policy and Development
> - Forum on Forests
> - Economic and Social Council (ECOSOC)
>
> ### UNFCCC – United Nations Framework Convention on Climate Change
>
> ### World Health Organization
>
> ### World Meteorological Organization

and climate and described vegetation zones for the first time. More recently Darwin's *Origin of Species* proposed a mechanism that in retrospect fits within the boundaries of modern ecology. Indeed, it was one of his followers, Ernst Haeckel, who in 1866 coined the term, whilst Eduard Suess developed the concept of the biosphere in 1875, although the term ecosystem wasn't coined until 1935, and biodiversity didn't fall into common parlance until the early 1980s.

In 1972 the Stockholm Declaration referred to safeguarding land, flora and fauna through careful planning and management. Twenty years later Agenda 21 (UNDESA, 1992) was heavily focused on ecological impacts and the conservation of biodiversity, with Chapter 15 dealing directly with the latter and other chapters dealing with planning and the management of land resources, deforestation, desertification, fragile mountain ecosystems, sustainable agriculture, biotechnology and the protection of salt water and fresh water resources. In the interim period there had been some important events taking the issues forward, including the Nairobi Desertification Conference in 1977 and the Bonn Convention on the Conservation of Migratory Species of Wild Animals (CMS); an intergovernmental treaty that was signed in 1979 and came into force in 1983. The World Charter for Nature was adopted by the UN General Assembly in 1982, a central principle of which was that 'Ecosystems and organisms, as well as the land, marine and atmospheric resources that are utilized by man, shall be managed to achieve and maintain optimum sustainable productivity ...' (UN, 1982).

As with all Agenda 21 issues, responsibility for nature conservation and protection of biodiversity has been passed on to national governments to implement, with various departments and divisions of the UN providing support, education and monitoring services. The European Commission for example introduced the Habitats Directive in 1992 (EC, 1992) that has become the cornerstone of Europe's nature conservation policy.

## Sustainable consumption and production (SCP)

SCP is a cross-cutting theme which includes issues relating to all of the headings above, but requires separate attention because of the need to address the operation and behaviour of industry and consumers. SCP first appeared as a separate issue in 1992 as Chapter 4 of Agenda 21 to the Rio Declaration: 'Changing Consumption Patterns'. The UN Commission on Sustainable Development (CSD), which was formed as a result of Rio, defined SCP in 1995 as 'the use of services and related products which respond to basic needs and bring a better quality of life while minimizing the use of natural resources and toxic materials as well as the emissions of waste and pollutants over the life-cycle so as not to jeopardize the needs of future generations'. This was followed in 2002 by Chapter III of the Johannesburg Plan of Implementation: 'Changing unsustainable patterns of consumption and production' which gave birth to the 2003 Marrakech Process: Supporting Sustainable Consumption and Production. This in turn produced a 10 year framework of programmes (10YFP), which is currently under development at the time of writing and due to be launched in 2011. This ambitious framework involves a trans-global regional structure coordinated by UNEP and United Nations Department of Economic and Social Affairs (UNDESA) establishing programmes covering such activities as education, training, institutional capacity building, strengthening legal frameworks, mobilization of financial resources, technology transfer, etc.

## European policy and directives

### Background

Much of the national law relating to sustainability and the environment across the European Community (EC) is instigated and modified through Directives devised under the auspices of the European Parliament and Council of the European Union. The EC also develops pan-European law which is directly regulated, normally through the implementation of existing EC regulations. Box 1.3 lists the main EC directives and regulations that relate to the contents of this book. Most of these have been or are being adopted into UK law and will be discussed further in the next section and the relevant chapter in Part 2.

There is currently no one EC directive that deals directly with sustainable development, however the EU Sustainable Development Strategy sets out guiding principles for Member States and is subject to periodic review, the most recent of which was published in July 2009 (Commission of the European Communities, 2009). The 2006 review set out a 'Renewed Strategy' in Note 10917/06, (EC, 2006) with the aim of sustainable development promoting 'a dynamic economy with full employment and a high level of education, health protection, social and territorial cohesion and environmental protection in a peaceful and secure world, respecting cultural diversity'. This is a very broad and ambitious aim which enshrines many of the challenges that the modern world faces. These are listed under seven headings and associated targets, which form the basis of the EU reviews:

1   Climate change and clean energy;
2   Sustainable transport;
3   Sustainable consumption and production;
4   Conservation and management of natural resources;
5   Public health;
6   Social inclusion, demography and migration;
7   Global poverty and sustainable development.

The targets against which performance is tested are derived from both UN initiatives and EC Directives.

---

### Box 1.3 European Community publications dealing with sustainability and environmental issues

*Energy and climate change*

- 2002/91/EC Directive on the energy performance of buildings (EPBD) – currently under review
- 2006/32/EC Directive on energy end-use efficiency and energy services
- 2009/28/EC Directive on the promotion of the use of energy from renewable sources (Renewable Energy Directive – RED)
- 2005/32/EC Directive establishing a framework for the setting of ecodesign requirements for energy-using products
- EC Regulation No 106/2008 on a Community energy efficiency labelling programme for office equipment (Green Star Regulations)

- EC Regulation No 1980/2000 on revised Community ecolabel award scheme
- 2003/96/EC Directive restructuring the Community framework for the taxation of energy products and electricity
- 2004/8/EC Directive on the promotion of cogeneration based on a useful heat demand in the internal energy market
- 2009/72/EC Directive concerning common rules for the internal market in electricity
- 2009/73/EC Directive concerning common rules for the internal market in natural gas
- 2009/29/EC Directive amending Directive 2003/87/EC so as to improve and extend the greenhouse gas emission allowance trading scheme of the Community
- 2009/31/EC Directive on the geological storage of carbon dioxide
- EC Regulation No 397/2009 amending Regulation (EC) No 1080/2006 on the European Regional Development Fund as regards the eligibility of energy efficiency and renewable energy investment in housing
- EU White Paper: Adapting to Climate Change: Towards a European framework for action

## Water, waste and wastewater

- 98/83/EC Directive on the quality of water intended for human consumption (Drinking Water Directive)
- 91/271/EEC Directive concerning urban wastewater treatment
- 2008/98/EC Directive on waste and repealing certain Directives (Revised Waste Directive)
- 91/689/EEC Directive on hazardous waste
- 99/31/EC Directive on the landfill of waste (Landfill Directive)
- 2000/76/EC Directive on incineration of waste (Incineration Directive)

## Transport

- EC Regulation No 443/2009 setting emissions performance standards for new passenger cars as part of the Community's integrated approach to reduce $CO_2$ emissions from light-duty vehicles.

## Air quality and pollution

- 2008/50/EC Directive on ambient air quality and cleaner air for Europe (CAFE)
- 2004/42/EC Directive on the limitation of emissions of volatile organic compounds due to the use of organic solvents in certain paints and varnishes and vehicle refinishing products (VOC Paints Directive)
- 96/61/EC Directive concerning integrated pollution prevention and control
- EC Regulation No 2037/2000 on substances that deplete the ozone layer

### Environmental noise

- 2002/49/EC Directive relating to the assessment and management of environmental noise

### Ecology and biodiversity

- 92/43/EEC Directive on the conservation of natural habitats and of wild flora and fauna (Habitats Directive)

### Flood risk

- 2007/60/EC Directive on the assessment and management of flood risk

### Workplace health & safety

- 89/654/EEC Directive concerning the minimum safety and health requirements for the workplace (Workplace Directive)
- 2003/18/EC Directive amending 83/477/EEC on the protection of workers from the risks related to exposure to asbestos at work
- 97/11/EC Directive on the assessment of the effects of certain public and private projects on the environment (Environmental Impact Assessment (EIA) Directive)
- 2001/42/EC Directive on the assessment of certain plans and programmes on the environment (Strategic Environmental Assessment (SEA) Directive)

## Climate change and clean energy

In December 2008 the EC Parliament and Council reached agreement on a 'Climate Action and Renewable Energy Package' (refer to http://ec.europa.eu/environment/climat/future_action.htm) which committed Member States to 'reducing overall emissions to at least 20% below 1990 levels by 2020' with the possibility of increasing this to 30 per cent if other major emitters in the developed and developing worlds take on their fair share of the mitigation effort under a global agreement as set out in the 'Copenhagen Accord' signed at the UNFCCC in December 2009. The package also included a commitment to 'increasing share of renewable energy use to 20% by 2020'. A central plank of achieving these targets is the strengthening and expansion of the EU Emissions Trading System (EU ETS).

It can be seen from Box 1.3 that there have been numerous new and amended directives and regulations enacted in 2009. In addition to those listed, a proposal is currently out for consultation for recasting the EPBD Directive (EC, 2008b) which includes a number of enhancements including a stimulus for 'Member States to develop frameworks for low and zero carbon buildings'.

The Commission has also recently produced a White Paper on Adapting to Climate Change (EC, 2009) which sets out actions for increasing the resilience of:

- health and social policies;
- agriculture and forests;
- biodiversity, ecosystems and water;
- coastal and marine areas;
- production systems and physical infrastructure.

A number of instruments are described for achieving these actions, most importantly for financing adaptation projects and coordinating between Member States and other parties.

## Sustainable transport

A European Commission White Paper from 2001 set out transport policy for 2010, reviewed in 2005. This has resulted in a range of directives and proposals as well as the Greening Transport Package (Commission of the European Communities, 2008a) and a proposed Directive for pan-European Intelligent Transport Systems (Commission of the European Communities, 2008d). Proposals in the Greening Transport Package cover climate change, local pollution, noise, congestion and a whole series of cross-cutting issues. The Climate Action and Renewable Energy Package also includes a commitment to supply at least 10 per cent of transport fuel consumption from renewable sources by 2020. The proposal for a directive 'laying down the framework for the deployment of Intelligent Transport Systems in the field of road transport and for interfaces with other transport nodes' refers to a range of technologies, including processing, control, communications and electronics, that impact on vehicle design, as well as advanced approaches to traffic management, all of which are in the early stages of development (Commission of the European Communities, 2008a).

## Sustainable consumption and production

The European Commission is committed to the Marrakech Process (see International Overview above) and has consequently developed a Sustainable Development and Production and Sustainable Industrial Policy (SCP/SIP) Action Plan COM(2008)397 (Commission of the European Communities, 2008b), which sets out a strategy for updating and expanding the scope of the existing eco-labelling schemes, eco-management and audit scheme (EMAS) and Ecodesign Directive (European Council, 2009).

A primary driver of sustainable consumption and production is the Environmental Technology Action Plan (ETAP), adopted by the European Commission in 2004 (Commission of the European Communities, 2004). ETAP promotes eco-innovation across the EU with the aim of expediting the transition of innovative environmental technologies from research to a global market, attracting private and public investment through mechanisms such as Environmental Technology Platforms (ETP) based on public/private partnerships.

Initiatives include verification, target setting, mobilization of finance, incentivization, awareness raising and training.

The public sector of Member States will be expected to set an example by meeting targets set by the Green Public Procurement (GPP) process (EC, 2008a) that the European Commission is proposing, requiring that 50 per cent of tendering processes from 2010 will have to meet targets for core criteria derived from new and enhanced directives covering:

- eco- and energy labelling;
- energy star ratings for office equipment;
- eco-design;
- clean and energy efficient vehicles;
- biofuels and bioliquids.

Priority will be given to procurement by the following sectors:

- construction and maintenance of buildings;
- food and catering services;
- transport and transport services;
- energy;
- office machinery and computers;
- clothing;
- paper and printing services;
- furniture;
- cleaning products and services;
- equipment used in the health sector.

The Eco-design Directive (2005/32/EC) provides a framework for the implementation of a series of EC Regulations which came into force between December 2008 and July 2009 specifying minimum energy efficiency criteria for a wide range of products including electric motors, circulating pumps, televisions, set-top boxes, fridges and freezers, external power supplies, lighting products and standby/off-mode power consumption. The EC requires manufacturers supplying these products to Member States to implement efficiency improvements phased over a period that varies between product types, with some improvements to be introduced as early as July 2010, stretching to 2017. The directive is being amended to cover a more extensive list of products including those that have an indirect impact on energy consumption such as water-using products and windows.

## Conservation and management of natural resources

The Commission issued a 'Thematic strategy for the sustainable use of natural resources' in 2005 that committed to a series of actions including the development of a knowledge base, developing performance indicators, establishing a forum for policy development and working with UNEP and others on international initiatives. The EU has also recently published a number of action plans including their Biodiversity Action Plan (Commission of the European Communities, 2008c) COM (2008) 864 and the Action Plan for

Forest Law Enforcement Governance and Trade (FLEGT) derived from a 2005 EC Regulation setting up a licensing scheme for timber importation.[1]

The objectives of the Biodiversity Action Plan fall under ten headings:

1   Safeguarding important habitats;
2   Conserving biodiversity in the countryside;
3   Conserving marine biodiversity;
4   Integrating biodiversity into land use planning and development;
5   Reducing the impact of invasive alien species;
6   Strengthening international governance;
7   Providing assistance to countries outside the EU;
8   Reducing the impact from international trade;
9   Climate change adaptation;
10  Improving the knowledge base.

Specific species of flora and fauna have been protected for some years under the 1992 Habitats Directive, which includes species of bird which were formerly protected under the 1979 Birds Directive. The Natura 2000 network of sites across Europe was established to encompass the Special Areas of Conservation (SACs) and Special Protection Areas (SPAs) identified for specific species in the Habitats and Birds Directives respectively.

## Socio-economics

The European Commission recognizes that there is a potential dichotomy between economic growth with job creation and environmental sustainability. Consequently the Lisbon Strategy for Growth was launched in 2000 with the ambitious aim of full employment by 2010. Renewed in 2005, it was built on the three pillars of economic growth and competitiveness, social inclusion and environmental concerns. The guidelines for the latest cycle, 2008–2010, COM (2007) 803 (Commission for the European Communities, 2007) refer to the need to decouple economic growth from environmental degradation whilst creating economic stability for sustainable growth. A more realistic target of 70 per cent employment by 2010 was established, along with a series of targets relating to education, training for the long-term unemployed, childcare provision and investment in research and development.

The European Union also has a Health Strategy which was published in 2007 and set out a whole series of European Union Health Indicators which are being tracked by WHO-Europe.

# UK policy and legislation
## History and background

Sustainability policy and legislation in the UK has its roots in medieval London, arising from the risks associated with increasing densities of population, especially from fire. After a major fire in 1212, for example, thatched roofs were banned. Records are available from the early 14th century that demonstrate that disputes between neighbours were settled in a court known as an Assize of Nuisance, advised by a coterie of masons and carpenters known as 'viewers'. This system was adopted elsewhere in England; for example Bristol had viewers as early as 1391. However, London continued to grow with little planning, narrow streets and the wholesale use of timber in construction, resulting in 80 per cent of the city being wiped out in the Great Fire of 1666. This led to Charles II introducing the first Building Act of 1667, specifying fire resistant materials to be used in the reconstruction of the city and a minimum distance allowed between buildings to create fire breaks. These methods were adopted in an ad hoc manner by other cities, but a National Building Act was not proposed until 1841, when Lord Normanby introduced his Bill for Regulating Buildings in Large Towns which became the Metropolitan Building Act in 1844. This was followed in 1848 by the Public Health Act creating boards of health at a local level; planting the seeds of modern local government across the UK. The Local Government Act of 1858 enabled local authorities to regulate building through bylaws, whilst the Public Health Act of 1875, together with a set of Model Bye-laws led to the first building control at a national level: the 1st Series being introduced in 1877. Local Planning Authorities were not formalized until the Town and Country Planning Act of 1934 which created local planning authorities. Model Building Bye-laws continued as the main vehicle for building regulation, gradually extending their scope until in 1953 they included for the first time a chapter on the thermal insulation of houses. Scotland was the first to standardize building standards in the Building (Scotland) Act of 1959, whilst England and Wales produced their first Building Regulations in 1965 under the Public Health Act, of 1961. However, it was the Building Act of 1984 (Building Act, 1984) which established building control at a local level and Building Regulations have been published under this Act ever since.

The 1965 Regulations included minimum requirements for thermal insulation for dwellings, sound insulation for dwellings, refuse disposal, ventilation and room height, chimneys and fireplaces, heat producing appliances, incinerators, drainage, sanitary conveniences, wells, tanks and cisterns; as well as materials, fire precautions, structural stability and moisture prevention. It wasn't until 1976 that a section on the 'conservation of fuel and power in buildings other than dwellings' appeared, recognizable today as Part L2.

Modern Building Regulations are currently going through a transition, driven by EC legislation and the sustainability agenda. The main vehicle for this change is the Code for Sustainable Homes, which, although not enshrined in total in the Building Regulations, is being used to set benchmarks for specific criteria such as carbon dioxide emissions associated with building energy consumption. A scheduled programme of amendments has been committed to by Government, leading to 'zero carbon' dwellings and schools by 2016 and other buildings by 2019, although the definition of 'zero carbon' is currently out for consultation (Department for Communities and Local Government, 2007a and 2007b). Part G of the Building Regulations, which now deals with sanitation, hot water safety and water efficiency, has recently been amended to give maximum water consumption requirements for dwellings, although the exact benchmark is currently out for consultation (HMSO, 2010).

Workplace legislation in the UK has a history dating back to 1802 when the Factories Act first appeared on the statute book and the Factories Inspectorate was created in 1833. The main purpose was to limit the hours worked by children, requiring factory owners to provide schooling and pay attention to infectious disease. Workrooms were to be well ventilated and whitewashed twice a year. Over the years the Act evolved until in 1961 it included provisions covering most aspects of health, safety and welfare, most of which have been subsequently incorporated into the Health and Safety at Work etc. Act 1974 and its daughter regulations such as the COSHH Regulations (COSHH, 1988) and the '6 Pack' (see Box 1.4).

In parallel with the development of building and health and safety regulations, public health legislation was introduced through the Public Health Act in 1848, which also paradoxically led to the system of local government and associated planning law and building control that we have today. It had its genesis in the cholera outbreaks that arose from poor sanitation in the rapidly growing towns and cities and it placed the supply of water, sewerage, cleansing and paving under local control, with oversight from a General Board of Health.

---

## Box 1.4 Key UK policy and legislation relating to sustainability and the environment

### 2009

- The UK Low Carbon Industry Strategy. Department of Business Innovation and Skills (DBIS), Department of Energy and Climate Change (DECC)
- The UK Low Carbon Transition Plan: National Strategy for Climate & Energy (DECC)
- Quality of Place: Improving the Planning and Design of the Built Environment, Cabinet Office Strategy Unit (COSU)
- A Sustainable New Deal: A Stimulus Package for Economic, Social and Ecological Recovery, Sustainable Development Commission (SDC)
- Prosperity without growth? The transition to a sustainable economy, SDC
- Sustainable Development Indicators in your Pocket, Department for Environment, Food and Rural Affairs (DEFRA)
- UK Renewable Energy Strategy, DECC. The Carbon Reduction Strategy for Transport: Low Carbon Transport, A Greener Future, Department for Transport (DfT)
- Low Carbon and Environmental Goods and Services: An Industrial Analysis, Innovas/Department for Business, Enterprise and Regulatory Reform (BERR)
- Draft Noise Action Plan: Major Roads, DEFRA

### 2008

- Planning Policy Statement 12: Local Spatial Planning
- Energy Act
- Climate Change Act

- Home Information Pack (Amendment) Regulations
- Strategy for Sustainable Construction, HMG
- Meeting the Energy Challenge: A White Paper on Nuclear Power, Department for Trade and Industry (DTI)
- Site Waste Management Plan Regulations
- Public Health Act
- Progress Report on Sustainable Products and Materials, DEFRA

## *2007*

- Meeting the Energy Challenge: A White Paper on Energy, DTI
- The UK Energy Efficiency Action Plan, DEFRA
- The Energy Performance of Buildings (Certificates and Inspections) (England and Wales) (Amendment) Regulations
- Sustainable Communities Act
- UK Government Sustainable Procurement Action Plan, DEFRA
- Building a Greener Future: Policy Statement, Department for Communities and Local Government (CLG)
- Air Quality Standard Regulations
- Local Government and Public Involvement in Health Act
- Planning and Climate Change: Supplement to Planning Policy Statement 1 (PPS1)
- Town and Country Planning (Environmental Impact Assessment) (England and Wales) (Amendment) Regulations
- Environmental Permitting (England and Wales) Regulations
- The Air Quality Strategy for England, Scotland, Wales and Northern Ireland. DEFRA
- The Construction (Design and Management) Regulations
- Waste Strategy for England, DEFRA
- The Water Supply (Water Quality) Regulations 2000 (Amendment) 2007
- The Ecodesign for Energy-using Products Regulations

## *2006*

- Strong and Prosperous Communities: The Local Government White Paper, CLG
- Control of Asbestos Regulation
- Water Resources (Environmental Impact Assessment) (Amendment) Regulations (England and Wales)
- Planning Policy Statement 3: Housing
- Planning Policy Statement 25: Flood Risk
- Environmental Noise (England) Regulations, as amended
- Quiet Homes and Home Zones (England) Regulations

## *2005*

- Securing the Future: UK Government Strategy for Sustainable Development, DEFRA
- Securing the Future: Delivering UK Sustainable Development Strategy, DEFRA

- Planning Policy Statement 1: Delivering Sustainable Development
- Clean Neighbourhoods and Environment Act
- Planning Policy Statement 9: Biodiversity and Geological Conservation
- Hazardous Waste Regulations

## 2004

- Planning and Compulsory Purchase Act
- Environmental Assessment of Plans and Programmes Regulations
- Planning Policy Statement 22: Renewable Energy

## 2003

- The Water Act
- Water Environment (Water Framework Directive) (England and Wales) Regulations

## 2002

- Working with the Grain of Nature: A Biodiversity Strategy for England, DEFRA
- Living Places: Cleaner, Safer, Greener, Office of the Deputy Prime Minister (ODPM)
- The Strategy for Sustainable Farming and Food: Facing the Future, DEFRA

## 2001

- Planning Policy Guidance 13: Transport

## 2000

- Local Government Act

## 1999

- A Better Quality of Life: Strategy for Sustainable Development in the UK
- Water Supply (Water Fittings) Regulations 1999 (England and Wales)

## 1995

- Environment Act
- Planning Policy Guidance 2: Green Belts
- Reporting of Injuries, Disasters and Dangerous Occurrences Regulations

## 1994

- Planning Policy Guidance 24: Planning and Noise
- Urban Waste Water Treatment Regulations

## 1992

- 6 Pack including:

- Workplace Regulations
- Display Screen Equipment Regulations

*1991*

- Planning and Compensation Act
- Water Resources Act
- Construction Products Regulations

*1990*

- Environmental Protection Act
- Clean Air Act
- This Common Inheritance: Britain's Environmental Strategy

*1989*

- Noise at Work Regulations
- Health and Safety Information for Employees Regulations

*1988*

- Control of Substances Hazardous to Health Regulations
- Town and Country Planning (Assessment of Environmental Effects) Regulations

*1981*

- Health and Safety (First Aid) Regulations

*1974*

- Control of Pollution Act
- Health and Safety at Work etc Act

*1959*

- Rights of Light Act

The Local Government Act of 1858 replaced the Board of Health with a Local Government Board, leading to a broadening of the remit of local boards to cover duties that had previously been covered by the Town Police Clauses Act of 1847, including a range of public safety measures, such as the removal of obstructions in streets and of dangerous buildings, and the prevention and fighting of fire. The Public Health Act continued to provide a mandate for local authorities and in a landmark version enacted in 1875 gave them power to regulate the sizes of rooms, the space around houses and street width.

Both of these acts have evolved over the years with the most recent incarnation of the Local Government Act coming into force in 2000. Following the 2006 Strong and Prosperous Communities White Paper (Department for Communities and Local Government, 2006a) it evolved into the Local Government and Public Involvement in Health Act 2007 (Local Government and Public Health Involvement in Health Act, 2007). The Public Health Act has recently been used as a vehicle for introducing smoking bans in public places and age limits on the purchase of tobacco through the Health Act 2006 and associated Regulations.

Planning did not become regulated until the Housing and Town Planning Act of 1909 which included the aim to 'secure proper sanitary conditions, amenity and convenience in connection with the laying out of the land itself as well as any neighbouring land', whilst allowing local authorities to define zones for specified use, limit the number of buildings on a site and control appearance. The scope was extended to development beyond towns and cities by the Town and Country Planning Act (TCPA) of 1932. However, centralized planning did not enter the statute book until 1947, enabling major regeneration following Second World War, with green belt legislation following in 1955. It is interesting to note that even during the darkest days of the War, authorities were planning for post-war development and a number of studies and reports were produced, whilst a TCPA was published in 1944 to provide protection to trees; this has largely been transposed into modern planning as Tree Protection Orders (TPOs). The TCPA evolved over the next 45 years into its 1990 incarnation. A sister act to the TCPA, the Planning and Compulsory Purchase Act, was enacted in 2004, introducing, along with a number of amendments to the TCPA, requirements for Regional Planning Bodies to produce Regional Spatial Strategies and for local authorities to produce Local Development Frameworks.

Subsequently the Planning Act 2008 launched the Infrastructure Planning Commission with responsibility for considering planning applications for nationally significant infrastructure projects, such as nuclear power stations, the process being funded from levies imposed by local authorities on projects. It also includes a new requirement for councils to take action on climate change.

TCPA has also provided a vehicle for implementing the requirements of the EIA Directive, originally published in 1985, with the latest amendments incorporated in 2003 (2003/35/EC), with the Town and Country Planning (Environmental Impact Assessment) (England and Wales) (Amendment) Regulations 2007 introducing the most recent changes to the 1999 Regulations in England and Wales.

As stated above, the UK Government has also set out its plans for local government in the Local Government White Paper 2006 (Department for Communities and Local Government, 2006a) enacted by the Local Government and Public Involvement in Health Act 2007. These have resulted in a raft of requirements for local authorities to action Sustainable Community Strategies in Local Strategic Partnerships through Local and Multi Area Agreements, along with initiatives such as Neighbourhood Charters, City Development Companies and Community Calls for Action.

Much of the above has informed UK Government policies in sustainable development and the environment. In particular the UN Stockholm Conference in 1972, which was followed shortly after by the oil crisis – precipitated by conflict in the Middle East – led to a lengthy period of re-evaluation of priorities and strategies, not only with respect to energy but across all environmental policy. Prime Minister Edward Heath had created the Central Policy Review Staff under Baron Rothschild in 1970, leading to the 1973 Energy Policy for Britain, from which grew the first Department of Energy. Rothschild was suspicious of nuclear power and advocated energy conservation and investment in alternative energy sources. However, during the Thatcher years of non-intervention much of policy development was left to the energy industries. So it wasn't until the lead-up to the 1992 Rio Conference that a coherent strategy was set out in the seminal White Paper 'This Common Inheritance: Britain's Environmental Strategy' (This Common Inheritance: Britain's Environmental Strategy, 1990). Published in 1990, this anticipated most of the issues discussed at Rio, but focused particularly on the greenhouse effect, with consequent global warming, and on pollution control.

Since Labour came to power in 1997 there has been a whole raft of initiatives and policy statements in the areas of sustainable development, energy and climate change (Box 1.4). The key policies arising from this activity include the development of a sustainability strategy and corresponding indicators and review programme. The Government's *Better Quality of Life* (DEFRA, 1999) strategy published in 1999 required local authorities to develop Local Agenda 21 strategies by 2000 and established a system of national reporting based on Quality of Life Indicators. Following the World Summit in Johannesburg in 2002 the Government revised the emphasis of their strategy to *Securing the Future* (DEFRA, 2005) in their 2005 Sustainable Development Strategy, with corresponding Sustainability Indicators published annually (DEFRA website). The most recent version of these (2009) compares 68 indicators against baseline values from 1990 and 2003. The UK Framework Sustainability Indicators are listed in Box 1.5 and examples of results are shown in Figure 1.4.

A key delivery vehicle for Government sustainability strategy is the planning regime, rooted in Planning Policy Statement 1: Delivering Sustainable Development (Department for Communities and Local Government, 2005a) published in 2005 and supplemented in 2007 by Planning and Climate Change (Department for Communities and Local Government, 2007b). These key documents require local authorities to integrate sustainable development and climate change mitigation into development plans and planning decisions, taking into account:

- social cohesion and inclusion;
- protection and enhancement of the environment;
- prudent use of natural resources;
- sustainable economic development.

## Energy and climate change

Energy and climate change has been a particular focus in recent years, resulting in a White Paper 'Meeting the Energy Challenge' published in 2007 followed in 2008 by the Energy Act (summarized in Box 1.6) and the Climate Change Act (summarized in Box 1.7) and the Department of Energy and Climate Change (DECC) being formed in October of that year. Government departments have subsequently come up with a raft of strategies including a White Paper from DECC giving a 'National Strategy for Climate and Energy' (DECC, 2009a), a UK

---

## Box 1.5 UK Framework Indicators

- Greenhouse gas emissions
- Resource use – energy, water and materials
- Waste arisings
- Bird populations
- Fish stocks sustainability
- Air pollutant emissions and impact on ecosystems
- River quality
- Economic output
- Active community participation
- Crime
- Employment
- Workless households
- Childhood poverty
- Pensioner poverty
- Educational attainment
- Health inequality
- Mobility
- Social justice
- Environmental equality
- Well-being

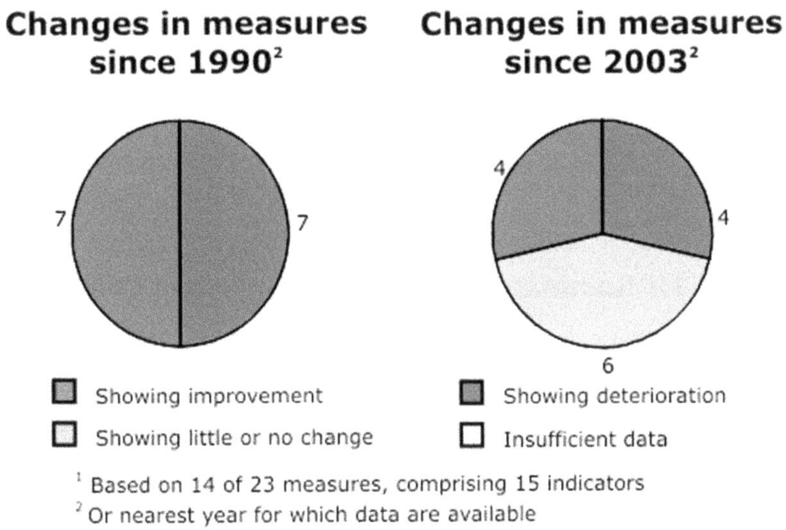

---

### Changes in measures since 1990[2]

### Changes in measures since 2003[2]

■ Showing improvement    ■ Showing deterioration

□ Showing little or no change    □ Insufficient data

[1] Based on 14 of 23 measures, comprising 15 indicators
[2] Or nearest year for which data are available

**Figure 1.4** *Sample of results from the 2009 Sustainable Development Indicators: Climate change and energy (CCE)[1]*

*Source:* Defra

Low Carbon Industrial Strategy (Department for Business Innovation and Skills and DECC, 2009), a Carbon Reduction Strategy for Transport (Department for Transport, 2009) and the UK Renewable Energy Strategy (DECC, 2009b).

A carbon budget was published in parallel with the Treasury budget in April 2009 setting out emission reduction targets of 22 per cent by the end of 2012, 28 per cent by 2017 and 34 per cent by 2022, promising that none of this would be achieved through offsetting via international credits.

The DECC White Paper sets out the Transition Plan to 2020 (DECC, 2009a) for transforming the UK power sector, homes and workplaces, transport, farming and the way land and waste are managed, to meet carbon budgets set for each Government department, secure energy supplies, maximize economic opportunities and protect the most vulnerable.

---

### Box 1.6 Summary of Energy Act 2008

The UK Energy Act became law on 26 November 2008. The Act includes a number of technical amendments to the 2004 Act, but summarized below are the key provisions that have a particular implication with respect to carbon mitigation and sustainability:

1 Provisions for licensing of offshore storage of $CO_2$ (Carbon Capture and Storage) and natural gas.
2 Enhancement of Renewables Obligation scheme.
3 Enables government to introduce feed-in tariffs for small-scale (<5MW) electricity generation.
4 Proposals to introduce mandatory smart metering by 2018–20.
5 Incentives for renewable heat installations through financial support mechanism.
6 Development of nuclear power plants must fully fund decommissioning and waste management.

---

### Box 1.7 Summary of Climate Change Act 2008

The UK Climate Change Act became law on 26 November 2008. The Act:

1 Established a UK national emissions reduction target of 80% by 2050, relative to a 1990 baseline for greenhouse gases (GHG).
2 Set an interim target of a 26% reduction by 2020 (for $CO_2$ only).
3 Required the Government to publish five yearly carbon budgets as from 2008 (for the period 2008–2012, 2013–2017 and so on) alongside a report on policies and programmes for meeting budget targets. Annual reports must set out net emissions for the UK and the role of international carbon credits in achieving those targets.

4  Legislated to include international aviation and shipping, with the Government to determine how they are to be included by 31 December 2012.

5  Created an independent Committee on Climate Change (CCC).

6  Required the CCC to advise the Government on carbon budgets, on the balance between domestic emissions reductions and the use of carbon credits and on adaptation strategies.

7  Placed a duty on the Government to assess the risk to the UK from the impacts of climate change (adaptation) in reports every five years.

8  Provided powers to establish trading schemes for the purpose of limiting GHG emissions – i.e. created the legal underpinning for the Carbon Reduction Commitment and future schemes.

9  Tasked the Government with making regulations under the Companies Act 2006 to require companies to report their GHG emissions as part of their directors' reports by 6 April 2012.

10  Obliged the Government to publish guidance for company GHG reporting by 1 October 2009.

11  Tasked the Government with reporting to Parliament on corporate GHG reporting and conducting a review of how it can contribute to UK national objectives.

12  Created powers to create waste reduction pilot schemes.

13  Obliged the Government to report annually on progress towards improving the efficiency and sustainability of the UK civil estate.

14  Enabled Government to make regulations about charges for single-use carrier bags.

15  Amended the provisions of the Energy Act 2004 on renewable transport fuel obligations.

The Renewable Energy Strategy responds to the 2009 EC RED (Renewables Directive) on how the UK is to deliver on its commitment to generate 15 per cent of energy supply from renewables by 2020, based on obtaining 30 per cent of electricity, 12 per cent of heat and 10 per cent of transport from renewable sources.

Planning policy for renewable energy was established in 2004 in PPS22: Renewable Energy (DECC, 2009b) which was introduced to ensure that regional planning bodies and local planning authorities use their powers to 'promote and encourage' renewable energy resources' by:

- setting out criteria for assessing applications;
- seeing wider environmental and economic benefits as material considerations in determining applications;
- encouraging small-scale installations where appropriate;
- fostering community involvement;
- considering the assessment of environmental, social and economic impacts.

Regional Spatial Strategies should include renewable energy targets to be achieved by 2010 and 2020. Local authorities may require a specified percentage of energy required for new developments to be obtained from renewable sources.

## Sustainable transport

The Department for Transport has published its 'Carbon Reduction Strategy for Transport' (Department for Transport, 2009) that sets out a raft of measures that depend largely on European and global initiatives, such as EC Regulations on emission performance standards for new passenger cars and global sectional targets for aircraft. It does include, however, proposals to subsidize ultra low emission cars and low emission bus technology, whilst using the Renewable Transport Fuel Obligation to increase the use of sustainable biofuels. The initiatives include:

- investing in 18 Cycling Demonstration Towns and Cities;
- a Sustainable Travel Towns programme;
- a National Cycle Plan;
- new guidance for Local Transport Plans;
- the 'Act on $CO_2$' Campaign;
- Green Fleet reviews;
- the Safe and Fuel Efficient Driving (SAFED) campaign;
- $CO_2$ emission trading to include aviation from 2012.

## Water and waste

Legislation covering water is divided between that dealing with water resources and measures to reduce consumption. Requirements governing the abstraction of water is covered by the Water Resources Act 1991 amended by the Water Act 2003, and include the specific duties of the Environment Agency. Requirements for the design of water installations to prevent leakage, excessive consumption, contamination, etc. are set out in the Water Fittings Regulations 1999 (Water Supply (Water Fittings) Regulations, 1999) whilst new requirements for maximum water consumption in dwellings are set out in Part G of the Building Regulations referred to above.

The Water Resources Act also covers requirements for discharging effluent and wastewater into natural water resources, whilst the 1994 Urban Waste Water Treatment Regulations (Urban Waste Water Treatment (England and Wales) Regulations, 1994) implement the eponymous EC Directive covering the collection, conveying and treatment of wastewater. Legislation covering municipal waste is complex and there are separate regulations for different types of waste, landfill and incineration. The primary legislation is the Control of Pollution Act 1974 augmented by the Environmental Protection Act 1990 modified by the EC Waste Framework to cover storage, treatment, recycling and transport. Separate Waste Strategies have been produced for England, Wales, Scotland and Northern Ireland and are reviewed annually. Local authorities are required to incorporate waste strategies into Regional Spatial Strategies and Local Development Frameworks.

The most recent Waste Strategy for England (DEFRA, 2007a) sets out a wide range of objectives and actions including:

- Decoupling waste growth from economic growth through prevention and re-use.
- Meeting or exceeding EC Framework targets for diverting waste from landfill, including investment in suitable infrastructure.
- Improving the integration of treatment for municipal and non-municipal wastes.
- Increasing recycling and energy from waste installations.
- Reducing greenhouse gas emissions by at least 9.3 million tonnes $CO_2$ equivalent compared to a 2006 baseline.
- Introducing a culture change through awareness raising campaigns.
- Extending Waste and Resources Action Programme (WRAP)'s Courtauld Commitment initiatives to reduce packaging, including single-use plastic bags.

The Climate Change Act 2008 (Box 1.7) allows for the making of regulations about charges for single-use bags, as well as the creation of waste reduction pilot schemes.

## Pollution and air quality

Modern regulation of air quality in the UK grew from a history of coal burning and smog in the large cities, resulting in the 1956 Clean Air Act (consolidated in 1993), which focused primarily on emissions from chimneys, and the Control of Pollution Act 1974, which implemented EC Directives from 1970 and 1972 that focused on emissions from road vehicles. Subsequent Directives and UK legislation dealt with the lead and other pollutants at source. It wasn't until 1989 that an EC Directive paid attention to the concentration of key pollutants in the air, setting limit values that were reflected in the Air Quality Standard Regulations that were published in the UK later the same year. The Environmental Protection Act 1990 gave local authorities responsibility for air pollution control, whilst the 1995 Environment Act established a statutory framework for air quality management requiring a National Air Quality Strategy, first published in 1997, the most recent of which was produced in 2007 and renamed 'The Air Quality Strategy for England, Scotland, Wales and Northern Ireland' (DEFRA et al, 2007) in concert with the latest version of the Air Quality Standard Regulations (Air Quality Standards Regulations, 64, 2007).

## Biodiversity and land use

Following the Second World War the UK Government not only focused on the regeneration of the blitzed towns and cities but produced landmark legislation for nature conservation and access to the countryside in the form of the 1949 National Parks and Access to the Countryside Act. This not only created the National Parks and opened up a network of footpaths but created the Nature Conservancy Council, which remains today as Natural England, Scottish Natural Heritage and the Countryside Council for Wales. Sites of Special Scientific Interest (SSSIs) were established by the National Nature Reserve, Wildlife and Countryside Act of 1981 which pre-empted the EC Habitats Directive by some 11 years. SSSIs were incorporated into the Special Areas of

Conservation (SACs) that comprise the Natura Network[2] and that has been established as British law by the Conservation (Natural Habitats, etc.) Regulations 1994 (as amended).

Green belt policy was first established for London by the Greater London Regional Planning Committee in 1935, with provision made in the 1947 TCPA for local authorities to include them in their development plans, but it wasn't until a landmark circular from Government to local authorities in 1955 that policy was implemented nationwide through a direction to local authorities. Current green belt policy and requirements are set out in Planning Policy Guidance, PPG2, (Department for Communities and Local Government, 1995) amended in 2001, which provides rules for limiting development in existing green belts. More recently in 2005 a Planning Policy Statement (PPS9) has set policy for biodiversity and the conservation of geology (Department for Communities and Local Government, 2005b) based on a strategy published by DEFRA in 2002 (DEFRA, 2002b) – *Working with the Grain of Nature: A Biodiversity Strategy for England*. This requires Regional Spatial Strategies and Local Development Frameworks to incorporate objectives to conserve and enhance biodiversity in both plan policies and decisions. DEFRA also produced a strategy for sustainable farming and food (DEFRA, 2002a) in 2002.

Government policy is to encourage urban regeneration through the development of brownfield i.e. previously developed sites. PPS1 requires local authorities to maximize outputs by, for example, giving preferential treatment to developments that include 'housing at higher densities on previously developed land' rather than low density greenfield development. PPS3: Housing (Department for Communities and Local Government, 2006b) sets out a Government target of a minimum of 60 per cent of housing development to be on previously developed (brownfield) land and for targets to be established in Regional Spatial Strategies and local development documents, indicating incentives that are to be provided to developers to achieve these targets. PPS3 also encourages local authorities to maximize the efficient use of land by setting density targets for specific zones, but with a minimum national target of 30 dwellings per hectare.

The potential for land to flood is an important limitation to development as well as being impacted on by climate change. The Government recognized this in 2006 with the publication of PPG25: Development and Flood Risk (Department for Communities and Local Government, 2006c) that requires local authorities to apply a sequential test when considering land for development 'to demonstrate that there are no reasonably available sites in areas with a lower probability of flooding'. For individual developments, flood risk assessments are to be provided by the developer and the Environment Agency is to be consulted.

## Socio-economic sustainability and communities

PPS1 also requires development plans to address social cohesion and inclusion through:

- the assessment of the impact of development on the social fabric of communities;

- measures to reduce social inequalities;
- access to jobs, health, housing, education, shops, leisure and community facilities;
- the assessment of and provision for community needs;
- the provision of safe, healthy and attractive places to live;
- facilities for physical activity.

A recent Cabinet Office report (HM Government, 2009) sets out current Government thinking on quality of place. The report defines their subject as 'the physical characteristics of a community that affect the quality of life and life chances of people living and working in it'. Their conclusions are summarized in Figure 1.5, which neatly encapsulates the key aspects of a sustainable community.

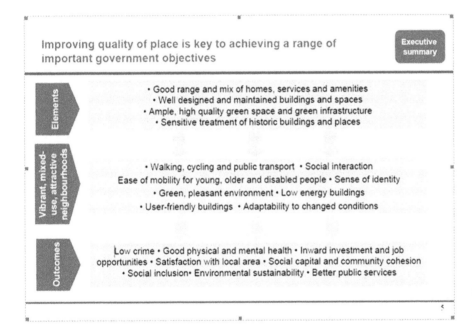

**Figure 1.5** *Quality of place – a summary*

*Source:* Cabinet office, Crown copyright

## Environmental noise

Exposure to noise is an important indicator of quality of life, particularly during the sleeping hours. This has been recognized by WHO Europe in their recently published Night Noise Guidelines for Europe (WHO Europe, 2009). These indicate that sleep disturbance occurs at lower levels than previously thought and DEFRA are currently reviewing Government planning guidance: PPG24 (Department for Communities and Local Government, 1994) with this in mind. The EC Environmental Noise Directive (END) 2002/49/EC required Member States to carry out noise mapping to identify existing hot spots and establish an Action Plan for mitigation. England has implemented END through the Environmental Noise (England) regulations (Environmental Noise (England) (Amendment) Regulations, 2009) and a public consultation process is under way on the Action Plan (DEFRA, 2009). It is interesting to note that the results from DEFRA's noise mapping exercise indicate that nearly 7 million people are exposed to noise levels at night which are 10dB above the target value for night-time noise levels ($L_{night}$) of 40dB recommended by WHO Europe, and some 2.7 million are exposed to levels above the interim value of 55dB. Planning applications for noise-sensitive developments such as dwellings are currently assessed under Planning Policy Guidance 24 (Department for Communities and Local Government, 1994) (which is under review at the time of writing). It sets out a series of Noise Exposure Categories (NEC A to D) and requires noise prediction for day- and night-time levels to be predicted. If noise levels exceed the NEC D level then planning permission would normally be refused, whilst below NEC A permission can be granted unconditionally. NEC B and C require some form of mitigation such as barriers or specialist windows and/or ventilation.

## Rights of light and overshadowing

The right to not having daylight into a dwelling obstructed by a neighbour dates back to the Prescription Act of 1832 and is currently covered by the Right to Light Act 1959, which includes relatively simple rules relating to the angle between the centre of a window and the proposed development, where this amenity has been available for at least 20 years. Right for Light legislation does not come within the scope of planning law, but local authorities will judge the impact of a proposed new development as part of an EIA. Most local authorities will accept an assessment using procedures set out in BRE's Site Layout Planning for Daylight and Sunlight: A Good Practice Guide (Littlefair, 1991).

## Wind environment

A wind impact assessment is similarly likely to form part of an EIA for developments that have a potential negative impact on the wind environment in the surrounding public domain, such as might apply to tall buildings. There is currently no UK wide planning policy or legislation that applies to this and the requirements for a planning application will be dependent upon legal precedent and the specific rules laid down by a local authority; for example the City of London Unitary Development Plan from 2002 required that developers are to 'produce evidence that unacceptable wind turbulence would not be created by substantial development schemes ...'. This will be carried over into the new Local Development Framework when it is adopted in 2011.

## Materials, construction and procurement

The UK Government is acting as a champion for the sustainable procurement of products and materials through the adoption of the EU Green Public Procurement criteria (see above) and promotion within public procurement of DEFRA's 'Buy Sustainable–Quick Wins' procurement specifications.[3]

A cross-departmental body, the Government Construction Client Panel (GCCP) was established in 1997 and produced a key set of actions and targets covering a range of sustainability criteria for public sector procurement, including the following to be achieved by 2003:

- All (100%) departments to use whole life costing in deciding between re-using existing premises or new build and in making decisions concerning sustainability.
- Meet good practice targets in designing to minimize waste.
- Meet 'Achieving Excellence' targets in lean construction.
- Minimize energy in construction and use.
- Meet good practice targets for minimizing pollution.
- Preserve and enhance biodiversity.
- Conserve water resources.
- All projects to be carried out under the Considerate Constructers' Scheme.
- Achieve BREEAM Excellent for new projects and Very Good for refurbishment.

These targets may be past their 'use-by date' but still have relevance and appear in some form in more recent Government policy on public sector procurement set out in the 2008 Strategy for Sustainable Construction (HM Government with Strategic Forum for Construction, 2008) and the Government's current Sustainable Procurement Action Plan (DEFRA, 2007b).

The Office of Government Commerce (OGC) includes within its terms of reference sustainable procurement and operation, and has adopted a series of targets for the Sustainable Operation of the Government Estate (SOGE),[4] including targets for carbon reduction, water consumption, timber procurement and waste arisings. These also appear in the Government's Sustainable Procurement Action Plan.

The EC Eco-design Regulations were enacted across Europe in 2008 and 2009 (see EC section above) and will apply to products supplied to consumers in the UK. As well as impacting on electronic and white goods, this will be extended to cover water-using products and building products such as windows that impact on energy consumption. The Ecodesign for Energy-Using Products Regulations 2007 (Ecodesign of Energy-Using Products Regulations, 2007) implement the framework from the Eco-design Directive but have excluded the product list and instead have been used to implement earlier EC Directives relating to the energy efficiency of boilers, refrigerators and ballasts for fluorescent lighting.

There is currently no legislation that requires the environmental impact or embodied energy of building materials to be considered in construction projects, however, the requirement for all dwellings to be assessed under the Code for Sustainable Homes (CSH) and the increasing use of the Building Research Establishment Environmental Assessment Method (BREEAM) for new non-residential buildings means that at least embodied impacts and the use of recycled materials are frequently being considered as part of the design process.

Similarly the drivers for sustainable construction methods such as the Considerate Constructers' Scheme (CCS) are primarily voluntary, in particular by developers attempting to achieve targets under CSH and BREEAM criteria. The CCS, or its equivalent local scheme, is sometimes a planning requirement, particularly in crowded city environments such as London where impacts on neighbours and the public are particularly important.

In 2008 Site Waste Management Plans became mandatory under the Regulations of the same name (Site Waste Management Plans Regulations, 2008). These require the principle contractor to prepare a plan which demonstrates how he is to reduce waste, maximize recycling and dispose of waste safely and legally.

# USA policy and legislation

## Sustainability, conservation and environmentalism

In the US the term 'sustainability' has not been widely used in the context of environmental, social or economic impact until recently. It is widely recognized that following the Second World War the US 'proposed their way of life, based on an unlimited consumption of land, goods and energy, as a model for the entire world' (Zaninetti, 2009). However, Roosevelt's 'New Deal' was in fact a recovery programme, responding to financial and economic calamity.

Historically, the terms 'green buildings' or 'green architecture' have been used as synonyms for sustainability, although this is usually applied to low energy design. However, other environmental issues were raised by pioneering environmentalists such G. P. Marsh in his book from 1864 *Man and Nature*. Conservation groups such as the Appalachian Mountain Club (AMC) and the Sierra Club were founded primarily to protect wilderness areas and establish national parks. The AMC was established in 1876 to explore and preserve the White Mountains in New Hampshire, whilst the Sierra Club was formed in 1892 by a group of professors from universities at Berkeley and Stanford to create national parks at Glacier and Mount Rainier. Conservation became a cause célèbre in the early 20th century, with some 49,000 people joining the National Audubon Society (NAS), dedicated to the conservation of endangered species of birds, in the year after it was formed in 1905.

The Inter-American Conference on the Conservation of Renewable National Resources met in the same year that the United Nations was formed – 1948 – and came up with the prescient principle that 'no generation can exclusively own the renewable resources by which it lives. We hold the commonwealth in trust for prosperity, and to lessen or destroy it is to commit treason against the future' (Fairchild, 1949). These words have resonance for subsequent definitions of sustainable development, such as those coined by Brundtland in 1987 (see International section above).

An active conservation movement blossomed in the 1950s and 1960s fuelled in part by widespread concerns about pollution, associated with some dramatic events such as oil spills, exposure to radiation from atom bomb testing and the widespread abuse of pesticides. Rachel Carson in her seminal book *Silent Spring*, published in 1962, highlighted the latter issue in particular. DDT had been widely used in the 1940s and 1950s to both eradicate mosquito populations and as an agricultural pesticide, however, it became evident that it was decimating bird populations and damaging human health. Rachel Carson's book became a best-seller and her thesis was accepted by the Kennedy presidency, and reinforced by the activities in Long Island of the Environmental Defense Fund, leading to an eventual DDT ban in the US in 1971.

Following the pattern established elsewhere in the developed world, a number of serious air pollution events occurred that led to a raft of legislation designed to establish the extent of the problem and set air quality criteria and control measures. In 1948 a five-day industrial smog in Donora, Pennsylvania left 20 dead and some 6000 with respiratory and related symptoms. In 1955 the Air Pollution Control Act mandated federal research programmes into the effects of air pollution and provided technical assistance to states. This was replaced in 1963 by the first Clean Air Act, which established funding for further study and the clean-up of air pollution, with the aim of defining air quality criteria and providing grants for air pollution control. The 1965 Motor Vehicle Air Pollution Control Act mandated auto emission standards, whilst the 1967 Federal Air Quality Act defined air quality control regions.

In 1970 the amended Clean Air Act consolidated the research carried out to that date and established a more robust set of criteria and controls, whilst in the same year the National Environmental Protection Act established the US Environmental Protection Agency (EPA) that was set the task of policing relevant legislation. This was augmented by enhancements to the Water Pollution Control Act in 1972 and the passing of the Endangered Species Act in 1973, the Safe Drinking Water Act in 1974, the Resources Conservation and Recovery Act in 1976 and the Clean Water Act in 1977.

All this legislative activity occurred during the Republican presidencies of Richard Nixon and Gerald Ford; however, after the brief period in office of Jimmy Carter, during which he produced an ambitious energy policy, expanded national parks and established a Department of Education, President Reagan presided over a period of anti-conservation and *laissez faire* government that left the environmental movement in the US divided and impotent.

However, George Bush Senior's presidency saw the launch of the Energy Star programme in 1992 and the first green building programme in Houston, Texas. The US Green Building Council was formed in 1993.

Although Bill Clinton's administration did tighten air pollution regulation, very little new environmental legislation was passed during the eight years of his incumbency, despite the Rio Declaration and Agenda 21 having been agreed the year before his election. However, the Partnership for Advancing Technology in Housing (PATH) was launched during the Clinton presidency. This set National Construction Goals 'to make American homes stronger, safer, more durable, more energy efficient and environmentally friendly'. The Leadership in Energy and Environmental Design (LEED) scheme, based on the UK BREEAM scheme, was launched by the US Green Building Council (USGBC) in 1998.

Since George W. Bush came to power in 2001, there have been a number of initiatives and Acts, some of which have divided opinion. For example, the Healthy Forests Restoration Act of 2003, designed to reduce the spread of forest fires by clearing undergrowth and requiring fire breaks to be created, was accused of providing free reign for logging companies. Another piece of legislation proposed during the Bush era was the Clean Skies Act 2003 that included many of the requirements for pollution reduction of the Clean Air Act but with lower targets, longer time spans and a cap and trade system that was intended to accelerate the process of granting permission for power plants.

An important component of the Clean Air Act is the requirement for a New Source Review (NSR), which in its original form in 1970, specified that the best available pollution control technology (BAPCT) be installed on new plant, but, in a 1977 revision, required that these controls be applied to existing plant when it is modified or expanded. However, it is alleged that this was widely ignored by the power companies and overlooked by the EPA until 1997 when a number of prosecutions were launched. However, the Bush administration in its 2001 review of National Energy Policy decided to introduce a threshold for the cost of modification that would trigger an NSR of 20 per cent of the value of the plant, or typically hundreds of millions of dollars. Hence the majority of ageing coal-fired power stations in the US largely remained unimproved, with emissions well above that stipulated in the Clean Air Act.

Early in the Bush presidency the Brownfields Revitalization and Environmental Restoration Act (BRERA) was passed with the intention of opening up contaminated sites for development by reducing the liability of developers under the earlier Comprehensive Environmental Response, Compensation and Litigation Act (CERCLA).

## Social and economic measures

The focus of the Roosevelt presidency's 'New Deal' following the Second World War was on recovery and regeneration, perpetuated by Truman's Fair Deal and the landmark Housing Act of 1949 that was designed to help eradicate slums and promote redevelopment.

John F. Kennedy's 'Frontier Program' included a set of proposals to create employment and improve housing provision. His Housing Bill of 1961 proposed an ambitious and complex housing programme to spur the economy, revitalize cities and provide affordable housing for middle- and low-income families.

Following Kennedy's tragic assassination in 1963 after less than two years in office, Lyndon Johnson continued many of the themes of his Frontier Program under the grandly titled Great Society programme that was intended as a vehicle for eliminating poverty and racial injustice. It launched spending programmes that addressed education, medical care (Medicare and Medicaid), urban problems and transportation.

Johnson created the Department of Housing and Urban Development (HUD) under the Housing and Community Development Act (HCDA) of 1965 and banned discrimination in the provision of social housing in the Fair Housing Act of 1968. The scope of the HCDA was extended by revisions in 1974, 1977, 1988 and 1992 to increase funding for community development block grants, urban development grants and housing for senior citizens and disabled persons.

## Energy and climate change

The US has benefited from an abundance of cheap fuel and hence for the greater part of the 20th century the energy consumed by buildings and industry, the efficiency of energy generation and the fuel used for transport were not high on the political agenda, at least not until the OPEC (Oil Producing and Exporting Countries) embargo and subsequent oil crisis in 1973.

President Ford responded with the Energy Policy and Conservation Act of 1975 which created a Strategic Petroleum Reserve and Corporate Average Fuel Economy (CAFE) target for automobile manufacturers. However, Carter's energy policy in 1977 went further than this by setting a series of goals to be achieved by 1985 including:

- reducing annual growth in demand to 2%;
- reducing petroleum consumption to 10% below 1977 levels;
- halving oil imports;
- establishing an oil reserve of 6 month's supply;
- increasing coal production by 1 billion tons/year;
- insulating all new buildings;
- using solar energy in two and a half million houses.

To help achieve these goals, the Carter administration enacted the National Energy Act in 1978 which included five sister Acts:

- The National Energy Conservation Policy Act (NECPA) required utilities to provide residential consumers with energy conservation advice.
- The Public Utility Regulatory Policies Act (PURPA) promoted the greater use of renewable energy by forcing utilities to buy power from non-utility power producers – resulting in an increase in cogeneration (combined heat and power).
- The Energy Tax Act promoted fuel efficiency and renewable energy through tax incentives and provided tax hikes for cars having fuel consumption less than 22.5 miles per gallon.
- The Power Plant and Industrial Fuel Use Act restricted oil and natural gas as a fuel for power generation and industry and encouraged the use of coal and alternatives, including nuclear energy.
- The Natural Gas Policy Act followed a period during which demand for gas had outstripped supply and attempted to correct this by price control at the wellhead initially, but intending that market forces would help match demand to supply.

Ronald Reagan continued Carter's push for coal, but, in spite of the Three Mile Island accident in 1979, targeted a major expansion of nuclear power generation, corresponding to 400 per cent by the year 2000.

During George Bush Senior's early years the Clean Air Act (1990) was amended to include a capping of sulphur emissions, an early template for carbon emissions trading, and in 1992 an Energy Policy Act was passed that included Production Tax Credits that provided further subsidy for renewables, although its uptake has proven to be cyclical. In 2005 the Clinton administration's Energy Policy Act opened up funding for renewables, including $150 million for biomass.

At the time of writing, the Obama administration's Climate Change Bill is yet to make it through the Senate. Box 1.8 provides a summary of the various

Acts that make up the Bill, which include numerous amendments to legislation referred to above. A key target is buried deep in the text at Title III, Sub-title A, where one can find the commitment to a 17 per cent reduction in GHG emissions by 2020. This compares with commitments of 25–30 per cent for the EC countries and 34 per cent for the UK (by 2022).

---

## Box 1.8 Summary of US Climate Change Bill

The Act is currently divided into five 'Titles' each with a series of 'Sub-titles', some of which are extremely complex. It includes amendments to a number of previous Acts, including the Clean Air Act 1990 (CAA), Energy Policy Act 2005 (EPA) and Energy Independence and Security Act 2007 (EISA). Below are listed the salient points distilled from a summary document available at the website given following this list:

### I. Clean Energy

A. Combined Efficiency and Renewable Electricity Standard
- Requires supply by large utility companies to be reduced through efficiency improvements and renewable energy generation by 6% by 2012, 9.5% by 2014, 13% by 2016, 16.5% by 2018 and 20% beyond 2020.

B. Carbon Capture and Sequestration
- Sets up an environment in which carbon capture and sequestration can be facilitated.
- Amends CAA to allow certification and permits for geological sequestration.
- Establishes a target of 50% carbon savings for utility companies that derive at least 30% of input from coal or coke, increasing to 65% after 2020.

C. Clean Transportation
- Electricity utilities to provide infrastructure for charging electric vehicles.
- Funding for Advanced Technology Vehicle Loan Programme to be doubled.

D. State Energy and Environmental Development (SEED) Accounts
- Establishment and funding of SEED to allow federal funding of local measures.

E. Smart Grid Advancement
- A major programme of enhancement of the electricity supply infrastructure across the US with the primary objective of more efficiently matching supply to demand through advanced monitoring and control techniques and promoting distributed generation, including on-site renewables.
- Explore the potential for Energy Star rated products to be integrated into Smart Grid.

F. Transmission Planning
- Facilitate the deployment of renewable energy including funding a grant programme.

G. Technical Correction to Energy Laws
- Amend EISA 2007 to set minimum luminous efficacy for general service lamps of 45 lumens/W by 2020.

H. Energy Efficiency Centres and Research

I. Nuclear and Advanced Technologies

- Funding for clean energy investments.
- National Bio-energy Partnership.
- Study on Thorium fuel cycle for nuclear reactors.

## II. *Energy Efficiency*

A. Building Energy Efficiency Programmes
- Immediate 30% reduction in energy use required by relevant building code, increasing to 50% by 2014 for dwellings and by 2015 for commercial buildings; and by a further 5% every 3 years until 2030.
- Develop and implement standards for retrofit – Retrofit for Energy and Environmental Performance (REEP) programme (Bill introduced to Congress May 2009).
- Make available grants for Energy Star rated dwellings.[5]
- Establish advanced energy labelling programme.
- Provide assistance to retail power providers for tree-planting programme.
- The Department of Housing and Urban Development (HUD) to facilitate local grants including solar energy schemes.

B. Lighting and Appliance Energy Efficiency Programmes
- Energy Policy and Conservation Act (EPCA) to be amended to improve efficacies of external luminaires and efficiencies of various appliances such as warm air furnaces under the Energy Conservation Program for Consumer Products other than Automobiles.
- Establish a new energy efficiency test for televisions and introduce a labelling system.
- Establish Best-in-Class Appliances Deployment programme for energy consuming consumer products and incentives for retailers to promote (Bill introduced to Congress May 2009).
- Extend WaterSense programme to include labelling.[6]
- Extend Energy Star system to include cost effectiveness and payback.

C. Transportation Efficiency
- Promulgate standards to achieve maximum possible reduction in emissions from heavy vehicles by the end of 2010.
- Develop targets for GHG emission reduction.
- Adapt EPA SmartWay programmes to promote and finance low GHG transport technologies.[7]

D. Industrial Energy Efficiency Programmes
- Extend voluntary American National Standards Institute (ANSI) industrial plant energy efficiency certification programme.[8]

- Award electricity generation that recovers thermal energy: Waste Energy Recovery Incentive grant programme.[9]
- Increase take-up of energy efficient electric motors.
- Initiate grants to finance GHG reduction by small and medium-sized manufacturers.

E. Improvements in Energy Savings Performance Contracts (ESPC)[10]
- Federal agencies to expand allowable types of energy transactions to include renewable thermal and on-site renewable.

F. Public Institutions
- Expand reach of energy sustainable and efficiency grants.
- Finance research into consumer behaviour in relation to energy use.

G. Miscellaneous
- Amend National Energy Conservation Policy Act 1978 (NECPA) that applies to energy management of federal buildings to require agencies to report to Congress on improvements in energy performance and uptake of innovative practices and technologies.
- Improve overall energy productivity of US by 2.5%/year 2012–2030 based on strategic plan with biennial reports to Congress.
- EPA to carry out feasibility study on national programme of labelling of products and materials.
- Promote within International Civil Aviation Organisation (ICAO) framework for regulation on GHG emissions from aircraft.

H. Green Resources for Energy Efficient Neighbourhoods Act 2009 (GREEN Act) (Submitted to Congress also as separate Bill in May 2009).
- Introduction of annual energy efficiency participation incentive through HUD.
- Introduction of new HUD energy efficiency standards.[11] Existing structures to achieve energy savings of 20% when rehabilitated or improved. New dwellings to be fitted with sockets for charging electric vehicles.
- Demonstration programme of 50,000 homes over a four-year period.
- Focus on under-served markets, including 'energy efficient' mortgages for very low to medium income families and provision of insurance for 50,000 mortgages under the National Housing Act by the end of 2012.
- Energy performance of dwellings to be certified by Home Energy Rating System Council accredited individuals.[12]
- Provide incentives for the planting of trees and native species.
- Ensure affordable housing has good access to public transport.
- Facilitate grants for community education programmes.
- Revitalization plans to comply with Green Communities criteria.[13]
- Alternate Energy Sources State Loan Fund to be established.

## III. Reducing Global Warming Pollution – The Safe Climate Act 2009

A. Reducing Global Warming Pollution
- Establish GHG emission targets compared to 2005 baseline of 97% by 2012, 83% by 2020, 58% by 2030, 17% by 2050, including allowances for deforestation in developing countries that have agreed to reduce deforestation.

- Defines reporting requirements for all major GHG emitters.
- Establishes allowances for offsetting.
- Target set to reduce cumulative emissions from deforestation of 6 billion tons by 2025.

B. Disposition of Allowances
- Amends CAA allowances for deforestation, carbon capture and storage and associated mechanism for distribution of funds.

C. Additional GHG Standards
- Refers to measures to eliminate ozone depleting substances.

D. Carbon Market Assurance
- Includes provisions for carbon market oversight.

E. Additional Market Assurance

## IV. Transitioning to a Clean Energy Economy

A. Ensuring Real Reductions in Industrial Emissions

B. Green Jobs and Worker Transition

C. Consumer Assistance
- Compensation to eligible low-income households.

D. Exporting Clean Technologies

E. Adapting to Climate Change: National Climate Change Adaptation Program: Global Change Research and Data Management Act 2009 and National Climate Services Act 2009 (both introduced to Congress in May 2009).
- Introduces funding programme to building resilience to climate change impacts 2011 to 2049.
- Public health preparedness.
- Establishes National Research Adaptation Panel required to develop research programme, implementation plan, reviewed periodically.
- Establishes International Climate Change Adaptation Program.

F. Deficit Neutral Budgetary Treatment

## V. Agricultural and Forestry Related Offsets

A. Establishes Offset Credit Program

B. US Department of Agriculture (USDA) GHG Emission Reduction and Sequestration Advisory Committee
- Offset programme for agriculture and forestry.

C. Miscellaneous
- Use of federal land for biomass production.

Note: Taken from the American Clean Energy and Security Act of 2009 as approved by House of Congress 26 June 2009. Available at www.govtrack.us/congress/bill.xpd?bill=h111-2454&tab= summary (accessed 19 November 2009).

# Notes

1   A summary of the Biodiversity Action Plan updated to 2008 is available at http://ec.europa. eu/environment/nature/info/pubs/docs/brochures/bio_brochure_en.pdf.
2   www.natura.org/.
3   www.DEFRA.gov.uk/sustainable/government/what/priority/consumption-production/ quickWins/, now known as Government Buying Standards.
4   www.ogc.gov.uk/estates_sustainability.asp.
5   www.energystar.gov/.
6   www.epa.gov/WaterSense/.
7   www.epa.gov/smartway/.
8   www.cee1.org/ind/industrial-program-planning/Plant%20Certification%20Strategic%20 Plan%205_18_08.pdf.
9   www.epa.gov/RDEE/registry/index.html.
10   www.epa.gov/epp/pubs/case/espc.htm.
11   www.internationalcodes.net/2009-international-energy-conservation-codes.shtml.
12   www.usahers.com/.
13   www.epa.gov/greenkit/index.htm.

# References

*Air Quality Standards Regulations 2007*. SI 2007/64, London: HMSO

BioRegional and WWF International (WWF), 2008. *One Planet Living*. Available at www.oneplanetliving.org/index.html (accessed 23 September 2009)

Brundtland Commission, 1987. *Report of the World Commission on Environment and Development: Our Common Future*. UN documents. Available at www.un-documents. net/wced-ocf.htm (accessed 16 January 2010)

*Building Act 1984*, London: HMSO

Commission of the European Communities (EU), 2004. *Communication from the Commission to the Council and the European Parliament – January 2004: Stimulating Technologies for Sustainable Development: An Environmental Technologies Action Plan for the European Union* (COM (2004) 38 final) Brussels: EU (published 2004). Available at http://ec.europa.eu/environment/etap/information/documents_en. html#002 (accessed 18 January 2010)

Commission of the European Communities, 2007. *Communication from the Commission to the Spring European Council – Integrated Guidelines for Growth and Jobs 2008– 2010*. (COM (2007) 803 final Part V) Brussels: EU. Available at http://ec.europa.eu/ growthandjobs/pdf/european-dimension-200712-annual-report-integrated-guidelines_ en.pdf (accessed 18 January 2010)

Commission of the European Communities, 2008a. *Communication From The Commission to The European Parliament and The Council – Greening Transport {SEC(2008)2006}*. (COM (2008) 433) Brussels: EU. Available at http://eur-lex.

europa.eu/LexUriServ/LexUriServ.do?uri=COM:2008:433:FIN:EN:PDF (accessed 18 January 2010, not available in English).

Commission of the European Communities, 2008b. *Communication from the Commission to the European Parliament, the Council, the European Economic and Social Committee and the Committee of the Regions on the Sustainable Consumption and Production and Sustainable Industrial Policy Action Plan.* (COM (2008) 397) Brussels: EU. Available at http://ec.europa.eu/environment/eussd/pdf/com_2008_397. pdf (accessed 18 January 2010)

Commission of the European Communities, 2008c. *Communication from the Commission to the Council, the European Parliament, the European Economic and Social Committee and Committee of the Regions – a mid-term assessment of implementng the EC Biodiversity Action Plan.* (COM (2008) 864) Brussels: EU. Available at http://ec.europa.eu/environment/nature/biodiversity/comm2006/pdf/ bap_2008_en.pdf (accessed 18 January 2010)

Commission of the European Communities, 2008d. *Proposal for a Directive of the European Parliament and of the Council laying down a framework for the deployment of Intelligent Transport Systems in the field of road transport and for interfaces with other transport modes {SEC (2008) 3083} {SEC (2008) 3084}* Brussels: EU. Available at http://eur-lex.europa.eu/LexUriServ/LexUriServ.do?uri=C OM:2008:0887:FIN:EN:HTML (accessed 18 January 2010)

Commission of the European Communites, 2009. *Mainstreaming Sustainable Development into EU policies: 2009 Review of the European Union Strategy for Sustainable Development.* COM (20090400) Brussels: Commission of the European Communities. Available at http://eur-lex.europa.eu/LexUriServ/LexUriServ. do?uri=COM:2009:0400:FIN:en:PDF (accessed 21 October 2009)

*COSHH (Control of Substances Hazardous to Health) 1988.* SI 1988/1657, London: HMSO

Department for Business Innovation and Skills and DECC, 2009. *The UK Low Carbon Industrial Strategy.* London: TSO. Available at www.berr.gov.uk/files/file52002.pdf (accessed 20 August 2010)

Department for Communities and Local Government, 1994. *Planning Policy Guidance 24: Planning and Noise.* London: HMSO. Available at www.communities.gov.uk/ documents/planningandbuilding/pdf/156558.pdf (accessed 20 August 2010)

Department for Communities and Local Government, 1995 (amended 2001). *Planning Policy Guidance 2: Green Belts.* London: HMSO. Available at www.communities. gov.uk/publications/planningandbuilding/ppg2 (accessed 26 January 2010)

Department for Communities and Local Government, 2005a. *Planning Policy Statement 1: Delivering Sustainable Development.* London: TPO. Available at www. communities.gov.uk/publications/planningandbuilding/planningpolicystatement1 (accessed 19 January 2010)

Department for Communities and Local Government, 2005b. *Planning Policy Statement 9: Biodiversity and Geological Conservation.* London: HMSO. Available at www. communities.gov.uk/planningandbuilding/planning/planningpolicyguidance/ historicenvironment/pps9 (accessed 19 January 2010)

Department for Communities and Local Government, 2006a. *Strong and Prosperous Communities – the Local Government White Paper.* London: HMSO

Department for Communities and Local Government, 2006b. *Planning Policy Statement 3, PPS3, Housing.* London: TSO. Available at www.communities.gov.uk/publications/ planningandbuilding/pps3housing (accessed 26 January 2010)

Department for Communities and Local Government, 2006c. *Planning Policy Statement 25: Development and Flood Risk.* London: TSO. Available at www.communities.gov. uk/documents/planningandbuilding/pdf/planningpolicystatement25.pdf (accessed 20 August 2010)

Department for Communities and Local Government, 2007a. *Building a Greener Future: Policy statement.* London: HMSO

Department for Communities and Local Government, 2007b. *Planning Policy Statement: Planning and Climate Change – Supplement to Planning Policy 1.* London: TSO. (Published 17 December 2007)

DEFRA (Department for Environment, Food and Rural Affairs), 1999. *A Better Quality of Life, a strategy for sustainable development in the UK.* Cm4345. London: HMSO, 1999.

DEFRA, 2002a. *Strategy for Sustainable Farming and Food – Facing the Future.* London, HMSO. Available at www.defra.gov.uk/foodfarm/policy/sustainfarmfood/documents/sffs.pdf (accessed 20 August 2010)

DEFRA, 2002b. *Working with the Grain of Nature – a Biodiversity Strategy for England.* London: HMSO. Available at www.defra.gov.uk/environment/biodiversity/documents/biostrategy.pdf (accessed 20 August 2010)

DEFRA, 2005. *Securing the Future. The UK Government Sustainable Development Strategy* (Cm 6467). London: TSO

DEFRA, 2007a. *Waste Strategy for England 2007. Cm 7086.* London: HMSO

DEFRA, 2007b. *UK Government Sustainable Procurement Action Plan Incorporating the Government response to the report of the Sustainable Procurement Task Force.* London: TSO. Available at www.DEFRA.gov.uk/sustainable/government/documents/SustainableProcurementActionPlan.pdf (accessed 26 January 2010)

DEFRA, 2009. *Consultation on a Draft Noise Action Plan, Environmental Noise (England) Regulations 2006, as amended July 2009.* London: HMSO. Available at www.DEFRA.gov.uk/corporate/consult/noise-action-plan/index.htm (accessed 26 January 2010)

DEFRA, Undated. *DEFRA>Sustainable>SD in Government>Reviewing Progress.* Available at www.DEFRA.gov.uk/sustainable/government/progress/index.htm (accessed 19 January 2010)

DEFRA, Scottish Executive, Welsh Assembly Government and Department of the Environment Northern Ireland, 2007. *The Air Quality Strategy for England, Scotland, Wales and Northern Ireland Vol 1 and Vol 2. Cm7169.* London: TSO. Available at www.defra.gov.uk/environment/quality/air/airquality/strategy/ (accessed 20 August 2010)

Department for Transport, 2009. *Low Carbon Transport: A Greener Future – A Carbon Reduction Strategy for Transport – Cm 7682.* London: TSO. (Published 26 June 2009)

DECC (Department for Energy and Climate Change), 2009a. *The UK Low Carbon Transition Plan – National Strategy for Climate Change.* London: The Stationery Office (TSO). (published 26 June 2009)

DECC, 2009b. *The UK Renewable Energy Strategy – Cm7686.* London: TSO. (Published 26 June 2009)

*Ecodesign of Energy-Using Products Regulations, 2007.* SI 2007/2037, London: HMSO. Available at www.opsi.gov.uk/si/si2007/uksi_20072037_en_1 (accessed 26 January 2010)

*Environmental Noise (England) (Amendment) Regulations 2009.* SI 2009/1610. London: HMSO. Available at www.DEFRA.gov.uk/corporate/consult/environoice-regs2006/regulations.pdf (accessed 26 January 2010)

EC (European Commission), 1992. *Council Directive 92/43/EEC of 2 May 1992 on the conservation of natural habitats and of wild fauna and flora.* (Directive COM) (Updated 21 November 2008) Brussels: European Commission. Available at http://ec.europa.eu/environment/nature/legislation/habitatsdirective/index_en.htm (accessed 16 January 2010)

EC, 2006. *The Renewed EU Sustainable Development Strategy.* (Council of The European Union (EU SDS) renewed Strategy 10917/06). Brussels: EU. (Published

2006). Available at http://register.consilium.europa.eu/pdf/en/06/st10/st10917.en06. pdf (accessed 17 January 2010)

EC, 2008a. *Communication from the Commission to the European Parliament, the Council, the European Economic and Social Committee and the Committee of the Regions – Public procurement for a better environment.* COM (2208) 400 final. Brussels: EU. Available at http://eur-lex.europa.eu/LexUriServ/LexUriServ.do?uri=COM:2008:0400:FIN:EN:HTML (accessed 19 January 2010)

EC, 2008b. *Proposal for a Directive of the European Parliament and of the Council on the energy performance of buildings (recast).* (COM (2008) 780 final). Brussels: EU. Available at http://eur-lex.europa.eu/LexUriServ/LexUriServ.do?uri=COM:2008:0780:FIN:EN:PDF (accessed 20 August 2010)

EC, 2009. *Adapting to Climate Change: Towards a European framework for action.* (White Paper COM (2009) 147 final). Brussels: EU (Published 2009). Available at http://eur-lex.europa.eu/LexUriServ/LexUriServ.do?uri=COM:2009:0147:FIN:EN:PDF (accessed 17 January 2010)

European Council, 2009. *Directive 2009/125/ec of the European Parliament and of the Council of 21 October 2009 establishing a framework for the setting of ecodesign requirements for the energy-related products (recast)* Official Journal of the European Union 31 October 2009. Brussels: EC. Available at http://eur-lex.europa.eu/LexUriServ/LexUriServ.do?uri=OJ:L:2009:285:0010:0035:en:PDF (accessed 18 January 2010)

Fairchild, W. B., 1949. 'Renewable Resources: A World Dilemma: Recent Publications on Conservation', *Geographical Review*, vol 39, no 1, pp86–98. Available at www.jstor.org/stable/211159 (accessed 26 January 2010)

HM Government with Strategic Forum for Construction, 2008. *Strategy for Sustainable Construction.* Department for Business Innovation and Skills. London: TPO. Available at www.berr.gov.uk/whatwedo/sectors/construction/sustainability/sustainablestrategy/page48779.html (accessed 26 January 2010)

HM Government, 2009. *World Class Places: The Government's strategy for improving quality of place.* London: TSO. Available at www.communities.gov.uk/documents/planningandbuilding/pdf/1229344.pdf (accessed 26 January 2010)

HMSO, 2010. Approved Document G. *Sanitation, Hot Water Safety and Water Efficiency.* London: HMSO. Available at www.planningportal.gov.uk/uploads/br/100312_app_doc_G_2010.pdf (accessed 29 October 2010)

IUCN (The World Conservation Union), 1980. *Caring for the Earth. (World Conservation Strategy).* Extract. Available at www.ciesin.columbia.edu/IC/iucn/CaringDS.html (accessed 16 January 2010)

IUCN, 1991. *Caring for the Earth: A Strategy for Sustainable Living.* Gland, Switzerland: IUCN. Available at www.iucn.org (accessed 16 January 2010)

Littlefair, P., 1991. *Site Layout Planning for Daylight and Sunlight: A guide to good practice.* (BRE Report 209) London: BRE (Reprinted 2005). Available at www.brebookshop.com/details.jsp?id=321450 (accessed 26 January 2010)

Local Government and Public Involvement in Health Act, 2007. London: HMSO. Available at www.opsi.gov.uk/acts/acts2007/ukpga_20070028_en_1 (accessed 19 January 2010)

Malthus, T., 1798. *An Essay on the Principle of Population.* [e-book] St Paul's Church Yard, London. Available at www.esp.org/books/malthus/population/malthus.pdf (accessed 16 January 2010)

Meadows, D. H., Meadows, D. W., Randers, J. and Behrens III, W. I., 1972. *The Limits to Growth.* London: Pan Books

Mesarovic, M. and Pestel, E., 1975. *Mankind at the Turning Point.* (The Second Report to the Club of Rome). London: Hutchinson

MEA (Millennium Ecosystem Assessment), 2005. *Overview of the Millennium Ecosystem Assessment*. 2005. Available at www.maweb.org/en/About.aspx#2 (accessed 28 October 2009)

*Site Waste Management Plans Regulations, 2008*. SI 2008/314, London: TSO. Available at www.legislation.gov.uk/uksi/2008/314/contents/made (accessed 20 August 2010)

Stern, N., 2006. *The Economics of Climate Change*. Available at www.hm-treasury.gov. uk/stern_review_report.htm (accessed on 30 March 2010)

*This Common Inheritance: Britain's Environmental Strategy*, 1990. (White Paper) London: HMSO

UN (United Nations General Assembly), 1982. *37/7 World Charter for Nature*, UN General Assembly. 48th Plenary Meeting 28 October 1982. Available at www.un.org/ documents/ga/res/37/a37r007.htm (accessed 20 August 2010)

UNDESA (UN Department of Economic and Social Affairs), Division of Sustainable Development, 1992. *Agenda 21 Annex 1 Rio Declaration on Environment and Development*. Available at www.un.org/documents/ga/conf151/aconf15126-1annex1. htm (accessed 26 January 2010)

UNEP (United Nations Environment Programme), 1972. *Declaration of the United Nations Conference on the Human Environment*. Stockholm, 5–16 June 1972. UNEP. Available at www.unep.org/Documents.Multilingual/Default.asp?DocumentI D=97&ArticleID=1503&I=en (accessed 18 September 2009)

UNEP, undated. *Medium Term Strategy 2010–2013: Environment for Development*. UNEP. Available at www.unep.org/PDF/FinalMTSGCSS-X-8.pdf (accessed 16 January 2010)

UNEP, Division of Technology, Industry and Economics, Unknown Date. *International Source Book on Environmentally Sound Technologies (ESTs) for Municipal Solid Waste Management (MSWM)*. (International and Environmental Technology Centre (IETC), IETC Technical Publication Series Number 6). Overview and contents at: www.unep.or.jp/ietc/estdir/pub/msw/ (accessed 19 October 2009)

*Urban Waste Water Treatment (England and Wales) Regulations, 1994*. SI 1994/2841, London: HMSO

*Water Supply (Water Fittings) Regulations, 1999*. SI 1999/1148, London: HMSO

WHO Europe, 2009. *Night Noise Guidelines for Europe*. Denmark: 2009. Available at www.euro.who.int/__data/assets/pdf_file/0017/43316/E92845.pdf (accessed 26 January 2010)

Zaninetti, J.M., 2009. *Sustainable Development in the USA*. Summary. Indianapolis, IN: Wiley.

# 1.3
# Assessment Methodologies

## Introduction

The ability to predict and measure the success of a building or development against pre-set environmental or sustainability criteria is a challenge that has been tackled with varying degrees of success. Before the UK Building Research Establishment (BRE) launched its Environmental Assessment Method (BREEAM) in 1990 the only methods available for predicting the performance of buildings were those designed to calculate specific criteria such as thermal loads, energy or water consumption or life cycle impacts of product manufacture. These models were designed primarily to size equipment, predict running costs or optimize the manufacturing process.

In the early 1960s concerns over the limitations of raw materials and energy resources led to the development of the first exercises in life cycle assessment, combining projections of cumulative energy and raw material use. One of the first papers on the subject was presented by Harold Smith on chemical intermediates and products at the World Energy Conference in 1963 (quoted in Vigon, 1994). In 1969 Coca Cola pioneered life cycle inventory analysis with a study on drinks cans, which became known as Resource and Environmental Profile Analysis (REPA) in the US and Ecobalance in Europe, that is the process of quantifying the resource use and environmental releases of products.

This process has been adapted by BRE and others in the development of an integrated measure of environmental impact of construction materials, such as the Ecopoint environmental profiling system used in the Green Guide to Specification[1] that is used as part of both BREEAM and the Code for Sustainable Homes (see Chapter 3.12 Materials Specification).

Over the years there has been a particular focus on the development of environmental assessment tools for the construction sector, catering either for different building types or generic development. Most are designed for particular countries, taking into account different legislative requirements, cultures and practices. Composition varies, with some combining primarily environmental criteria, whilst others bring in socio-economic impacts.

There are a number of protocols that can be used to assess complex projects at the masterplan or concept design stage. In the UK the BRE has recently launched BREEAM Communities which has evolved from and integrates with regional tools such as the South East of England Development Agency's (SEEDA) Sustainability Checklist (see Other UK protocols below), designed for use during the planning stage of a project and including socio-economic criteria.

In Europe larger construction projects are required by law to submit an Environmental Impact Assessment (EIA) as an Environmental Statement (ES) with planning applications (refer to Chapter 1.4 Environmental Impact Assessment). These tend to include both qualitative and quantitative measurements of impact, the scope of which is normally agreed with the local planning authority in advance. Traditionally these have included chapters on design quality, visual impact/townscape, transportation, archaeology, built heritage, wind environment, sunlight/daylight, light pollution, electronic interference, flight paths, air quality, noise and vibration, soil and groundwater, ecology, construction and socio-economic impacts (see Box 1.12 in Chapter 1.4 Environmental Impact Assessment). Recently sustainability has typically been incorporated as a separate chapter in an ES or, when specified by the local authority, as a separate statement with the planning application. Similarly most local authorities require separate Energy Statements. These are frequently looking for a level of performance in excess of that required by Building Regulations.

Although an EIA provides a useful framework for recording predictions of the environmental and socio-economic performance of a development, its primary function is to enable the planning authority to judge whether a development presents unacceptable risks to the environment, whether it is compatible with local and regional planning frameworks and what benefits it brings to the community.

There are similar requirements in the US and elsewhere which will be discussed in Chapter 1.4.

## Building Research Establishment Environmental Assessment Method (BREEAM)

In order to provide a method by which environmental impacts can be combined in a single measure, BRE developed BREEAM in 1990. BREEAM provides a protocol for integrating a diverse range of environmental parameters into a single measurement tool. Originally designed for offices only, BREEAM now covers a variety of different scenarios as set out in Box 1.9.

---

### Box 1.9 BREEAM assessment protocols

*For new buildings in UK:*

BREEAM Offices
BREEAM Retail (e.g. shopping centres, supermarkets, banks, shops)
BREEAM Education (e.g. primary and secondary schools, further education colleges)
BREEAM Higher Education
BREEAM Healthcare:

- Hospitals
- GP surgeries
- Health centres and clinics

BREEAM Fire Stations

BREEAM Multi-residential (e.g. care homes, student accommodation)

Code for Sustainable Homes (England, Wales and Northern Ireland)

Ecohomes (Scotland)[a]

BREEAM Other Buildings:

- Bespoke for Single Building
- Bespoke for Multiple Buildings
- BREEAM Prisons
- BREEAM Courts
- BREEAM Data Centres
- Tailored protocols for specific clients, e.g:
  - Whitbread Premier Inn
  - Forestry Commission visitor centres

## *New Communities in the UK:*

BREEAM Communities (Residential and Multi-use in England):

- Bespoke Communities

## *Existing buildings in UK:*

BREEAM In Use

EcohomesXB

BREEAM Healthcare XB

BREEAM International:

- BREEAM Europe[b]
  - Offices
  - Retail
  - Industrial
- BREEAM Gulf
- BREEAM-NL (Netherlands) (see www.breeam.nl/)
- Bespoke International, e.g. Toyota Showrooms
- Bespoke Communities International[c]

Notes: Refer to the BREEAM website at www.breeam.org/index.jsp

[a] BRE is currently developing a protocol for refurbishing dwellings.

[b] 'Europe' for this purpose is defined as the European Union, the EFTA countries (Switzerland, Iceland), current candidate countries (Turkey, Croatia, Macedonia) and Albania, Belarus, Bosnia and Herzegovina, Moldovia, Montenegro, Serbia and Ukraine.

[c] BREEAM-ES (Spain) is under development at the time of writing with Residential and Commercial (combined offices, retail and industrial) the first to be launched and Communities and In-Use versions to follow.

BREEAM was not designed specifically as a measure of sustainability, nor as a way of combining all of the issues covered by an EIA. In fact, although it includes many of the elements of both, it does not cover either comprehensively. It does not include many of the socio-economic impacts associated with long-term sustainability nor does it include factors such as wind environment or overshadowing of neighbouring buildings that may form part of an EIA. It also includes factors which are arguably only loosely related to sustainability, many of which appear under the heading of 'Health and Wellbeing', such as zoning of temperature controls and sound transmission through walls and floors.

BREEAM was designed initially to provide a measure of the quality of the environmental design of buildings with a view to increasing standards across a range of issues from the global to the local. So as well as tackling carbon dioxide emissions from operational energy, transport, construction and (indirectly) materials, it deals with greenhouse gases from refrigerants, insulation and other materials. It also covers local pollution emissions into the air from boilers and into groundwater due to run-off. Local issues are also tackled through assessment of waste recycling strategies, recycling of aggregates and a series of credits that reward the use of brownfield land and enhancement of ecological value.

It could be argued that a building that is not pleasant and comfortable to work in is not sustainable because it may result in low productivity and become unpopular with occupants and even be unusable. BREEAM and its offshoots reward good design through a series of credits dealing with health, well-being and comfort. For non-residential buildings in moderate climates, a well-designed naturally ventilated building that makes maximum use of daylighting will gain credits under a variety of headings. It is no accident that in 2006 the Scottish Natural Heritage headquarters in Inverness achieved the highest BREEAM score to that date (refer to case history in Chapter 3.7 Computer Simulation of Building Environments). It is a fully naturally ventilated office building with excellent daylighting and extensive use of sustainable and recycled materials.

The materials used in the construction of a building are assessed in terms of their environmental impact, through a process that has been developed by BRE and published online as the Green Guide to Specification.[2] This assigns ratings to a large database of composite building elements based on a review of their impacts under headings such as embodied greenhouse gases, water consumption, emissions, etc. Also assessed is the environmental management of the supply chain, including such established certification systems as the FSC (Forest Stewardship Council).

BREEAM does not assess operational impacts alone but also construction issues, including a commitment to 'considerate' contracting[3] and recording and reporting on construction impacts, including waste management and recycling, pollution control and resource use.

Although each of the BREEAM protocols has been tailored for a specific building type, there are key credits, such as land use and ecology that they have in common. Assessors have to be licensed for specific protocols – BRE provide lists of licensed assessors on their website.

BRE introduced major changes to the UK protocols in 2008. Following the example set by the Code for Sustainable Homes (CSH) a number of changes were made which borrow from the US scheme Leadership in Energy and Environmental Design (LEED). First, a requirement for post construction assessment was introduced, with formal certificates not being awarded until the completed building is validated. Second, a series of mandatory credits were set; which means that the assessment fails if these are not achieved. Third, a link was established between the performance in energy and water credits and the final rating. Also 10 per cent of the final score is allocated to 'innovation', comprising measures that are not covered by other BREEAM Credits and that have a measurable environmental benefit that can be justified to a judging panel appointed by BRE. Finally, a new rating of 'Outstanding' was devised for BREEAM that sits above Excellent, Very Good, Good and Pass; certain minimum scores are required for specific credits to achieve an Outstanding rating. Outstanding corresponds to a score of 85 per cent or greater compared to 70 per cent for Excellent, 55 per cent for Very Good, 45 per cent for Good and 30 per cent for Pass.

BREEAM Communities was launched in 2009 as the first UK certification scheme designed for the planning application stage of multi-building developments. The protocol is designed for use with either residential or mixed-use developments in England. Other developments, including non-residential projects, developments outside England or those which are defined as neither new nor regeneration, require a tailored bespoke protocol.

The scope of BREEAM Communities is different from other BREEAM protocols in that it is designed for use with a development comprising different types of building and includes socio-economic issues as well as environmental issues that are to be considered at the planning application stage of a project. All issues are to be assessed for the development as a whole, although there is also a requirement for each building to be assessed under the appropriate BREEAM protocol (CSH for dwellings). Issues covered include climate change mitigation and adaptation, place shaping, community consultation and involvement, ecological enhancement and biodiversity action plan, options for sustainable transport, sustainable use of resources and addressing local and regional economies. The scheme is designed to integrate with Sustainability Checklists developed by the Regional Development Agencies (see Other UK protocols) and include weightings to reflect the importance of key criteria in different regions. BREEAM Communities allows two stages of certification, with an Intermediate Certificate being awarded at the outline planning stage and a Final Certificate at the detailed planning stage.

BREEAM in Use replaced the little-used BREEAM Existing Offices in 2009. The protocol is divided into an Asset Rating (inherent issues), Building Management Rating and Organizational Effectiveness (operational issues). These cover many of the same issues as a design assessment, such as energy and greenhouse gas emissions, water consumption, waste management, air quality, noise and lighting, but also property and fire protection measures. A commitment to a BREEAM in Use assessment is a requirement for an Outstanding BREEAM rating at the design and construction stage.

# Code for Sustainable Homes (CSH)

CSH is a 'spin-off' from the BREEAM stable that started life as EcoHomes but is now owned by the Department for Communities and Local Government (CLG).[4] Since 31 March 2008 it has been a requirement for all new dwellings to have a CSH certificate under the Home Information Pack (Amendment) Regulations 2008. However, there is no requirement to achieve a given rating under the scheme, unless one is specified by the funder or planning authority. For example, for public funding for social housing in England, Wales and Northern Ireland a level 3 rating is required. This does not apply to Scotland where, at the time of writing, Ecohomes still applies.

CSH covers many of the same issues as BREEAM but uses a different rating system that runs from level 1 to 6, rather than Pass to Outstanding.

Table 1.1 is extracted from the Technical Guide for Version 2 of the CSH (CLG, 2009b) setting out the scope for the assessment process. The mandatory

**Table 1.1** *Summary of environmental categories and issues*

| Categories | Issues |
| --- | --- |
| Energy and $CO_2$ emissions | Dwelling emission rate (M) Building fabric<br>Internal lighting<br>Drying space<br>Energy labelled white goods<br>External lighting<br>Low or zero carbon (LZC) technologies<br>Cycle storage<br>Home office |
| Water | Indoor water use (M)<br>External water use |
| Materials | Environmental impact of materials (M)<br>Responsible sourcing of materials - basic building elements<br>Responsible sourcing of materials - finishing elements |
| Surface Water Run-off | Management of Surface Water Run-off from developments (M)<br>Flood risk |
| Waste | Storage of non-recyclable waste and recyclable household waste (M)<br>Construction waste management (M)<br>Composting |
| Pollution | Global warming potential (GWP) of insulants<br>$NO_x$ emissions |
| Health and Well-being | Daylighting<br>Sound insulation<br>Private space<br>Lifetime homes (M) |
| Management | Home user guide<br>Considerate constructors scheme<br>Construction site impacts<br>Security |
| Ecology | Ecological value of site<br>Ecological enhancement<br>Protection of ecological features<br>Change in ecological value of site<br>Building footprint |

(M) denotes issues with mandatory elements.

**Table 1.2** *Code levels for mandatory minimum standards in $CO_2$ emissions*

| Code level | Minimum percentage reduction in dwelling emission rate over target emission rate |
| --- | --- |
| Level 1 (*) | 10 |
| Level 2 (**) | 18 |
| Level 3 (***) | 25 |
| Level 4 (****) | 44 |
| Level 5 (*****) | 100 |
| Level 6 (******) | 'Zero Carbon Home' |

credits must be achieved as a minimum to obtain a Code Level 1 score. The energy and water consumption credits are awarded on the basis of standards for Code Levels as set out in Tables 1.2 and 1.3.

The dwelling and target emission rates (DER, TER) are those determined for Building Regulations compliance as set out in Approved Document L1A: Conservation of fuel and power in new dwellings (ADL1A).[5] A 'zero carbon home' is defined as a dwelling that not only has a DER of zero (regulated emissions) but also has no carbon emissions associated with appliances and cooking (unregulated emissions), averaged over the year. The Standard Assessment Procedure (SAP) is currently being modified to incorporate the necessary calculations for zero carbon. According to a recent impact assessment on zero carbon homes (CLG, 2009a) developers will be expected to achieve zero carbon from 2016 by a combination of measures, at least 70 per cent of which will be through savings in 'regulated emissions' and the remainder, plus unregulated emissions, will be from a prescribed list of 'allowable solutions'. Consultation on amendments to Parts L and F of the Building Regulations closed in September 2009, including a commitment to match the Code Level 3 requirement for a DER of 25 per cent below the 2006 TER.

Amendments to Part G of the Building Regulations coming into force in 2010 require a maximum water consumption of 125 litres per person per day, including 5 litres for outdoor use and hence match the Code Level 1 and 2 requirements in the table below.

**Table 1.3** *Code levels for mandatory maximum standards in indoor water consumption*

| Code level | Maximum indoor water consumption in litres per person per day |
| --- | --- |
| Level 1 (*) | 120 |
| Level 2 (**) | 120 |
| Level 3 (***) | 105 |
| Level 4 (****) | 105 |
| Level 5 (*****) | 80 |
| Level 6 (******) | 80 |

# Other UK protocols

## Regional sustainability checklists

BREEAM Communities links with regional sustainability checklists, the latest versions of which can be found at the following websites, (relevant to English Regional Development Agencies only):

- www.southeast.sustainability-checklist.co.uk/
- www.sustainabilitywestmidlands.org.uk/projects/?/Sustainability+Planning+Checklist/128
- www.sdchecklist-northwest.org.uk/
- www.checklistsouthwest.co.uk/
- www.eastofenglandchecklist.co.uk/checklist

The checklist for the northeast was still under development at the time of writing. More information on this can be found at www.strategyintegrationne.co.uk/page.asp?id=133.

These checklists have evolved with support from BRE from the pioneering South East of England Development Agency (SEEDA) online checklist launched in 2003. They now follow approximately the same structure as BREEAM Communities (see above).

## Rating schemes

### Lifestyle Indicator

UK consultants Best Foot Forward (BFF) has developed a Lifestyle Indicator available for self-assessment free at: www.old.bestfootforward.com/tools/. Using broad brush lifestyle questions, it gives an approximate indication of personal carbon footprints in tonnes of $CO_2$ per annum and an ecological footprint in global hectares.

### Envirowise Indicator

BFF have also developed Envirowise Indicator, which is available at: www.envirowise.gov.uk/uk/Our-Services/Tools/Envirowise-Indicator.249257.html. This is an online assessment tool for use by UK businesses, enabling the input of price and consumption data for different types of premises. Outputs give totals of carbon and cost and enable the impacts of various measures to be checked.

### CEEQUAL

The Civil Engineering Environmental Quality Assessment and Award Scheme (CEEQUAL)[6] was developed by Crane Environmental for the Institution of Civil Engineers (ICE) for use in assessing projects such as roads, railways, airports, coast and river works, water treatment works, power stations, etc. Version 4 was released in December 2008 and a further version for term projects, such as highway maintenance commissions, is under development at the time of writing. The scope is similar to BREEAM and other environmental rating tools covering:

- project management;
- land use;
- landscape;
- ecology and biodiversity;
- historic environment;
- water resources and environment;
- energy and carbon in use;
- material use (including embodied impacts);
- waste management;
- transport;
- effects on neighbours;
- relations with local community and other stakeholders.

Assessments have to be carried out by suitably trained assessors and verified by independent third parties. There are a number of award types that currently include:

- Whole Project Award (with or without an Interim Client and Design Award);
- Design only Award;
- Client and Design Award;
- Construction only Award;
- Design and Construction Award.

Ratings are assigned as Pass, Good, Very Good or Excellent.

## International protocols

### Rating tools

#### LEED (US)

LEED (Leadership in Energy and Environmental Design) has the same roots as BREEAM but is administered by the US Green Building Council and has fewer versions than BREEAM.[7] Although the headings are different to BREEAM the issues covered are similar, but with obviously a strong bias toward US standards and legislation. There are versions of LEED for New Construction commercial buildings (LEED NC), Existing Buildings, Commercial Interiors, Core and Shell, Schools and Retail. LEED for Homes, which was launched in 2008, is designed for application to housing developments of all sizes, ranging from individual houses to high density apartment schemes. Unlike CSH, however, LEED for Homes is purely voluntary and, because of its cost, has not so far been extensively adopted. There is also a version for Neighbourhood Development that was launched in 2009 and has similarities to the BREEAM Communities scheme referred to above.[8]

The LEED NC protocol awards points against the following environmental performance issues:

- Sustainable sites: Erosion and sedimentation control (pre-requisite), site selection, density, connectivity, use of brownfield sites, public transport, alternative fuels, parking, site disturbance, provision of open space, efficient use of land, storm water design, heat island effect and light pollution.
- Water efficiency: Landscaping, air conditioning, wastewater technologies and demand reduction.
- Energy and atmosphere: Commissioning, minimum performance and CFC reduction (pre-requisites), optimum performance, renewables, additional commissioning, ODP emissions, measurement and verification, green power.
- Materials and resources: Storage and collection of recyclables (pre-requisite), building re-use, construction waste, resource re-use, recycled content, local sourcing, renewable materials and certified wood.
- Indoor environmental quality: Minimum indoor air quality (IAQ) and environmental tobacco smoke control (pre-requisites), outdoor air monitoring, increased ventilation, construction air quality, low emitting materials, indoor pollutants, controllability, thermal comfort, daylight, views.
- Innovations: innovative measures to enhance the environmental performance of the building.

Ratings are awarded on a scale from Certified to Platinum through Silver and Gold. Certification is based on an assessment on the completed building.

Versions of LEED have been introduced for use in Canada,[9] India,[10] Mexico[11] and Brazil.[12]

## SpEAR (UK)

Arup's Sustainable Project Appraisal Routine (SpEAR) is a method of visualizing potential sustainability performance under four headings portrayed on a rose diagram.[13]

- Environmental: air quality, project siting, water quality, ecological conservation, infrastructure and site works, transport options, heritage and regulations and guidance.
- Natural resources: materials, water use, energy use, land use and waste hierarchy.
- Societal: building user satisfaction, health and safety, form and space, access, amenity, cultural heritage.
- Economy: Local development, affordability, employment/skills base, competition effects and viability.

The protocol can be applied to masterplans and building designs for any kind of development anywhere in the world, but it is not in the public domain and can only be used by Arup personnel.

## NAHB tool (US)

The US National Association of House Builders launched a set of Model Green Home Building Guidelines in 2008 and, in partnership with the International Code Council, a National Green Building Standard (ANSI ICC 700-2008) including a Green Scoring Tool which can be used to rate the design of new and refurbished dwellings.[14] The tool covers energy, water, resource efficiency, lot and site development, indoor environmental quality and home owner education. The ratings awarded are bronze, silver, gold and emerald. The energy points are allocated on the basis of enhanced performance in comparison with the Energy Star[15] target and vary from a 15 per cent improvement for bronze to a 60 per cent improvement for emerald.

## Green Globes (Canada/US)

Not to be confused with Green Globe 21 (see EC3 Global below), Green Globes grew out of a version of BREEAM that was developed for existing buildings in Canada by BRE.[16] There are a number of different versions of Green Globes that have evolved separately for Canada and the US. In Canada the version for new buildings is known as Green Globes Design and is owned and operated by ECD Jones Lang Lasalle. These all involve online data input and third party verification for certification. The protocol covers the same headings as the US version, that is project management, site ecology and flood risk, energy, water, resources (materials and waste), emissions and effluents and indoor environment (ventilation, pollutants, heat, light and noise). The US scheme for new buildings is known as Green Globes NC (new construction) and is owned and operated by the Green Buildings Initiative (GBI).

The schemes for existing buildings have also gone their separate ways, with the scheme in Canada having dropped the Green Globes branding entirely and adopted the name of its current owner and operator: the Building Owners and Managers Association (BOMA). BOMA Building Environmental Standards (BESt) are available for office buildings, shopping centres, open retail and light industrial units. Instead of Globes these protocols have adopted four levels of award. The scheme for apartment buildings has retained the Green Globes title however. In the US the scheme is known as the Green Globes Continual Improvement Assessment for Existing Buildings (CIEB). It can only be used for commercial buildings, is run by GBI and retains the Globe rating system, but with a maximum of four Globes available. There is also a version for existing office buildings in the UK, run by GEM UK. The scope of all of these schemes for existing buildings covers energy, indoor environment (air quality, lighting and noise), resources (waste reduction and recycling), emissions and effluents, environmental management and water.

Green Globes Fit-Up covers similar territory to the Canadian NC scheme but adapted to the interior fit-out of new or existing buildings; covering project management, energy, water, resources (materials and waste), emissions and effluents and indoor environment (ventilation, pollutants and comfort).

Green Globes provides a separate protocol for Building Emergency Management Assessment (BEMA) which covers hazard identification, risk assessment, mitigation, preparedness, response plan and availability of resources and recovery plan.

A scheme is currently being developed under the Continental Automated Buildings Association (CABA) banner 'powered by Green Globes' for the evaluation of intelligent buildings in terms of a 'Building Intelligence Quotient' (BIQ™).

### GBTool (Canada)

GBTool was developed initially by Natural Resources Canada as part of the Green Building Challenge 2000 process. A number of versions have been trialled by national teams globally and compared at a series of conferences. The project is now administered by the International Initiative for a Sustainable Building Environment (iiSBE),[17] however, at the time of writing the download site was not accessible. The tool is intended primarily for research purposes and the spreadsheets have been adapted by national collaborators to suit local legislation and practice.

### SICES (Mexico)

The Mexico Green Building Council (MGBC) is currently developing an environmental assessment tool that will focus initially on commercial buildings and low income housing.[18]

### AQUA (Brazil)

The Brazilian *Alta Qualidade Ambientale* (AQUA) system is operated by the Vanzolini Foundation at the Polytechnic University of Sao Paulo.[19]

AQUA is based in part on the French HQE scheme (see below) and covers eco-construction, comfort, eco-consumption and quality of indoor environment and water supply. Schemes are available for offices, school buildings and hotels.

### EC3 Global (Australia)

EC3 Global is an Australia-based environmental consultancy which has launched a number of rating systems that can be used to support organizations attempting to reduce their environmental impact.[20] Their Green Globe 21 product is based on the UN Agenda 21 initiative adapted for the tourism sector and can be used for organizations or resorts/buildings. There are four 'Standards': a Company Standard, a Standard for Communities, an International Ecotourism Standard and a Design and Construct Standard. Organizations can become affiliates by expressing an interest in Green Globe 21, go through benchmarking only or become fully certified.

EC3 Global's Earthcheck Smart Buildings 3i comprises an environmental compliance checking module called EnviroCheck and greenhouse gas, water and waste benchmarking systems, with third party certification.

### Green Star (Australia)

The Australian Green Star rating system shares a lot of the features with BREEAM and LEED, on which it was based.[21] It was developed by the Green

Building Council of Australia (GBCA) and the Office Design and As Built protocol is currently in its third version. Recent additions to the suite of protocols can be used for education facilities, multi-unit residential buildings, healthcare facilities, retail centres and office interiors. Protocols for industrial units, mixed-use developments and convention centres are currently being piloted. As one might expect, the scope is similar to BREEAM, covering management issues, indoor environmental quality, energy, transport, water, materials, land use and ecology, emissions and innovation. Ratings are allocated on a star system, with up to six stars available. Only projects awarded four stars or more can apply for formal certification, with four stars being considered to represent 'best practice', five stars 'Australian excellence' and six stars 'world leadership'.

Green Star schemes are also available in New Zealand[22] – with separate versions for new office, industrial and education buildings and a version for existing buildings on the way – and South Africa[23] – new office buildings only, with a pilot version for retail now available.

## NABERS

The National Australian Building Environmental Rating Scheme (NABERS) is a performance-based rating system for existing offices, homes, hotels and retail buildings using measured operational parameters.[24] NABERS was launched in 2000 and is managed by the New South Wales Department of Environment, Climate Change and Water and has separate protocols for energy use, water consumption, waste and indoor environment, with further protocols being rolled out in due course for refrigerants, stormwater run-off, sewage, landscape diversity, transport and occupant satisfaction. NABERS Energy for offices has replaced the Australian Building Greenhouse Rating (ABGR) and can be used for new office buildings. NABERS Energy for data centres was under development at the time of writing.

### BASIX (Australia)

The Building Sustainability Index (BASIX) has been developed by the New South Wales (NSW) Government to provide a simple tool for targeting thermal comfort, energy and water saving for homes.[25] Meeting these targets is mandatory for all development applications in NSW. It is an online self-assessment tool that enables anyone to print out a certificate for submission to the Council if it can be demonstrated that targets are met. This sets out the features that have been committed to and is followed through by an inspection of the completed dwelling.

### HK-BEAM (Hong Kong)

The Hong Kong Building Environmental Assessment Method (HK-BEAM),[26] currently operated by the BEAM Society, was launched in 1996 for new and existing office buildings. Since that time some 200 buildings have been certified under the scheme, not only in Hong Kong but also Beijing, Shanghai and

Shenzhen. A scheme for high rise residential buildings was launched in 1999 and in 2003 the schemes were revised so that a wide range of building types and mixed-use developments could be assessed from flexible protocols for new and existing buildings. HK-BEAM Plus was published in November 2009 and has been modified in line with other protocols such as BREEAM and LEED including identifying certain criteria as pre-requisites for certification.

The scope covers site issues, materials and waste, energy, water and effluent, indoor environment (health and safety, indoor air quality, ventilation, comfort, amenities) and enhancements/innovations. The scheme has been designed to take account of the humid sub-tropical climate prevailing in Hong Kong. The classifications range from bronze to platinum. Currently all assessments are carried out through the Business Environment Council, although consideration is being given to setting up a network of suitably trained licensed BEAM Assessors.

## CEPAS (Hong Kong)

The Comprehensive Environmental Performance Assessment Scheme (CEPAS) was launched in 2006 by the Hong Kong Special Administrative Region (HKSAR) Buildings Department.[27] Developed by Arup HK, points are awarded under a series of headings:

- indoor environmental quality;
- building amenities;
- resource use;
- loadings;
- site amenities;
- neighbourhood amenities;
- site impacts;
- neighbourhood impacts.

Bonus points are awarded for innovation under each heading. Each heading covers a number of criteria, which depend upon the stage of the assessment. Different protocols are available for pre-design, design, construction and five-yearly operation stage assessments. Overall ratings are determined as bronze, silver, gold or platinum awards.

## Evaluation Standard for Green Building (China)

This standard was first implemented in June 2006 under the National Project Construction Standardisation Information Network. It can be used to evaluate dwellings, public buildings, offices, shopping malls and hotels and covers energy-saving and use, water saving and water resource utilization, materials and resource use, indoor environmental quality and operations management.[28]

## BCA Green Mark (Singapore)

The Building and Construction Authority (BCA) Green Mark Scheme was launched in 2005 and has separate versions for residential, non-residential, existing, overseas, office interiors, landed houses, infrastructure, new and existing parks and district projects.[29] The protocol for non-residential buildings was revised in 2008 and covers energy, water, environmental protection, indoor environmental quality and 'other green features and innovations'. The energy issues covered have separate criteria for air-conditioned and non air-conditioned buildings and a series of criteria covering lighting, car park ventilation, lifts and escalators, other energy efficient features resulting in predicted savings and bonus points for renewable energy. Water efficiency issues include points for water efficient fittings, irrigation and cooling towers, metering and leak detection. Environmental protection covers sustainable construction (including low impact materials), 'greenery', environmental management (including the use of a Construction Quality Assessment System – CONQUAS), access to public transport and environmentally friendly refrigerants. Indoor environmental quality covers thermal comfort, noise, indoor pollution and use of high frequency ballasts. Points are also awarded for the use of innovative green features such as rainwater harvesting and pneumatic waste conveying. Assessments have to be carried out by BCA Assessors, with awards ranging from certified, gold, gold[Plus] to platinum, with the last two qualifying for the Green Mark Incentive Scheme (GMIS), with associated Government funding designed to promote sustainable design.

## CASBEE (Japan)

The Comprehensive Assessment System for Built Environmental Efficiency (CASBEE) commenced development in 2001 and is currently operated by the Japan GreenBuild Council (JaGBC).[30] Different schemes are available for:

- new construction;
- existing buildings;
- renovation;
- heat island;
- urban development;
- urban area and buildings;
- home (detached house).

The protocol for new construction is divided into issues that broadly fall under the headings of Quality (Q) and Load Reduction (LR); with Quality covering indoor environment issues, such as noise, temperature, light and air quality; 'quality of service' covering functionality, amenity, maintenance, earthquake resistance, service life, reliability, flexibility and adaptability; whilst 'outdoor environment on site' includes ecology, landscape, townscape, local character and comfort issues.

Load Reduction includes: energy issues such as loads, use of natural energy, efficiency of building services and efficient operation; resources, including water conservation, sustainable and low pollution materials; off-site environment including consideration of global warming, local environment (heat island, pollution, load on local infrastructure); and surrounding environment (noise, vibration, odour, wind impact, overshadowing and light pollution).

### GRIHA (India)

The Green Rating for Integrated Habitat Assessment (GRIHA) has been developed by The Energy Resources Institute (TERI) of India.[31] It has been designed to be adaptable for new commercial (offices, retail, hotels, hospital and healthcare buildings) covering pre-construction, planning and construction issues. The assessment protocol is divided into four stages:

1   Site selection and planning: including conservation and efficient utilization of resource and health and well-being during construction.
2   Building planning and construction stage: comprising water consumption, energy (end-use, embodied and construction, renewable), recycle, recharge and re-use of water, waste management, health and well-being (including pollution, ozone depletion, air and water quality).
3   Building operation and maintenance: validation of 'green' performance.
4   Innovation: such as alternative transportation, lifecycle cost analysis, or as suggested by client.

Scoring is on a scale from one to six stars. Assessments are carried out by TERI and evaluated by an expert panel.

### TGBRS (India)

TERI has also developed the Teri Green Building Rating System (TGBRS) which is suitable for both new and existing commercial and residential buildings and was launched in 2003, although no information beyond a press release is available on this system on the TERI website.[32] It addresses 'the issue of inefficient use and wastage of precious energy resources'. It also uses a star rating system and is based on site planning, building envelope, building systems, water and waste management and green design practices.

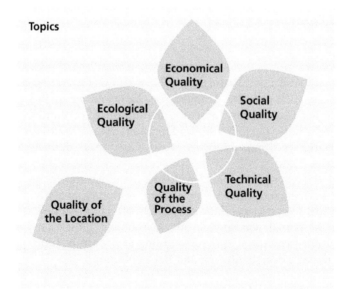

**Topics**

Economical Quality

Social Quality

Ecological Quality

Technical Quality

Quality of the Process

Quality of the Location

**Figure 1.6** *DGNB certification scheme schematic*

*Source:* Image copyright DGNB

## German Sustainable Building Certification

The German Sustainable Building Council (*Deutsche Gesellschaft für Nachhaltiges Bauen* – DGNB) launched a new certification system in 2009 with a programme of auditor training, following piloting on a number of office buildings across Germany.[33] These were assessed under the 'New Construction of Office and Administration Buildings – Version 2008'. The protocol is designed to be easily adapted to other building types and a rolling programme of end-use-specific protocols is planned. The protocol covers some 49 issues under the general headings shown in Figure 1.6.

The rating of bronze, silver or gold is based on the criteria attached to the circle in the figure, whilst the 'Quality of Location' is assessed for information only.

## HQE (France)

*La demarche HQE* (*Haute Qualite Environnementale*) was launched in 2005 and administered by *l'Association HQE*.[34] Current versions are available for offices and educational buildings and hospitals and healthcare facilities. Versions are being developed for commercial centres, hotels, logistics centres and existing buildings. The protocols are divided into two sections:

1 Managing the impacts on the outdoor environment, including:
   • relationship between buildings and immediate environment;
   • integrated choice of construction methods and materials;
   • avoidance of nuisance from construction activities;
   • minimizing energy, water, construction waste, maintenance and repair.

2 Creating a pleasant indoor environment, including:
   • control of temperature, humidity, noise, air quality, odour and water quality;
   • visual attractiveness;
   • hygiene and cleanliness indoors.

Separate protocols are also available for individual dwellings: NF *Maison Individuelle* – HQE[35] and for real estate transactions of individual or groups of homes: NC Housing – HQE.[36]

## qualigreen (France)

*Evaluations* qualigreen is an alternative scheme to HQE, but it avoids the certification process and is therefore claimed to be cheaper and more useful in supporting the design process.[37] It is available through greenlogic, who also provide a tool for assessing embodied carbon which they sell under the *Chantier Carbone* brand.

## Eco-Quantum (The Netherlands)

The Dutch Eco-Quantum life cycle assessment tool is used to determine the environmental performance of construction projects and new energy systems and calculates the efficiency of environmental improvements.[38] It is owned and operated by IVAM, the inter-faculty Department of Environmental Sciences at the University of Amsterdam. It is currently only available in Dutch.

### EcoEffect (Sweden)

The EcoEffect system is an environmental profiling system developed jointly by the Royal Institute of Technology, Stockholm and the University of Gavle.[39] Impacts are evaluated in terms of energy use, material use, indoor environment, outdoor environment and life cycle costs. Instead of using established benchmarks, the system compares various indicators against other buildings in its database and in various formats; for example a plot of internal against external load numbers. The internal load number represents the quality of the internal environment, which could be represented by occupant satisfaction in an existing building; whilst the external load number would be an aggregate of energy and material use. Although this system was first developed in 1999, it has yet to be fully employed as a commercially available tool for construction professionals.

### Ecoprofile (Norway)

Økoprofil (Ecoprofile) is the main Norwegian environmental assessment scheme.[40] It is currently operated by the Norwegian Building Research Institute (SINTEF Byggforsk) and was launched initially in 1994 as 'Environmental Profile'. It is based on predicting performance and impacts under the headings of external environment, resource consumption and indoor air quality.

### BEAT (Denmark)

BEAT is a tool designed to calculate environmental profiles for buildings, building products and materials.[41] It is owned and operated by the Danish Building Research Institute at Aalborg University. The tool uses an editable database comprising energy sources, means of transport, products, building elements and a baseline building, with an output in the form of various environmental profiles, either in the form of tables or graphs. The database can be edited to suit individual requirements, but the default information is for Danish products and buildings.

### PromisE (Finland)

The Finnish Environmental Assessment and Classification System for Buildings (PromisE) is owned and operated by the VTT Technical Research Station.[42] Separate schemes are available for new and existing office, retail and residential buildings, with a similar scope for all buildings, but different weightings applied for each criterion. There are four broad categories of issue covered: the health of users, consumption of natural resources, environmental loadings and environmental risks. The criteria covered recognize the specific issues associated with Finnish building design including chemical emissions from materials (outgassing), moisture control and ventilation. Issues such as management, documentation, service life and monitoring provision are given more prominence than in other protocols.

## Protocollo Itaca (Italy)

*Protocollo Itaca* (Ithaca Protocol) was published in 2009 for the environmental assessment of new or refurbished dwellings.[43] Criteria are assessed within a range from –1 to +5 where 0 represents legal/current practice and 5 represents best practice. Headings covered are site quality, consumption of resources, indoor environmental quality (IEQ) and service quality as follows:

- Site quality: contamination, re-use of existing structures, accessibility to public transport, amenities and infrastructure.
- Resource consumption: embodied energy of materials, thermal properties of fabric, solar protection, heating and cooling energy, renewables; re-use, recycled and local materials, water use, $CO_2$ from materials and operation, wastewater to sewer, rainwater harvesting, heat island effect.
- IEQ: ventilation, radon, temperature, daylight, noise and electromagnetic pollution.
- Quality of service: Building energy management system, operation and maintenance documentation, cycle storage, waste storage, recreation space, home automation and security.

## LiderA (Portugal)

The *Sistema de Avaliacao da Sustentabilidade* (Certification System of Environmentally Sustainable Construction) – LiderA – was developed in 2008 for the evaluation and certification of residential, commercial and tourism developments at the concept, design, construction, operation and renewal stages.[44] The main headings cover local resources, loads, environmental comfort, adaptability, socio-economic issues, environmental management and innovation.

## VERDE (Spain)

*Herramienta VERDE* is owned and operated by the Green Building Council España (GBCE) which provides a certification service for dwellings and office buildings only.[45] Certification can be carried out through accredited assessors only, although preliminary self-assessment is possible. The following issues are covered:

- Place selection: strategies for waste recycling, light pollution.
- Energy and atmosphere: embodied energy, operational energy, renewables, emissions of $NO_x$ and ozone depleting substances.
- Natural resources: water use, rainwater harvesting, grey water, materials impacts, use of recycled materials and minimization of waste.
- Interior space quality: outgassing, ventilation, draught, temperature and humidity, daylighting, glare, quality of light, noise.
- Service quality: efficiency of space planning, efficient and localized control of lighting and thermal systems, adaptability, maintenance plans, BEMS.
- Socio-economics: disabled access, sunlight, private space, view, costs.

Ratings are awarded from 0 to 5 compared to a reference building, where 0 represents the minimum legal requirements/standard practice.

### SI-5281 (Israel)

Israeli Standard 5281: Buildings with Reduced Environmental Impact  was developed by the Standards Institute of Israel and adopted in 2005 as a voluntary green building standard for new residential and office buildings.[46] It uses a points system to assess designs for measures to reduce impacts associated with energy, land use, water use, wastewater, drainage, waste management, air quality and ventilation, radon and noise. Credit is given for the provision of cycle storage and green-labelled materials, with bonus points at the assessor's discretion, which might include such measures as renewables and roof gardens.

### *Estidama* (Abu Dhabi)

The *Estidama* (Sustainability) Pearl Rating system dates from 2007, but is currently being extensively revised since it is intended to build it into local codes and regulations.[47] It has been developed for the Abu Dhabi Urban Planning Council as part of their Plan Abu Dhabi 2030 sustainable development programme. The intention is to integrate it with BREEAM, LEED or Green Star depending on the preference of the developer or design team, although with a core requirement to adapt to the climate and culture of Abu Dhabi.

## Planning and masterplanning tools

### PLACE³S (US)

Planning for Community Energy, Environmental and Economic Sustainability (PLACE³S) is a California Energy Commission tool developed in the early 1990s and designed to assist communities 'retain dollars in the local economy, save energy, attract jobs and development, reduce pollution and traffic congestion and conserve open space'.[48] It relies on bespoke software to model energy, traffic, pollution, etc. and has been used extensively in the western states of America to assist in neighbourhood planning in particular.

### SCALDS (US)

The Social Cost of Alternative Land Development Scenarios (SCALDS) is an Excel spreadsheet based model that estimates the monetary and physical costs of urban land development sponsored by the US Department of Transportation Federal Highway Administration and developed by Parsons Brinkerhoff in 1998.[49]

### SPARTACUS (EU)

The System for Planning and Research in Towns and Cities for Urban Sustainability (SPARTACUS) was developed by a European consortium under the 4th Framework Programme for Research and Technology Development between 1996 and 1998 based on land use modelling programmes (MEPLAN) that used information on population, employment and traffic movement prediction to generate predictions of pollutant emissions (particulates, oxides of nitrogen and carbon monoxide) and noise levels through a module called Raster.[50] This information was analysed through a tool called the Urban Sustainability Evaluation and Interpretation Tool (USE-IT) in terms of:

- air pollution;
- resource consumption;
- exposure to pollutants and noise in homes;
- traffic deaths and injuries;
- equity of exposure to pollutants;
- time wasted in traffic jams;
- impact on public transport punctuality;
- city centre vitality;
- access to city centre and services;
- economic impacts.

There is no evidence that it has been applied commercially in town or city planning since its initial development as a research project.

**Table 1.4** *Summary of assessment schemes*

| Country/Title | Type | Versions | Link |
|---|---|---|---|
| **Abu Dhabi** | | | |
| Estidama | Env assessment | Under development | www.estidama.org/Default_en_gb.aspx |
| **Australia** | | | |
| Green Globe 21 | Benchmarking tool | Companies, communities, ecotourism, design and construct | www.ec3global.com/about/who-is-ec3/Default.aspx# |
| EC3 Earthcheck | Environmental compliance check | | ditto |
| Green Star | Env assessment | Office design and as built, education, residential, healthcare, retail, office interiors | www.gbca.org.au/green-star/rating-tools/ |
| NABERS | Env assessment | Existing offices, homes, hotels, retail | www.nabers.com.au/ |
| BASIX | Energy/water rating | New dwellings | www.basix.nsw.gov.au/information/about.jsp |
| Brazil | | | |
| AQUA | Env assessment | New offices, school buildings, hotels | www.vanzolini.org.br/conteudo.asp?cod_site=0&id_menu=493 |
| LEED Brasil | Env assessment | As LEED US | www.gbcbrasil.org.br/in/index.php?pag=certificacao.php |
| Canada | | | |
| Green Globes | Env assessment | New and existing buildings, fit-out, emergency management, automation | www.greenglobes.com/ |

**Table 1.4** *Summary of assessment schemes (Cont'd)*

| Country/Title | Type | Versions | Link |
|---|---|---|---|
| GBTool | Research tool | | www.iisbe.org/ |
| LEED Canada | Env assessment | As LEED USA | www.cagbc.org/leed/what/index.php |
| **China** | | | |
| Evaluation Standard for Green Buildings | Evaluation tool | New dwellings, public buildings, offices, shopping malls, hotels | www.risn.org.cn/Norm/xxbz/ShowCalib1.aspx?CalibID=60043&IsEdit=False). |
| **Denmark** | | | |
| BEAT | Env profile tool | New buildings, products, materials | www.en.sbi.dk/publications/programs_models/beat-2002 |
| **Finland** | | | |
| PromisE | Env assessment | New and existing offices, retail, dwellings | http://virtual.vtt.fi/virtual/proj6/environ/ympluok_e.html |
| **France** | | | |
| HQE | Env assessment | New offices, education, healthcare | www.assohqe.org/documents_certifications_hqe.php |
| qualigreen | Env assessment | New buildings | www.greenlogic.fr/qualigreen.php |
| **Germany** | | | |
| DGNB Certificate | Env and socio-economic assessment | New offices and admin buildings | www.dgnb.de/en/certification/methodical-principle-certification-system/index.php |
| **Gulf States** | | | |
| BREEAM Gulf | Env assessment | New multi-use buildings | www.breeam.org/page.jsp?id=196 |
| **Hong Kong** | | | |
| HK-BEAM | Env assessment | New and existing commercial buildings | www.hk-beam.org.hk/general/home.php |
| CEPAS | Env assessment | New buildings | www.bd.gov.hk/english/documents/index_CEPAS.html |
| **India** | | | |
| GRIHA | Env assessment | New offices, retail, hotels, healthcare | www.grihaindia.org/index.php?option=com_content&task=view&id=13 |
| TGBRS | Env assessment | New and existing commercial and residential | http://teriin.org/ |
| LEED India | Env assessment | As LEED US | www.igbc.in:9080/site/igbc/tests.jsp?event=22869 |
| **Italy** | | | |
| Protocollo Itaca | Env profile | New and refurb residential | www.itaca.org/valutazione_sostenibilita.asp |

**Table 1.4** *Summary of assessment schemes (Cont'd)*

| Country/Title | Type | Versions | Link |
|---|---|---|---|
| **Mexico** | | | |
| SICES | Env assessment | New commercial and low-income housing | www.mexicogbc.org/certificacion.php |
| LEED Mexico | Env assessment | New construction, interiors, existing, pilots for shell and core, new homes and housing developments | www.cadmexico.com.mx/fundacion/noticias/01/diseno/03/dis_01_03.htm |
| **Netherlands** | | | |
| Eco-Quantum | Env profile | New buildings and energy systems | www.ivam.uva.nl/?18 |
| **New Zealand** | | | |
| Green Star NZ | Env assessment | New offices, interiors, industrial education | www.nzgbc.org.nz/main/greenstar |
| **Norway** | | | |
| Ecoprofile | Env profile | New and existing buildings | www.byggsertifisering.no/PortalPage.aspx?pageid=142 |
| **Portugal** | | | |
| LiderA | Env and socio-economic assessment | New and existing commercial, tourism and residential | www.lidera.info/?p=MenuContPage&MenuId=15&ContId=29 |
| **Singapore** | | | |
| BCA Green Mark | Env assessment | New and existing residential, non-residential, interiors, infrastructure, district projects, parks | www.bca.gov.sg/GreenMark/green_mark_buildings.html |
| **South Africa** | | | |
| Green Star SA | Env assessment | New offices, retail pilot | www.gbcsa.org.za/greenstar/ratingtools.php |
| | Env and socio-economic assessment | New and existing commercial, tourism and residential | www.lidera.info/?p=MenuContPage&MenuId=15&ContId=29 |
| **Spain** | | | |
| BREEAM-ES | Env assessment | Residential and commercial, communities, existing offices – all under development | www.breeam.org/index.jsp |
| **Sweden** | | | |
| Ecoeffect | Env profile | New buildings | www.ecoeffect.se/ |

**Table 1.4** *Summary of assessment schemes (Cont'd)*

| Country/Title | Type | Versions | Link |
|---|---|---|---|
| **UK** | | | |
| BREEAM | Env Assessment | New offices, retail, HE, education, industrial, healthcare, fire stations, multi-residential, bespoke, prisions, courts, data centres, communities, existing offices, healthcare and residential | www.breeam.org/index.jsp |
| Code for Sustainable Homes | Env assessment | Residential | www.communities.gov.uk/planningandbuilding/ buildingregulations/legislation/codesustainable/ |
| SpEAR | Env and socio-economic assessment | New buildings | www.arup.com/Services/Sustainability_ Consulting.aspx |
| CEEQUAL | Env assessment | New civil engineering projects | www.ceequal.co.uk/ |
| BFF | Lifestyle indicator | Personal carbon footprint | http://old.bestfootforward.com/tools/ |
| Envirowise Indicator | Assessment tool | Businesses | www.envirowise.gov.uk/uk/Our-Services/Tools/ Envirowise-Indicator.249257.html |
| **US** | | | |
| LEED US | Env assessment | New construction, commercial interiors, core and shell, schools, retail, healthcare, neighbourhood, residential | www.usgbc.org/DisplayPage. aspx?CMSPageID=222 |
| Green Globes | Env assessment | New construction, existing commercial buildings | www.greenglobes.com/ |
| NAHB | Env assessment/ standard | New homes | www.nahbgreen.org/Guidelines/ansistandard. aspx |
| PLACE³S | Planning tool | New development | www.energy.ca.gov/places/ |
| SCALDS | Planning tool | New development | www.fhwa.dot.gov/scalds/scalds.html |
| **Europe** | | | |
| BREEAM Europe | Env Assessment | Offices, retail and industrial | www.breeam.org/index.jsp |
| SPARTACUS | Planning tool | New development | http://virtual.vtt.fi/virtual/proj6/yki4/spartacus.htm |

# Notes

1    www.thegreenguide.org.uk/.
2    See Chapter 3.12 Materials Specification and www.thegreenguide.org.uk.
3    www.ccscheme.org.uk/.
4    www.communities.gov.uk/planningandbuilding/buildingregulations/legislation/
     codesustainable/.
5    A pdf is available at www.planningportal.gov.uk/uploads/br/BR_PDF_ADL1A_2006.pdf
     (accessed 2 February 2010).
6    www.ceequal.co.uk/.
7    www.usgbc.org/DisplayPage.aspx?CMSPageID=222.
8    www.usgbc.org/DisplayPage.aspx?CMSPageID=148.
9    www.cagbc.org/leed/what/index.php.
10   www.igbc.in:9080/site/igbc/tests.jsp?event=22869.
11   www.cadmexico.com.mx/fundacion/noticias/01/diseno/03/dis_01_03.htm.
12   www.gbcbrasil.org.br/in/index.php?pag=certificacao.php.
13   www.arup.com/Services/Sustainability_Consulting.aspx.
14   www.nahbgreen.org/Guidelines/ansistandard.aspx.
15   www.energystar.gov/.
16   www.greenglobes.com/.
17   www.iisbe.org/.
18   www.mexicogbc.org/certificacion.php.
19   www.vanzolini.org.br/conteudo.asp?cod_site=0&id_menu=493.
20   www.ec3global.com/Default.aspx.
21   www.gbca.org.au/green-star/rating-tools/.
22   www.nzgbc.org.nz/main/greenstar.
23   www.gbcsa.org.za/greenstar/ratingtools.php.
24   www.nabers.com.au/.
25   www.basix.nsw.gov.au/information/about.jsp.
26   www.hk-beam.org.hk/general/home.php.
27   www.bd.gov.hk/english/documents/index_CEPAS.html.
28   www.risn.org.cn/Norm/xxbz/ShowCalib1.aspx?CalibID=60043&IsEdit=False.
29   www.bca.gov.sg/GreenMark/green_mark_buildings.html.
30   http://ibec.or.jp/CASBEE/english/index.htm.
31   www.grihaindia.org/index.php?option=com_content&task=view&id=13.
32   http://teriin.org/.
33   www.dgnb.de/en/certification/methodical-principle-certification-system/index.php.
34   www.assohqe.org/documents_certifications_hqe.php.
35   www.mamaisoncertifiee.com.
36   www.cerqual.fr/.
37   www.greenlogic.fr/qualigreen.php.
38   www.ivam.uva.nl/?18.
39   www.ecoeffect.se/.
40   www.byggsertifisering.no/PortalPage.aspx?pageid=142.
41   www.en.sbi.dk/publications/programs_models/beat-2002.
42   http://virtual.vtt.fi/virtual/proj6/environ/ympluok_e.html.
43   www.itaca.org/valutazione_sostenibilita.asp.
44   www.lidera.info/?p=MenuContPage&MenuId=15&ContId=29.
45   www.gbce.es/tools/general-information.
46   www.heschel.org.il/fellows/files/Israeli%20Standard%205281.pdf.
47   www.estidama.org/Default_en_gb.aspx.
48   www.energy.ca.gov/places/.
49   www.fhwa.dot.gov/scalds/scalds.html.
50   http://virtual.vtt.fi/virtual/proj6/yki4/spartacus.htm.

## References

CLG (Department for Communities and Local Government), 2009a. *Zero Carbon Homes: Impact Assessment*. London: CLG. Available at www.communities.gov.uk/publications/planningandbuilding/impactzerocarbon (accessed 2 February 2010)

CLG, 2009b. *Code for Sustainable Homes Technical Guide for Version 2*. CLG. Available at www.communities.gov.uk/planningandbuilding/buildingregulations/legislation/codesustainable (accessed 2 February 2010)

Vigon, B. W., 1994. *Life Cycle Assessment: Inventory Guidelines and Principles*. Florida: CRC Press

# 1.4
# Environmental Impact Assessment

## Introduction

Environmental impact assessment (EIA) first appeared as a mandate on federal agencies in the US in 1969. A requirement was introduced for a detailed statement to be produced on the environmental impact of 'proposals for legislation and other major federal actions significantly affecting the quality of the human environment'. This requirement appeared in the 1969 National Environmental Policy Act (NEPA) and applied to actions by federal agencies including policy development and the permitting of construction projects requiring federal funding. It was left to individual states to implement legislation to cover local development. For example, the California Environmental Quality Act (CEQA) was enacted in 1970 requiring state and local agencies to follow a protocol of analysis and public disclosure of the potential environmental impacts of development projects. The European Union (EU) began drafting an EIA Directive in 1977, although it was not adopted wholesale until 1988. This delay was primarily due to the opposition of the Conservative UK Government, which at that time considered its existing planning laws to be adequate.

Globally, the requirement for environmental assessment as part of the process for permitting development remains patchy, particularly across the developing world. Australia, New Zealand and Canada have extant legislation, whilst there was limited use of EIAs in China as early as 1973, included in the Environmental Protection Law of 1979, augmented by a series of measures relating to construction projects and consolidated in the 2003 EIA Law. This allows retrospective Environmental Statements and there is little evidence of projects being delayed due to the absence of an Environmental Statement.

Participating countries signed up to the UN Convention on Environmental Impact Assessment in a Transboundary Context, held in Finland in 1991, requiring consultation between neighbouring countries where there is a significant risk of a construction project in one country impacting on its neighbours. This was followed by meetings in Kazakhstan and Kiev, resulting in the Protocol on Strategic Environmental Assessment being adopted in 2003.

The Aarhus Convention in 1998 required the involvement of all stakeholders in decisions impacting on sustainable development and focused on the interactions between the public and public authorities, which in turn led to amendments to EIA law globally.

## European Union

With Directive 85/33/EEC the European Union introduced a mix of mandatory and discretionary procedures for assessing environmental impact that Member States implemented by 1988. Article 3 required the direct and indirect effects of a development on humans, fauna, flora, soil, water, air, climate, landscape, material assets and cultural heritage to be assessed, along with interactions between any of the above.

The EIA Directive, as it became known, included a number of Annexes that set out the categories of projects to which the Directive applied. For Annex I projects an EIA was mandatory, whilst for Annex II projects Member States were allowed to specify 'certain types of projects as being subject to an assessment or may establish the criteria and/or thresholds necessary to determine which ... are to be subject to an assessment ...' (EIA Directive Article 4).

Following the first review of the 1985 Directive, it was amended in 1997, strengthening elements relating to screening, scoping and alternatives, and again in 2003 (2003/35/EC) (see Box 1.3) to reflect the commitments made by the EC under the Aarhus Convention relating to public participation. Following rulings by the European Court of Justice in 2006 in cases involving the London Borough of Bromley, as well as the UK Government, an amendment was forced on British law that requires a further EIA on reserved matters at the detailed design stage of a project if one has been carried out on the concept but significant changes to risk have resulted from design development. This resulted in amendments to the EIA legislation in England enacted in 2008 (see the next Section).

The Strategic Environmental Assessment Directive 2001 (2001/42/EC) grew out of the EC Fifth Environment Action Programme which affirmed the importance of assessing the environmental effects of plans and programmes.

Following the UNECE transboundary EIA Convention in 1991, a meeting in Sofia in 2001 resulted in an agreement by Member States to establish a minimum environmental assessment framework through the assessment of plans and programmes, including situations where there could be resultant impacts in neighbouring states.

## UK

Prior to the 1985 Directive, planning legislation in the UK included measures for the protection of the environment through the Town and Country Planning Act and legislation protecting green belts, securing rights to light, etc. For this reason the UK Government fought hard to water down the initial proposals for the EIA Directive, delaying publishing by some six years. The Directive was not implemented through primary legislation, but instead under the European Communities Act 1972 using a series of regulations in a number of policy areas, including the Town and Country Planning (Assessment of Environmental Effects) Regulations 1987 (see Box 1.4) leaving a number of sectors, such as transport, agriculture, forestry and offshore development uncovered. These loopholes were eventually plugged by the Planning and Compensation Act of 1991.

The Town and Country Planning (Environmental Impact Assessment) Regulations 1999 (T&CP EIA Regs) responded to the amendments incorporated into the EC Directive 97/11/EC (see above), whilst the 2007 amendment (see Box 1.4) responded to the judgement of the European Court of 2006 referred to above.

Box 1.10 lists the regulations that were introduced to fill the gaps not covered adequately by the general regulations referred to above. Most apply to England and Wales and, unless indicated otherwise, they are mirrored by corresponding legislation in Scotland and Northern Ireland.

The SEA Directive (see above) has been transposed into English law through the Environmental Assessment of Plans and Programmes Regulations 2004 (see Box 1.4), with similar provisions for Wales, Scotland and Northern Ireland. It has become normal practice for authorities to combine SEAs with the Sustainability Appraisal process, required under the Planning and Compulsory

---

## Box 1.10 Environmental Impact Assessment Regulations in England and Wales

- Environmental Impact Assessment (Agriculture) (England) (No 2) Regulations 2006;
- Water Resources (Environmental Impact Assessment) (England and Wales) Regulations 2003;
- Environmental Impact Assessment (Forestry) England and Wales Regulations 1999;
- Environmental Impact Assessment (Land Drainage Improvement Works) (Amendment) Regulations 2006;
- Environmental Impact Assessment (Fish Farming in Marine Waters) Regulations – England and Wales and Scotland;
- Marine Works (Environmental Impact Assessment) Regulations 2007;
- The Harbour Works (Environmental Impact Assessment) Regulations 2000 – England and Wales and Scotland;
- Highways (Environmental Impact Assessment) Regulations 2007 (England and Wales);
- Transport and Works (Environmental Impact Assessment) Regulations 2000;
- Electricity Works (Environmental Impact Assessment) (England and Wales) Regulations 2000;
- Nuclear Reactors (Environmental Impact Assessment for Decommissioning) Regulations 1999 – England and Wales and Scotland;
- Offshore Petroleum Production and Pipe-lines (Assessment of Environmental Effects) (Amendment) Regulations 2007 – England, Wales, Scotland, Northern Ireland;
- The Public Gas Transporter Pipe-line Works (Environmental Impact Assessment) (Amendment) Regulations 2005 England and Wales and Scotland;
- The Pipe-line Works (Environmental Impact Assessment) Regulations 2000 England, Wales, Scotland, Northern Ireland.

Purchase Act 2004 (see Box 1.4) where a Sustainability Appraisal promotes 'sustainable development through the integration of social, environmental and economic considerations into the preparation of ... Regional Spatial Strategies ... Development Plan Documents and Supplementary Planning Documents.'[1]

The T&CP EIA Regs incorporate Schedules that are broadly similar to the Annexes in the EIA Directive, listing applications for which EIAs are mandatory and discretionary in Schedules 1 and 2 respectively. For most developments for which this book is relevant the 'urban development' heading in Schedule 2 will apply, limited to sites over 0.5ha in total area within the boundary or that fall within a 'sensitive' area. The Regulations define sensitive sites as those that fall within certain classifications, such as having a neighbouring site of special scientific interest, falling within a National Park, World Heritage Site, etc. The Multi-Agency Geographical Information database has a consolidated map of sensitive sites for a given development.[2]

## EIA Process

The Department for Communities and Local Government (CLG) has produced a guide to procedures for an EIA (CLG, 2000).[3] This strongly recommends that developers consult with the relevant authorities early in the design process in order to establish whether an EIA is required, known as a 'screening opinion' and, if in the affirmative, agree a scope, known as a 'scoping opinion'. If the developer disagrees with the screening opinion then he can appeal to the Secretary of State (Welsh Assembly in Wales). Special arrangements are required for projects with Permitted Development Rights in a Simplified Planning Zone or an Enterprise Zone.

Appendix 5 of the CLG Guide provides a checklist of matters to be considered for inclusion in an Environmental Statement (summarized in Box 1.11).

Although the scope of an EIA varies, the contents of a typical Environmental Statement for a mixed use development in the Greater London area are given in Box 1.12.

---

### Box 1.11 Summary of requirement for inclusion in an Environmental Impact Assessment

- Description of development during construction and operation: physical characteristics; land use; processes; residues and emissions to water, air and soil; noise and vibration; light; heat; radiation.
- Alternatives and reasons for choice.
- Direct, indirect, secondary, cumulative, short-, medium- and long-term, permanent, temporary and positive environmental effects on humans, fauna, flora, soil, water, air, climate, landscape, material assets and cultural heritage through the existence of the development and use of natural resources.
- Measures to prevent, reduce and/or offset significant adverse effects.
- Non-technical summary.
- Any deficiencies in assessing risk.

**Box 1.12 Typical scope of an Environmental Statement
for mixed-use development in London**

1  Introduction
2  EIA methodology
3  Alternatives and design evolution
4  The proposed development
5  Site preparation and construction
6  Planning policy context
7  Socio-economics
8  Transportation and access
9  Wind microclimate
10  Daylight, sunlight and overshadowing – external
11  Daylight, sunlight and overshadowing – internal
12  Light pollution and solar glare
13  Electronic interference
14  Archaeology
15  Built heritage
16  Ground conditions
17  Water resources, drainage and flood risk
18  Air quality
19  Noise and vibration
20  Ecology
21  Aviation
22  Cumulative impact assessment
23  Residual impact assessment and conclusions

Separate statements may also be included with the planning application dealing
with townscape, conservation and visual impact assessment; sustainability; and
energy strategy.

## US

The US has no standardized national approach to environmental impact
assessment for planning applications and the permitting of built structures.
Each state and major city, such as New York, has its own procedure. For
example New York State introduced a requirement for environmental
assessment through the State Environmental Quality Review Act (SEQRA) in
1975 and New York City has introduced the City Environmental Quality
Review (CEQR).[4] 'The CEQR process requires City agencies to assess, disclose
and mitigate the environmental consequences of their decisions to fund, directly
undertake or approve an action.' The process requires the agency to undertake
an initial evaluation (similar to a screening opinion under UK law) which tests
the scheme against the criteria set out in the New York Department of Energy
Conservation (DEC) regulations known as '6 NYCRR Part 617' which appear
as Appendix 1 to the CEQR Technical Manual.[5] Type I actions are those for
which it is likely that an Environmental Statement will be required, although

there is scope for agencies to add their own requirements and adjust the threshold criteria to suit local circumstances. In summary Type I actions include:

- adoption of land use plans, etc.;
- changes in allowable uses within district zones;
- state or local agency transactions involving sites of more than 100 acres;
- residential development meeting defined thresholds, e.g. more than 2500 units in a large city;
- non-residential sites of either 10 acres or more, surface water volume of more than 2 million gallons/day, parking for more than 1000 vehicles, gross floor area of 240,000 sq ft or more (large city);
- structures taller than 100 feet;
- developments having a non-agricultural use to be built in a district defined as agricultural;
- work on structures defined as 'historic'.

Thresholds are reduced if the development will impact on public parkland, recreation areas or open space.

Type II actions are also defined, including such things as repair, replacement and maintenance.

Once it has been decided that an Environmental Assessment (EA) is required then a pro forma is provided for submission to the 'Lead Agency'. For example in New York City the Lead Agency for a new private development would typically be the responsibility of the Office of the Deputy Mayor for Economic Development and Rebuilding.

The EA has to be based on the 'Criteria of Significance' obtained from the following:

- Substantial adverse change to air quality, ground or surface water quality or quantity, traffic volumes, noise levels, solid waste quantities, erosion, flooding, leaching and drainage problems.
- Destruction of vegetation or fauna, interference in migration of fish or wildlife, impact on habitat area, threats to endangered species, adverse effects on natural resources, impairment of the legally defined 'Critical Environment Area'.
- Conflict with community plans or goals.
- Impairment of character or quality of historical, archaeological, architectural or aesthetic resource, or community/neighbourhood character.
- Energy use; health hazard; change in use/intensities; population increase.
- Cumulative impacts.

The CEQR Manual requires that impacts be considered in terms of physical setting, probability of occurrence, duration, irreversibility, geographic scope (local, regional, global), magnitude and number of people affected.

Box 1.13 shows a typical contents list for an EA for a large mixed use development in New York City.

---

**Box 1.13 Environmental Assessment scope for mixed-use development in New York City**

- Land use, zoning and public policy;
- Socio-economic conditions;
- Community facilities;
- Open space;
- Shadows;
- Historic resources;
- Urban design and visual resources;
- Neighbourhood character;
- Natural resources;
- Hazardous materials;
- Waterfront revitalization programme;
- Infrastructure;
- Solid waste and sanitation services;
- Energy;
- Traffic and parking;
- Transit and pedestrians;
- Air quality;
- Noise;
- Construction impacts;
- Public health.

---

# Notes

1  www.communities.gov.uk/planningandbuilding/planning/sustainabilityenvironmental/sustainabilityappraisalsa/.
2  www.magic.gov.uk/.
3  www.communications.gov.uk/publications/planningandbuilding/environmental impactassessment.
4  www.nyc.gov/html/oec/html/ceqr/ceqrpub.shtml.
5  www.nyc.gov/html/oec/downloads/pdf/CEQR_Technical_Manual_Appendices.pdf.

# Reference

Department for Communities and Local Government (CLG), 2000. *Environmental Impact Assessment: A Guide to Procedures*. London: Thomas Telford

# Part 2

# Sustainability and Masterplanning

# 2.1
# Sustainable Communities

## Introduction

A definition for sustainable development is given in Part 1, referring to the seminal report from Brundtland dating from 1987. The aim of Part 2 is to set out an approach to masterplanning that achieves the underlying objectives behind this definition, that is:

1   To provide a design pattern book for development that is based on the principles of sustainability.
2   To provide a framework for the design of sustainable communities.

There are a number of definitions for a sustainable community. The US based National Resources Defense Council (NRDC) refers to sustainable communities as those that are '...capable of maintaining their present levels of growth without damaging effects',[1] whilst the UK Department for Communities and Local Government (CLG) has a definition that focuses particularly on social sustainability:

> Sustainable communities are places where people want to live and work, now and in the future. They meet the diverse needs of existing and future residents, are sensitive to their environment, and contribute to a high quality of life. They are safe and inclusive, well planned, built and run, and offer equality of opportunity and good services for all.'[2]

CLG suggests that a sustainable community should offer decent homes at prices people can afford, good public transport, schools, hospitals, shops and a clean, safe environment. It also refers to the need for open public space where people can relax and interact and that residents should be offered a say on the way their neighbourhood is run.

CLG also sets out what it sees as the principal components of a sustainable community as being active, inclusive, safe, well run, environmentally sensitive, well designed, well built, well connected, thriving, well served (with schools, hospitals, etc.) and fair for everyone.

'Environmentally sensitive' is a catch-all phrase that includes most of the issues that would be covered by an EIA and a BREEAM or CSH assessment (refer to Part 1 above).

Decisions that impact on the sustainability of a community occur at various stages of the planning process: from the development of Government policy through Regional Spatial Strategies, Local Development Frameworks to masterplanning and the detailed design of individual developments. A sustainable community may be the result of the successful implementation of some or all of these measures. However, success is more likely if policy and legislation provide drivers that require developers to fund sustainability measures both for new development and improving the existing surrounding community.

The 2007 Sustainable Communities Act requires central government to consult with local communities via the local authorities on what measures are required to improve their communities.

The Act is designed to strengthen the role of communities. It provides a simple process by which the ideas generated by local communities are fed through their local authority and a body known as the 'selector' to central government. The government will consult the selector and try to reach agreement on which of the short-listed proposals should be implemented. Government will publish an action plan setting out how it will take forward the suggestions that it adopts.

Through the publication of 'Local Spending Reports', communities are better informed about the public funding that is spent in their area. This should enable local authorities, their partners and communities to take better informed decisions about the priorities they choose to pursue to promote the sustainability of their local community (CLG, 2007).

## Composition of a Sustainable Community

The recently published BREEAM Communities scheme, with regional checklists, and the LEED Neighbourhood Development protocol referred to in Part 1 both provide useful templates for planning sustainable communities globally. There is significant overlap between the two schemes and the CLG definition referred to above, but there are gaps in all of them. The intention of Table 2.1 is to provide a comprehensive list of the issues by which a sustainable community can be judged, indicating the source, be it expressed as a design feature (as in BREEAM or LEED) or a desired outcome (as in the CLG components of a sustainable community).

Clearly not all of the outcomes set out by CLG are within the ambit of the developer or masterplanning design team, although their influence will depend on the specific constraints associated with the development and the scale and ambition of the project. A large project such as the Battersea Power Station development (see case history, page 93) has sufficient scale and scope to impact on the sustainability of a large area, in this case part of Vauxhall, Nine Elms and Battersea and, once built, will provide a catalyst for the economic growth of an area that is currently decaying, poorly served and poorly connected. Large projects have the 'muscle' to influence local development plans and can have a significant impact on the vitality of a community, whereas smaller projects will have smaller impacts and will tend to respond to planning drivers rather than lead them.

**Table 2.1** *Composition of a sustainable community*

| BREEAM Communities 2009 | LEED Neighbourhood Development 2009 | CLG components of a sustainable community (based on outcomes) |
|---|---|---|
| Climate and energy:<br>• flood management<br>• air, ground and water pollution<br>• heat island<br>• energy efficient design<br>• water efficient design<br>• renewables<br>• access to services infrastructure | Green infrastructure and buildings<br>• stormwater management<br>• building energy efficiency<br>• solar orientation<br>• heat island reduction<br>• building water efficiency<br>• water efficient landscaping<br>• on-site renewable energy resources<br>• district heating and cooling<br>• infrastructure energy efficiency<br>• light pollution reduction<br>• construction activity pollution prevention | Environmentally sensitive<br>• minimize climate change through energy efficiency and renewables<br>• reduce noise pollution<br>• minimize pollution to land, water and air |
| Resources<br>• materials re-use and procurement<br>• waste recycling and composting<br>• water resources and pollution | Green infrastructure and buildings<br>• existing building re-use<br>• historic resource preservation and adaptive use<br>• minimize site disturbance<br>• recycled content in infrastructure<br>• solid waste management infrastructure<br>• wastewater management | Environmentally sensitive<br>• minimize waste and dispose to good practice<br>• make efficient use of natural resources<br>• encourage sustainable production and consumption<br>Well designed and built<br>• high quality, durable buildings using materials which minimize negative environmental impacts |
| Transport<br>• walkable neighbourhoods<br>• cycle networks<br>• public transport<br>• green transport plans<br>• car clubs, flexible parking<br>• home zones | Smart location and linkage<br>• locations with reduced automobile dependence<br>• bicycle network and storage<br>• housing and jobs proximity<br>Neighbourhood pattern and design<br>• walkable streets<br>• compact development<br>• connected and open community<br>• reduced parking footprint<br>• street network<br>• transit facilities<br>• transportation demand management<br>• tree-lined and shaded streets | Well connected<br>• appropriate level of local parking<br>• good connectivity through telecoms and internet access<br>• good access to regional, national and international communications networks<br>Environmentally sensitive<br>• create opportunities for safe walking and cycling<br>• reduce dependence on cars<br>Well designed and built<br>• accessible jobs, key services and facilities by public transport, walking and cycling |
| Ecology and biodiversity<br>• enhancing ecological value<br>• biodiversity action plan<br>• use of native species | Smart location and linkage<br>• (protection of) imperilled species and ecological communities<br>• wetland and water body conservation<br>• agricultural land conservation<br>• steep slope protection<br>• site design for habitat or wetland and water body conservation/restoration<br>• long-term conservation management of habitat or wetland and water bodies | Environmentally sensitive<br>• protect and improve biodiversity |

**Table 2.1** *Composition of a sustainable community* (Contd)

| BREEAM Communities 2009 | LEED Neighborhood Development 2009 | CLG components of a sustainable community (based on outcomes) |
|---|---|---|
| Business and economy<br>• inward investment<br>• prioritized business sectors<br>• job creation<br>• local employment<br>• complement existing businesses | Neighbourhood pattern and design<br>• local food production<br>Smart location and linkage<br>• housing and jobs proximity | Thriving<br>• wide range of jobs and training opportunities<br>• sufficient land and buildings to support economic prosperity<br>• job and business creation with benefits for local community<br>• strong business community with links to wider economy<br>• economically viable and attractive town centre<br>Well designed and built<br>• accessibility of jobs by public transport, walking and cycling |
| Community<br>• inclusive design<br>• mixed-use<br>• affordable units<br>• stakeholder consultation<br>• development user guide<br>• sustainable facility management | Neighborhood pattern and design<br>• visitability and universal design<br>• mixed-use neighbourhood centres<br>• mixed-income diverse communities<br>• community outreach and involvement<br>• access to civic and public spaces<br>• access to recreation facilities<br>• neighbourhood schools | Active, inclusive, safe and fair<br>• social inclusion and good life chances for all<br>• have due regard for future generations<br>• create a sense of community identity and belonging<br>• create tolerance, respect and engagement for all<br>• engender friendly, cooperative and helpful behaviours<br>• create opportunities for cultural, leisure, community, sport, etc.<br>Well run<br>• representative accountable governance<br>• effective community engagement<br>• strong and informed partnerships<br>• sense of civic values, responsibility and pride<br>• high quality, mixed-use, durable, flexible and adaptable buildings<br>Well served<br>• well-performing local schools, FE and HE institutions<br>• high quality integrated local healthcare and social services<br>• high quality services for families and children<br>• good range of shops and community facilities |
| Place shaping<br>• effective land use<br>• defensible space<br>• Secure by design<br>• Active frontages<br>• Access to green space<br>• Affordability | Smart location and linkage<br>• Smart/preferred location – infill sites and adjacent to existing development<br>• brownfield redevelopment | Well-designed and built<br>• sense of place and local distinctiveness<br>• user-friendly public and green spaces<br>• diversity, affordability and accessibility of housing |

**Table 2.1** *Composition of a sustainable community* (Contd)

| BREEAM Communities 2009 | LEED Neighborhood Development 2009 | CLG components of a sustainable community (based on outcomes) |
|---|---|---|
| | | • appropriate size, scale, density, design and layout to suit locality<br>• buildings and public spaces which promote health, reduce crime and make people feel safe<br>Environmentally sensitive<br>• create cleaner, safer and greener neighbourhoods<br>Active, inclusive and safe<br>• low levels of crime, etc. with visible community-friendly policing |
| Buildings BREEAM/CSH assessed | Green infrastructure and buildings<br>• certified green buildings | Well-designed and built |

The remaining chapters of Part 2 deal with the strategies and decisions that are in the hands of the masterplanning design team.

# Case history

## Battersea Power Station

The proposed development of the site currently occupied by the iconic Battersea Power Station is on an under-utilized brownfield site that has lain dormant since the early 1980s.[3] Being located within the Vauxhall/Nine Elms/Battersea Opportunity Area (OA), the site has been recognized as an area in need of redevelopment and renewal.

The developer is proposing a new riverside community with thousands of homes, shopping, a hotel, cafes, and office floor space. It will be a new urban quarter with a balance between homes and offices, so people can live within walking distance of their workplace. The power station building will provide culture and entertainment amenities, and will be a destination for Londoners and tourists to visit.

The developer is working with Transport for London and London Underground to deliver a new extension of the Northern Line from Kennington station, to link the site to the underground network.

The site is dominated by the power station building, one of the world's largest brick buildings. It forms the cornerstone of the development proposals. The developer has established an aspiration to achieve a 'zero carbon' power station in use.

The primary energy for the building will be derived from a biofuel-fired combined cooling heat and power (CCHP) plant located in the Energy Centre beneath the Power Station Park. 'Zero carbon' status will be achieved by:

**Figure 2.1** *Architect's impression of proposed redevelopment of Battersea Power Station site*
*Source:* Image courtesy of Treasury Holdings/Raphael Vinoly

- generating low carbon electricity, heating and cooling serving site-wide district heating, cooling and power networks on a phased basis;
- the export of low carbon electricity to offset high carbon grid supplied electricity;
- district heating connected to the proposed OA district heating network.

The CCHP installation will initially utilise a mix of biofuels and natural gas but with the flexibility to incorporate waste-derived fuel (e.g. methane from bio-digestion) should this become available and viable for the latter phases of the development;

The development will offer the following general sustainability features:

- a high density mix of uses on a brownfield site;
- a high quality, inclusive design and enhanced public realm;
- respect for the natural environment, with enhanced biodiversity and open space;
- minimum possible carbon emissions from both operation and materials to mitigate against climate change;
- accessible, usable and permeable for all users;

- sustainable, durable and adaptable in terms of design, construction and use;
- secure, safe and accessible environments where crime and disorder, including terrorism and fear of crime, do not undermine the quality of life or community cohesion;
- practical and legible;

As well as the extension of the Northern Line from Kennington station to link the site to the underground network, the following transport initiatives are currently envisaged:

- improvements to bus services;
- high quality pedestrian and cycle networks, including improvements to the Thames Path to improve connectivity to the river and the strengthening of east–west links to Battersea Park and Queenstown Road Station;
- potential extension of the river bus service.

Site-wide grey water recycling and the use of borehole abstraction will meet part of the potable water demand, landscape irrigation and fire hydrant services for the development. The construction will conform to current best practice standards for both site operation and waste management. Reused, recycled and prefabricated construction materials will be used, where practicable.

The development will form a new sustainable and creative district in Wandsworth, and will act as 'a catalyst for [the] social and economic growth' of the wider OA. The area surrounding the site is a historically deprived area where environmental exclusion has brought down aspirations and achievements of the local population for decades. The two wards closest to the development, Queenstown and Latchmere have some of the highest unemployment rates and lowest qualifications in the borough. The Index of Multiple Deprivation classifies these neighbourhoods as severely deprived in the dimensions of employment, education, training and skills development.

It is estimated that the development will generate in the region of 25,000 net direct construction related jobs during the anticipated 14-year construction programme. Once the development is complete and operational, the opportunities for on-site inward investment through offices, retail, food and beverage, cultural and arts business represent an overall uplift of the OA area. These new businesses should create new employment for some 17,000 people in a diverse range of sectors, including services, catering, public services, with opportunities in sport, arts and cultural facilities.

A key aspect of building a sustainable community will be to ensure that the on-site public services are accessible and integrated to provide the best service for the local population. The development will include space for new social infrastructure, culture and community facilities, which is likely to include a health clinic, to expand and improve healthcare provision for all OA residents.

The development is likely to cater for all children requiring nursery school spaces through the provision of conveniently located nursery facilities. Play spaces across the development will be located close to residential areas with enhanced facilities closest to the larger family dwellings.

The Power Station Park to the north of the power station will link to the open space/public realm provided on the jetty. This will create a new 'local park', which will help to mitigate the known existing deficiency in this area.

(Based on information provided by URS Corporation Ltd and Treasury Holdings.)

## Notes

1   www.nrdc.org/reference/glossary/s.asp.
2   http://webarchive.nationalarchives.gov.uk/+/http://www.communities.gov.uk/archived/ general-content/communities/whatis/.
3   The discussion of the Battersea Power Station development in this Case History is based on information provided by URS Corporation Ltd and Treasury Holdings.

## Reference

CLG (Department for Communities and Local Government), 2007. Sustainable Communities Act 2007: A Guide. Department for Communities and Local Government

# 2.2
# Land Use and Density

## Introduction

The sustainable allocation of land for development and the determination of optimum density may be primarily the responsibility of local planning authorities, but land owners and developers will frequently contribute to the final allocation.

In the UK, Planning Policy Statement 1: Sustainable Development PPS1 – 2005[1] requires planning authorities to 'set a clear vision for the future pattern of development' and, to paraphrase para. 27(iv), bring forward sufficient land of suitable quality in appropriate locations to meet the expected needs for housing, industrial, retail, commercial, leisure, and recreation development, and exploitation of raw materials, accounting for accessibility, transport, infrastructure, waste management, flood risk and hazards.

Elsewhere PPS1 refers to planning authorities giving preferential treatment to developments that include 'housing at higher densities on previously developed land' (brownfield development) and it is interesting to note that the percentage of residential development on previously developed land in England has risen from 54 per cent in 1990 to 78 per cent in 2008 (DEFRA, 2009), although for '*all* new development' the increase was much smaller: that is from 47 per cent in 1990 to 52 per cent in 2006. Government targets for the percentage of housing development on brownfield land set out in PPS3 (CLG, 2006a) is 60 per cent. Densities for housing increased from 25 dwellings per hectare (dw/ha) in 1990 to 46 dw/ha in 2008.

The approach to land use planning and zoning in the US has been decentralized, with policy left to local states, counties and cities. The US Smart Communities Network considers that most local zoning ordinances have isolated employment locations, shopping, services and housing from each other and that low density development catering for motor vehicle access has led to unsustainable urban sprawl.[2]

This section examines what might constitute sustainable land use and the main factors that drive an appropriate density of development.

## Sustainable land use

When deciding on a site for development, developers and planning authorities have a very different agenda. The land buyer for a developer will be interested in purchasing a site as an investment that may not be realized for a number of years, depending on the market. Some developers have extensive 'land banks' that have grown through strategic purchases with a view to maximizing return whilst meeting local authority allocations. On the other hand PPS12 (CLG, 2008) requires planning authorities to identify zones for particular uses, based on a strategic view of the long-term needs for the area. The ideal situation is when the developer's and planning authority's objectives are in harmony. The key issues that dictate a site's suitability for development are discussed in the following sub-section.

## Greenfield versus brownfield

Development of sites that have not previously been built on (greenfield) is controversial, particularly if that site is in a green belt. Although the UK Government clearly promotes brownfield development (see above) nearly 50 per cent of all development in 2006 was on greenfield sites. In the UK the pressure on these sites will become greater as house building programmes catch up with Government targets. However, Planning Policy Statement 4 (CLG, 2009) makes it clear that local planning authorities 'should ensure that the countryside is protected for the sake of its intrinsic character and beauty, the diversity of its landscapes, heritage and wildlife, the wealth of its natural resources and to ensure that it may be enjoyed by all' with new development restricted to the edge of existing settlements.

Historically, in the US there has not been so much pressure on the countryside as in the UK, although the US has introduced grants for cleaning up 'brownfields' under the Small Business Liability Relief and Brownfields Revitalization Act 2002[3] and in recent years there has been more attention paid to the gradual creep of urban sprawl.

Sometimes the most sustainable option would be to refurbish existing buildings rather than demolish and build new. The decision needs to be based on a life cycle assessment of carbon emissions and costs. Nick Baker provides a comprehensive view of the decision-making process in his recently published 'Handbook of Sustainable Refurbishment' (Baker, 2009).

Development in built-up areas that is primarily comprised of hard surfaces will have little impact on the heat island effect in summer, although if the site has no operational buildings on it prior to development there will be some impact on winter temperatures associated with the heat transmission from heated buildings. Significant development on a greenfield site will increase both summer and winter temperatures around the buildings. This can be mitigated to some extent by extensive use of soft landscaping and green roofs. Water bodies and shading from trees will also reduce external temperatures close to the ground and can be used to reduce summer overheating, improving the viability of natural ventilation or reducing air conditioning loads.

Each of the following will have a bearing on both the sustainability of the site and hence the ultimate development, the type of use that the site is suitable for, and the economic feasibility of the project.

## Condition and nature of site

A brownfield site will normally have a history of development that could include some contaminated land and underground obstructions. If there are existing buildings on the site then these may contain asbestos or contamination. Specialist installations involving the storage or processing of radioactive materials, explosives or biological hazards, for example, will require special attention.

At the other end of the spectrum a greenfield site that has high ecological value, for example with protected trees, flora or fauna, is on a migratory route or has recognized national or international importance will present major planning obstacles and/or costs. For example in the UK a site harbouring great crested newts will require significant mitigation, such as having to move all the newts to a suitable neighbouring site with extensive use of fencing to prevent their return. Even brownfield sites may present ecological challenges; for example if bats are nesting in existing buildings.

The topography of a site will obviously impact on construction costs; building on a slope is more expensive than on the flat, however, there may be advantages to a sloping site in that a southerly slope in particular maximizes the potential for exploiting passive solar radiation with south-facing windows, conservatories and solar panels whilst, for a low rise development in particular, a slope maximizes the number of units that have a view.

Similarly the topography and exposure of a site might predispose it to the installation of wind turbines. Sites that sit over an accessible aquifer might be able to economically utilize geothermal energy, borehole cooling or ground source heat pumps.

Ground condition, including the load bearing ability and structural stability of soil and underlying rocks, the level of the water table and underground obstructions will primarily impact on the design and cost of foundations and below grade structures.

## Flood risk

Flood risk has become one of the most critical issues for planners, developers and, of course, occupiers of properties in flood-prone areas. Flooding is not a new phenomenon of course, but it has become a focus in the global warming debate. As referred to in Part 1, in 2006 the UK Government introduced new requirements for regional and local planning authorities with respect to the sustainability appraisal of their plans and the risk assessments required to accompany planning applications in PPS25, (CLG, 2006b). In the US flood risk of development is tied in with property insurance through the National Flood Insurance Reform Act 1994.[4] In the UK maps of flood risk are available from the Environment Agency, whilst in the US they are published by the Federal Emergency Management Agency (FEMA).

Although a brownfield development may be prone to flooding, in most cases the area of permeable land will not be significantly reduced. In fact with a combination of soft landscaping and green roofs it is usually possible to increase the attenuation of flood water. On the other hand, a greenfield site will usually end up with a greater area of impermeable surfaces and therefore contribute to a greater flood risk in the area. It is the uncontrolled growth of development in river catchment areas that is a major contributor to an increase in flood events over the last century.

## Accessibility, transport and services

A poorly connected residential development results in an enclave that is separate from the community, with a heavy reliance on motor vehicles and the old and disadvantaged left isolated. A development can be isolated in a city centre, if dislocated by a busy road, or in the middle of the countryside, if there is not the critical mass or services required to form a viable community. The ideal would be for the entire community to be able to walk or cycle safely to the amenities they need on a regular basis, such as schools, healthcare facilities, shops, post offices and community facilities. Good links to centres of employment and transport hubs would minimize the reliance on cars, whilst providing good access for delivery and service vehicles that do not create conflict between pedestrians, cyclists and vehicles.

These issues are covered in more detail in Chapters 2.4 Social Sustainability and 2.7 Integrated Sustainable Transportation Planning.

## Infrastructure and capacity

A brownfield development will normally be able to connect into an existing infrastructure of utilities and roads. An assessment of demand and capacity will be required to determine whether that infrastructure is capable of serving the new use. A greenfield development on the other hand may require an entirely new utility infrastructure and road connections to match the new demand. In either case, reinforcement (increasing the capacity) of existing electricity, gas, water, sewers and roads can be very costly. An imaginative approach to demand reduction and on-site renewables can significantly reduce the amount of reinforcement required (refer to Chapter 2.6 Energy Strategy and Infrastructure).

# Density of development

The relationship between density and sustainability is a complex one. It involves such factors as the efficient use of land, the sustainability of tall buildings and issues surrounding the social impact of high rise development. Density generally refers to the number of dwellings per unit land area; however, in a mixed-use development the amount of usable floor space also comes into the equation. In city planning, high density development is a fact of life. In the publisher's summary of Edward Ng's recently published book *Designing High-Density Cities* (2009) it is claimed that 'high density cities can support closer amenities, encourage reduced trip lengths and the use of public transport and therefore reduce transport energy costs and carbon emissions', reducing urban sprawl and incursions into the green belt.[5] This applies equally to residential, commercial or mixed-use development.

In the report on *The economic impact of high density development and tall buildings in central business districts* published by the British Property Federation in 2008 (Colin Buchanan & Partners, 2008) significant economic advantages are identified due to the 'agglomeration effect' through businesses clustering together to create critical mass, economies of scale and an increased labour pool to draw from, resulting in better value for money from increased competition.

There is also the question of whether the resultant tall buildings are more resource and carbon efficient to build and operate than the corresponding low rise development. This is explored in some detail in the report by Pank and others for the Corporation of London in 2002 (Pank et al, 2002) and is examined further in Chapter 3.2 Operational Energy and Carbon. Embodied energy in high rise buildings is significantly higher than in low rise, for example a study of five office buildings in Melbourne published in 2001 reports that the two highest buildings (42 and 52 storeys) have embodied energy some 60 per cent higher than the three low rise (3, 7 and 15 storeys). It is more difficult to identify a similar trend in the relationship between operational energy and height. Because of the smaller external surface area, a high rise residential building that is not air conditioned should have a lower energy consumption per gross floor area than a detached house built to the same insulation standards. However, it is more difficult to achieve very high insulation standards, particularly low air leakage for the upper storeys. Office buildings, on the other hand, are a challenge to naturally ventilate if they have more than seven or eight storeys.

## Notes

1    www.communities.gov.uk/publications/planningandbuilding/planning/planning policystatement1.
2    www.smartcommunities.ncat.org/landuse/luintro.shtml.
3    www.epa.gov/brownfields/laws/sblrbra.htm.
4    www.fema.gov/pdf/nfip/riegle/pdf.
5    see www.earthscan.co.uk/?tabid=21001.

## References

Baker, N. V., 2009. *The Handbook of Sustainable Refurbishment: Non-Domestic Buildings*. London: Earthscan

CLG (Department for Communities and Local Government), 2006a. *Planning Policy Statement 3: Housing*. London: Communities and Local Government. Available at www.communities.gov.uk/publications/planningandbuilding/pps3housing (accessed 12 August 2010)

CLG, 2006b. *Planning Policy Statement 25: Development and Flood Risk*. London: CLG. Available at www.communities.gov.uk/documents/planningandbuilding/pdf/planningpolicystatement25.pdf (accessed 12 August 2010)

CLG, 2008. *Planning Policy Statement 12: Creating strong safe and prosperous communities through Local Spatial Planning*. London: CLG. Available at www.communities.gov.uk/documents/planningandbuilding/pdf/pps12lsp.pdf (accessed 12 February 2010)

CLG, 2009. *Planning Policy Statement 4: Planning for Sustainable Economic Growth*. December, London: CLG. Available at www.communities.gov.uk/documents/planning andbuilding/pdf/planningpolicystatement4.pdf (accessed 12 February 2010)

Colin Buchanan & Partners, 2008. *The Economic Impact of High Density Development and Tall Buildings in Central Business Districts*. London: British Property Federation. Available at www.ctbuh.org/Portals/0/People/WorkingGroups/legal/legal/WG_BPF_Report.pdf (accessed 12 February 2010)

DEFRA (Department for Environment, Food and Rural Affairs), 2009. *Sustainable Development Indicators*. DEFRA. Available at www.defra.gov.uk/sustainability/government/progress/national/index.htm (accessed 12 February 2010)

Ng, E. (ed.) 2009. *Designing High-Density Cities*. London: Earthscan

Pank, W. Girardet, H. and Cox, G., 2002. *Tall Buildings and Sustainability*. London: City of London. Available at www.cityoflondon.gov.uk/nr/rdonlyres/de3830fe-d52d-4b10-b8b6-ab8eeb001404/0/bc_rs_tallbuild_0202_fr.pdf (accessed 12 February 2010)

# 2.3
# Massing and Microclimate

## Introduction

The amount of development that can be incorporated into a given site will depend on a number of potentially conflicting factors. The developer will calculate the floor area required for various uses from a prediction of the market demand on completion. In a volatile property market this is likely to be reviewed at a number of stages during the development process. On the other side of the equation there will be a whole series of constraints that may limit the possible floor area that the site can accommodate. If this falls below the developer's economic limit, then it may be necessary to vary the size of phases, delay development entirely until the market picks up, or sell the site (preferably with planning permission).

This section deals with those issues that impact on the massing of the buildings on the site. Most of these apply to high density urban development, rather than low rise buildings such as housing or business park type development.

Once it is decided that an architectural solution to a brief requires tall buildings, the following issues have to be addressed:

- impact on the wind environment for pedestrians using entrances, landscaped areas and public realm;
- overshadowing and impact on daylight and sunlight experienced by neighbours and in landscaped areas;
- visual impact;
- constraints from flight path envelopes;
- noise and air quality.

English Heritage (EH) and the Commission for Architecture and the Built Environment (CABE) in their 2003 publication Guidance on Tall Buildings (EH/CABE, 2003) recommend that consideration be given to 'the effect on the local environment (of tall buildings), including microclimate (wind), overshadowing, night-time appearance, vehicle movements ... and those in the vicinity of the buildings'. Similarly, para 4B.9 of the London Plan (Greater London Authority, 2008) states that 'all large scale buildings, including tall buildings should be of the highest quality design and in particular be sensitive to their impact on microclimates in terms of wind, sun, reflection and overshadowing'.

# Wind environment

Every new building or structure will interfere with wind flows locally compared to an empty site. For low rise developments these changes will generally not cause any problems for pedestrians using the resulting landscaped areas, neighbouring pavements and public realm. However, tall buildings can create downdraughts, eddies or vortices that can accelerate prevailing winds to an extent that is uncomfortable for an unacceptable percentage of the time, or even dangerous. Wind can also impact on the safe and comfortable use of balconies, terraces and entrances.

Of course wind can also be useful in driving natural ventilation or wind turbines: the higher winds experienced at the roof level of tall buildings may create viable conditions for wind turbines that would otherwise perform badly in the sheltered regions of city centre roofscapes.

There is no widely used rule of thumb available to decide when a wind impact assessment is required. This will not only depend on the height of the building or buildings, but on the exposure of the site and the height of buildings that already exist on the site, if any. For example, if the planned building is lower than one to be demolished to make way for it then it is likely that the wind assessment can be scoped out of the EIA.

A decision has to be made at the masterplanning stage of a project on whether a wind impact assessment is likely to be required as part of the detailed

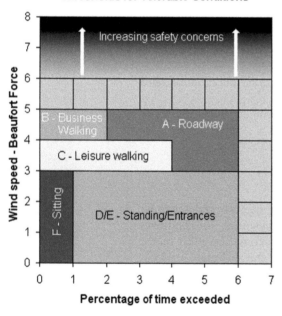

| Beaufort Force | Wind speed m/s |
| --- | --- |
| 0 | <0.3 |
| 1 | 0.3 – 1.5 |
| 2 | 1.6 – 3.4 |
| 3 | 3.5 – 5.4 |
| 4 | 5.5 – 7.9 |
| 5 | 8.0 – 10.7 |
| 6 | 10.8 – 13.8 |
| 7 | 13.9 – 17.1 |
| 8 | 17.2 – 20.7 |
| 9 | 20.8 – 24.4 |
| 10 | 24.5 – 28.4 |

Beaufort Force 8 is a gale and 10 is a storm.

**Figure 2.2** *Lawson Criteria and conversion from wind speed to Beaufort scale*

*Source:* Image courtesy of RWDI Consulting Engineers

EIA and, if so, whether sufficient accuracy can be achieved using a computer simulation model or whether a wind tunnel test is required. Generally speaking a project in the London area will require a wind impact assessment if the proposed building is ten storeys or more, although an initial desk study by an experienced wind engineering consultant might be considered for lower buildings.

The criteria widely adopted by wind environment specialists in the UK are the Lawson Comfort Criteria (Lawson and Penwarden, 1977). Lawson developed thresholds of tolerability for six usage categories representing the full range of activities that pedestrians using the space around a building might undertake: from roads and car parks through to sitting. (Figure 2.2).

It can be seen from Figure 2.2 that the Lawson Comfort Criteria are based on the percentage of the year that specific wind speeds are likely to occur. For example if a wind speed at an entrance to a building of 5.4m/s (Force 3) is exceeded for 6 per cent of the year or more then mitigation will be required. It is also necessary to establish whether gusts of wind are likely to create wind speeds for short periods that could blow a vulnerable pedestrian over.

In order to predict the wind speeds around a proposed building or buildings and design suitable mitigation measures for when the Lawson Comfort Criteria are exceeded, a specialist assessment is required. Air flow outdoors is extremely complex and the presence of multi-faceted buildings surrounding a site is very difficult to simulate. Although a number of consultants do offer computer simulation using computational fluid dynamics (CFD) the current generation of models does not model gusts and complex building geometries that accurately, with 'Large Eddy Simulation' the only technique that gets close to the right

**Figure 2.3** *Typical wind tunnel set up for wind environment modelling*

*Source:* Image courtesy of RWDI Consulting Engineers

behaviour, but requiring very long processing times and high costs to do so. Wind tunnel modelling is therefore the favoured option for detailed planning applications, whilst the author advocates the use of wind tunnel tests as part of the design process, particularly for complex projects. Where there are multiple tall buildings proposed, it may be of assistance during the masterplanning to determine the optimum positions for buildings and landscaping features to minimize wind impacts and potentially costly mitigation. Figure 2.3 shows a typical wind tunnel rig set up to measure scaled wind velocities in pedestrian areas. The turntable is rotated and wind tunnel fan adjusted to simulate wind speeds experienced in a given location through a full 360° of wind directions.

Wind directions, including eddies and downdraughts can be visualized using a smoke generator. This is useful in establishing positions for baffles, canopies, etc. in order to reduce wind speeds in critical locations, such as entrances and seating areas.

## Daylight, sunlight, overshadowing and glare

The right to daylight for the occupants of dwellings is enshrined in property law and is not considered as part of a planning application. It is set out in the Rights of Light Act 1959 and comes into play only for windows that have received uninterrupted daylight for 20 years, granted by deed or registered under the Act.

Where a proposed development is likely to overshadow neighbouring dwellings to an extent greater than existing buildings on the site, then the planning authority will require an assessment. In the UK this will normally require the tests of overshadowing set out in the BRE guidance (Littlefair, 1991). For example, the City of London in their Unitary Development Plan state that they would 'resist development which would reduce noticeably the daylight and sunlight available to nearby dwellings and open spaces to levels which would be contrary to BRE guidelines' (London, 2002). Although the BRE guidance has no statutory authority, it is recognized by planning authorities throughout the UK as the authoritative method for assessing daylight, sunlight and overshadowing impacts.

Although the BRE analysis can be carried out by hand, for a complex development it is normal practice to use a validated computer simulation model to determine the impact. The BRE tests differentiate between daylight, sunlight and overshadowing, where 'daylight' refers to the diffuse light received from an overcast sky and the unobstructed view of the sky and resultant daylighting in the building; 'sunlight' is the exposure to the sun's rays during summer and winter in habitable rooms; and 'overshadowing' refers to the shadows cast on gardens, amenity areas and other open spaces of existing buildings.

The planning authority will require these assessments to be carried out on the impacts on neighbouring dwellings, but they may also require analysis if there are dwellings within the proposed development. Even if this is not required by the planning authority, a daylight factor analysis will be required for a CSH assessment and to meet the requirements of the Housing Corporations Scheme Development Standards (Chapter 3.6 Light and Lighting).

## Daylight

BRE provides two methods for assessing daylight impact: Vertical Sky Component (VSC) and Average Daylight Factor (ADF). VSC modelling predicts the amount of light available from a standard overcast sky, usually referred to as a Commission Internationale l'Eclarage (CIE) Standard Overcast Sky. The BRE VSC test determines the amount of unobstructed sky viewable compared to the situation without the new development. A VSC below 27 per cent is poor, whilst a reduction of 20 per cent compared with the existing situation is considered to be an adverse impact.

The VSC test gives only an approximate indication of daylight impact, so the ADF method is to be preferred. However, the latter requires information on the room sizes, use, layout and internal surface reflectances that may be difficult to obtain. The average daylight factor is the illuminance predicted at the 'working plane' (a horizontal plane 0.7m above the floor) expressed as a percentage of the CIE overcast sky illuminance on an unshaded horizontal surface outside, which is 5000 lux for England (compared with 18,000 lux for the equator). Values for dwellings as set out in BS8206 Part 2: 1992 (BSI, 1992) should be no less than 1.5 per cent for living rooms, 2 per cent for kitchens and 0.5 per cent for bedrooms. If use is not known, it may be acceptable to use a default value of 1.5 per cent, adopting typical room sizes and surface finishes assumed.

## Sunlight

Potential impacts on available sunlight are assessed using the BRE's Annual Probable Sunlight Hours (APSH) method which is used to predict the sunlight available during both summer and winter for the main window of each habitable room that faces within 90° of due south. The winter period is taken to be from 21 September to 21 March and the summer comprises the remaining months. A window is considered adversely impacted if a point at the centre of the window receives 20 per cent lower sunlight hours (APSH) during either summer or winter. An APSH of 25 per cent or less is considered to be poor.

## Overshadowing

Overshadowing of neighbouring buildings affects both daylight and sunlight, but the overshadowing test in the BRE guidance assesses the amount of permanent shading resulting from the new development in neighbouring gardens and other open spaces used for leisure and amenity. The assessment is carried out for 21 March and the impact is considered to be adverse if the area in permanent shadow on that date is increased by 20 per cent or more. Ideally no more than 25 per cent of the assessed area should be in permanent shadow, with a maximum value of 40 per cent recommended.

It is good practice to model shadow paths on an hourly basis for open spaces and neighbouring buildings for 21 March, 21 June and 21 December to give a visual indication of transient overshadowing, although there are no criteria applied for this test.

## Glare

The main issue with glare is to assess the risk of drivers being temporarily unsighted by either an image of the sun reflected in a shiny surface of a proposed building or by external lighting. The risk of the latter occurring is easily overcome by ensuring external lighting is directed away from line of sight – a particular issue for security lighting controlled from presence detectors. Reflection of the sun will generally only be a risk at low solar altitude angles and can be predicted by modelling the solar geometry and obstructions around the building in relation to surrounding roads and location of reflective surfaces such as windows and polished metal cladding.

## Visual impact

The issues surrounding visual impact are of fundamental importance to planning applications and involve the greatest amount of subjectivity on the lines of 'beauty being in the eyes of the beholder'. As with other elements of

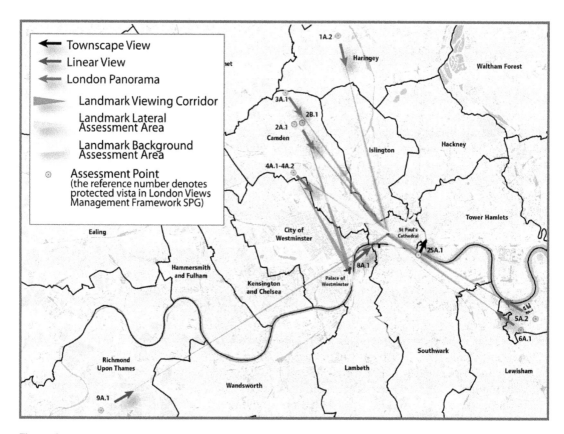

**Figure 2.4** *London protected views*

*Source:* Reproduced from the London Plan website (see note 1). Contains Ordnance Survey data © Crown copyright and database rights 2010

an EIA, context is very important and the issues are very different for greenfield and inner city development. The authoritative guidance for landscape and visual impact assessment is provided by the Landscape Institute and Institute of Environmental Management and Assessment (LI/IEMA, 2002). In the UK, county and local authorities have drawn up landscape character assessments which are to be used as a baseline for any development landscape and visual impact assessment. Cities such as London, San Francisco and Vancouver, where views of historic importance have to be protected, have developed maps of protected views, an example of which is shown for London in Figure 2.4.[1]

## Flight path envelopes

Tall buildings must be assessed to ensure that they do not enter into flight path safeguarding zones for airports in the area. The height restriction will be based on the distances between the proposed building and the runway, the horizontal distance from its centreline and the difference in height above sea level between the runway and the site. Although there are consultants who can advise on this, the ultimate arbiters are the national Civil Aviation Authorities – the CAA in the UK or the Federal Aviation Authority (FAA) in the US. Rules are produced locally, for example the CAA of New Zealand publish guidance on 'Objects and Activities Affecting Navigable Airspace', whilst the London Borough of Newham has published supplementary planning guidance on London City Airport Safeguarding (Newham, 2005) which gives rules for the dimensions of 'physical safeguarding areas – obstacle limitation surfaces' a plan of which is shown in Figure 2.5.

The information in this document is very useful for an initial assessment for projects close to London City Airport but it will be necessary to approach the Environment and Planning Manager at the Airport and the Safety Regulation Group at the CAA early in the planning process. In general, for all tall building projects proposed within, typically, 10km of any airport, designers should seek out information from the airport authorities on safeguarding surfaces in order to establish height limits for their proposed development. For very tall buildings, rules for airspace serving a number of airports might apply, in which case the CAA Safety Regulation Group should be consulted directly.

## Noise and air quality

The relationship between a building and surrounding roads and railways fundamentally impacts on decisions relating to the location of noise-sensitive activities, openings and ventilation openings. Preliminary mapping of noise and pollution levels that are likely to result from the development should be carried out when deciding massing and site layout. More detail is provided in Chapters 3.5 Air Quality, Hygiene and Ventilation and 3.8 Noise and Vibration.

## LONDON CITY SAFEGUARDED ASSESSMENT SURFACES – PLAN VIEW

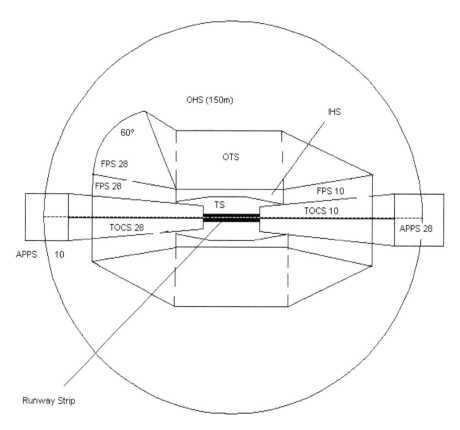

**Figure 2.5** London City Airport Safeguarding envelopes

Source: London City Airport and London Borough of Newham

## Note

1   Detailed maps and diagrams of protected views in London can be downloaded at www.london.gov.uk/thelondonplan/maps-diagrams/map-4b-02.jsp.

## References

BSI, 1992. *Lighting for buildings. Code of practice for daylighting.* BS8206-2: 1992. London: BSI

EH/CABE (English Heritage/Commission for Architecture and the Built Environment), 2003. *Guidance on Tall Buildings*. London: CABE. Available at www.cabe.org.uk/files/guidance-on-tall-buildings-2003.pdf (accessed 18 February 2010)

Greater London Authority, 2008. *The London Plan – Spatial Development Strategy for Greater London. Consolidated with alterations since 2004*. London: GLA. Available at www.london.gov.uk/thelondonplan/docs/londonplan08.pdf (accessed 18 February 2010)

Lawson, T. and Penwarden, A. D., 1977. The effects of wind on people in the vicinity of buildings. In: K. Eaton (ed.) *Procedings of the Fourth International Conference on Wind Effects on Buildings and Environment*. Available at http://books.google.co.uk/books?id=pRY9AAAAIAAJ&pg=RA2-PA605&lpg=RA2-PA605&dq=wind+Lawson+criteria&source=bl&ots=4f01Xr4mAd&sig=J9epO3oRI7CzWQnyLc5-I_7vgbs&hl=en&ei=NPhJS9C1JY214Qb5is2GAw&sa=X&oi=book_result&ct=result&resnum=7&ved=0CBsQ6AEwBjgU#v=onepage&q=wind%20Lawson%20criteria&f=false (accessed 18 February 2010)

LI/IEMA (Landscape Institute and Institute of Environmental Management & Assessment), Wilson, S., (ed.), 2002. *Guidelines for Landscape and Visual Impact Assessment*. 2nd ed. London and New York: Spoon Press

Littlefair, P. J., 1991. *Site Layout Planning for Daylight and Sunlight – A guide to good practice*. (BRE Report 209) London: BREPress. Available at www.brebookshop.com/details.jsp?id=287485 (accessed 18 February 2010)

London, City of, 2002. *City of London Unitary Development Plan: Policy EN35*. London: City of London Corporation

Newham, 2005. *Supplementary Planning Guidance Notes: London City Airport Safeguarding*. London: London Borough of Newnham. Available at www.newham.gov.uk/NR/rdonlyres/DDCBDF78-DF87-48DA-BD60-AF77F8043679/0/LondonCityAirport.pdf (accessed 19 February 2010)

# 2.4
# Social Sustainability

## Introduction

Definitions for social sustainability are included within those for sustainable communities given in Chapter 2.1 Sustainable Communities. The components of social sustainability are set out in Table 2.1 under the headings of 'Community' and 'Place shaping', along with certain aspects of 'Transport'. Social sustainability also has strong links with economic sustainability which will be dealt with in the next chapter.

A more dynamic delineation of social sustainability is given by Colantonio and Dixon in their recent report 'Measuring Socially Sustainable Urban Regeneration in Europe' as follows:

> Social sustainability concerns how individuals, communities and societies live with each other and set out to achieve the objectives of development models which they have chosen for themselves, also taking into account the physical boundaries of their places and planet earth as a whole. At a more operational level, social sustainability stems from actions in key thematic areas, encompassing the social realm of individuals and societies, which ranges from capacity building and skills development to environmental and spatial inequalities. In this sense, social sustainability blends traditional social policy areas and principles, such as equity and health, with emerging issues concerning participation, needs, social capital, the economy, the environment, and more recently, with the notions of happiness, well being and quality of life. (Colantonio and Dixon, 2009)

It could be argued that a development that fails in meeting the objectives of social sustainability results in a failed community, even if it performs extremely well environmentally.

This chapter will look at what contribution a developer and masterplanning design team can make to social sustainability during the development of a concept design.

## Amenities and facilities

A community does not simply comprise a group of homes but requires good access to the amenities and facilities required by the occupants of those homes on a day-to-day basis. Adults and children alike should be able to safely access shops, schools, healthcare facilities, community, social and recreational facilities, as well as transport nodes that connect them easily to town and city centres or places of employment.

Provision should be made for facilities such as crèches, primary schools and doctor's surgery; a range of retail facilities, such as a post office, pharmacy and corner shop; and open space with play areas, ideally within 500m safe walking distance of the development, or 1km at the most. In addition, facilities such as a bank or cash point, leisure centre, community centre, place of worship, public house and open access public area should be available or provided within 1km of the development.

It is worth referring to Ecohomes Credits Tra 1 (Public Transport) and Tra 3 (Local Amenities) (BRE, 2006) for benchmarks against which to test a site location and development proposal. The CSH has omitted these credits so that the rating has become an assessment of the dwelling performance only, independent of site location.

To obtain a maximum score for Ecohomes Tra 3 the requirement is for at least 80 per cent of the development to be within 500m safe walking distance of a food shop and post box and within 1km of at least five of the following:

- post office, bank/cash point, pharmacy, primary school, medical centre, leisure centre, community centre, place of worship, public house, children's play area and outdoor open access public area.

The 80 per cent criterion is calculated by measuring the distance from each external front door to the relevant amenities, with 80 per cent of the doors being within the specified distance to the amenity.

A safe pedestrian route is defined in Tra 1 and paraphrased as follows:

- Any pavement or path that is an official right of way dedicated to pedestrians 900mm or wider and provided with good artificial lighting where it passes through a built-up area or there is on-street parking. Note: in rural areas this could be a grass verge provided it is level, continuous and meets the above criteria; or:
- The carriageway of low-traffic roads with a speed limit of 20 mph or lower; or:
- In rural areas only, the carriageway of well-lit roads with traffic calming measures designed to reduce speeds to no more than 30mph, or where the speed limit is 50mph, the road is well-lit and there are no bends or significant junctions and a clear line of sight for at least 300m in either direction from any crossing points.
- Where the speed limit is greater than 20mph crossing points should be via a subway, footbridge, traffic island, zebra, pelican, toucan or puffin crossing; or:
- Exceptionally, the crossing of a road having an official speed limit of 30mph is allowed if traffic calming is designed to reduce speeds to 20mph at the crossing point, including chicanes, traffic humps or similar, provided there is a clear line of sight for at least 300m in either direction from the crossing point.

The extent to which a development should incorporate these facilities will depend on what exists locally and negotiations with the local planners on

developer contributions to community projects under a 'Section 106' Agreement
or planning obligation (from S106 of the Town & Country Planning Act 1990).
The scope of this agreement will depend on the size of the development and
may include allowances for improvements to local roads, schools, healthcare
facilities, community facilities, other amenities and infrastructure projects.
Guidance on determination of planning obligations is to be found in the Office
of the Deputy Prime Minister (ODPM) Circular 05/2005 (ODPM, 2005).

During the development of a masterplan for a residential or mixed-use
development it is necessary to establish the size of facilities that are required
either within the development or that require enhancement locally. These can
be sized in liaison with the local planning authority and based on the number
of residential units and residents, along with any temporary population
working or visiting the completed development. A number of toolkits have
been developed to enable developers and planning authorities to assess Section
106 planning obligations. Some are applied by local authorities across the UK,
others designed for use within specified boundaries:

- The Three Dragons Affordable Housing Development Control Toolkit,
  developed by the Three Dragons consultancy and Nottingham Trent
  University for the Greater London Authority (GLA) and Housing
  Corporation.
- The Economic Appraisal Tool (EAT), developed by GVA Grimley and
  Bespoke Property Group for the Housing Corporation (renamed the Homes
  & Communities Agency).[1]
- The Healthy Urban Development Unit (HUDU) Model can be used to
  determine the number of hospital beds or floor space required to cater for
  an increase in housing units and corresponding population in the London
  area divided between acute elective, acute non-elective, intermediate care,
  mental health and primary care facilities. It is owned and operated by the
  NHS London HUDU and is available under licence to selected developers
  and consultants.[2] This model also can be used to determine capital and
  revenue outlay associated with the enhanced facilities.
- Sport England's Planning Contributions Kitbag is a tool that allows local
  authorities to put forward sports facilty needs arising from new development
  across England.[3]
- The Southwark Toolkit allows planning obligations for projects in the
  London Borough of Southwark to be estimated covering on-site affordable
  housing, education, employment during construction and within the
  development, open space for children and sports, transport, public realm,
  archaeology, healthcare and community facilities.[4]

Other local authorities have similar toolkits that should be obtained early in the
masterplanning process.

## Secured by design

The Office of the Deputy Prime Minister (ODPM) published *Safer Places: The
Planning System and Crime Prevention in 2003* (ODPM, 2004) in which they

stated that 'safety and security are essential to successful, sustainable communities. Not only are such places well-designed, attractive environments to live and work in, but they are also places where freedom from crime, and from the fear of crime, improves the quality of life.' In PPS3: Housing (CLG, 2006) planning authorities are tasked to create safe environments, particularly for children.

Security and crime prevention needs to start at the masterplanning stage of a project. The principles of the Association of Chief Police Officers' Secured by Design (SBD) initiative should be applied, ideally in liaison with the local Architectural Liaison Officer (ALO) or Crime Prevention Design Adviser (CPDA).[5]

The SBD Principles document can be downloaded from the SBD website along with a series of guidance notes on specific applications such as new homes, schools and hospitals. It sets out a series of key principles based on addressing both the social and physical environments and designing out 'crime features', which it categorizes under the headings of 'movement generators', such as paths that create links between major centres through a quiet residential area; 'out of scale features', such as large supermarkets or football grounds that serve the wider community; and 'fear generators', including areas that are poorly lit and/or are hidden from sight.

The 'Design Key Points' given in the SBD document are reproduced in Box 2.1. These encapsulate many of the components of a sustainable community referred to earlier in Table 2.1.

---

### Box 2.1 Principles of Secured by Design Key Design Points

- Sensitive design that takes full account of the social and environmental context and encourages positive community interaction can help foster community spirit and a sense of shared ownership and responsibility. Where possible, the local community should be involved in the planning and design process.
- Provision of high quality landscape settings for new development and refurbishment, where external spaces are well-designed and well-integrated with the buildings, can help create a sense of place and strengthen community identity.
- Well-designed public spaces which are responsive to community needs will tend to be well used and will offer fewer opportunities for crime.
- Long-term maintenance and management arrangements must be considered at an early stage, with ownerships, responsibilities and resources clearly identified.
- Public and semi-private areas should be readily visible from nearby buildings or from well used rights of way.
- Natural surveillance is to be strongly encouraged, but care is needed particularly in residential development to ensure that privacy is not infringed.
- For residential development, parking should be provided close to and visible from the buildings where the owners live.

- Superfluous and unduly secluded access points and routes should be avoided.
- Access points to the rear of buildings should be controlled, for example by means of lockable gates (see also The Alleygater's Guide to Gating Alleys, link from SBD website).
- Roads to groups of buildings should be designed to create a sense of identity, privacy and shared ownership.
- Footpaths and cycleways should only be provided if they are likely to be well used.
- Footpaths and cycleways should be of generous width and have a suitable landscape setting to avoid creating narrow corridors which could be perceived as threatening.
- In terms of security, the design of the footpath is of equal importance to the design of the building. Where possible, the footpath should be at least 3m wide with a 2m wide verge on either side. Any shrub planting should start at the back of the verges.
- The position of planting and choice of species should be such that hiding places are not created. Thorny species of shrub can help to deter intruders.
- Good visibility should be maintained from either end, and along the route of footpaths and cycleways. Sharp changes in direction should be avoided.
- Footpaths and cycleways should not generally be routed to the rear of buildings, but if this is unavoidable a substantial buffer should be planted between a secure boundary fence and the footpath's margins, with planting designed so as to discourage intruders.
- Where developments adjoin waterways or rivers with towpath/footpath access, the buildings should 'face both ways', i.e. overlook the watercourse as well as the street.
- Footpaths and cycleways should be lit in built-up areas, except where the route is passing through woodland or an ecologically sensitive area, in which case an alternative lit route should be made available, such as a footway alongside a road.
- Alternative routes to important destinations may be beneficial, although a balance has to be struck between the advantages of greater choice and perceived security against the disadvantage of providing additional means of escape or of encouraging inappropriate movement of people.
- In the urban setting, open space, footpaths and cycleways should preferably be overlooked from buildings or traffic routes. Buildings should preferably face onto these areas, provided always that acceptable security for rear elevations can still be ensured.
- Property boundaries, particularly those at the side and rear, which adjoin public land, need to be secure. Windows should not provide easy access from public land. A substantial buffer planted on the outside of the fence line may help to discourage intruders.
- Improved lighting can be effective in reducing fear of crime, and in certain circumstances reducing the incidence of crime.

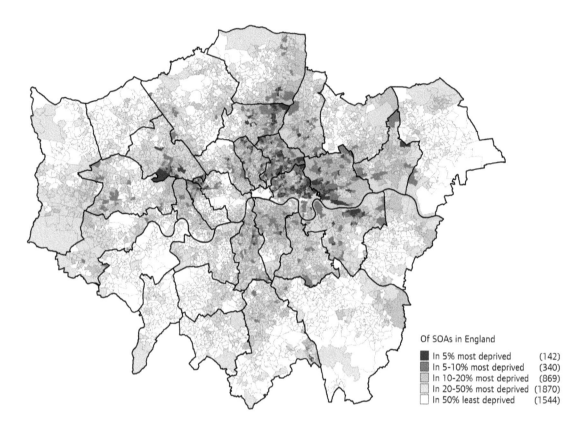

**Figure 2.6** *Indices of Multiple Deprivation for London 2007*
*Source:* CLG (2007)

## Deprivation

In the UK the CLG has developed a method of indicating the social and economic status of communities through the 'Index of Multiple Deprivation' (IMD). This uses census information under seven domains covering income, employment, health deprivation and disability, education, skills and training, barriers to housing and services, crime and living environment. Figure 2.6 plots the IMD values in 2007 for London.

## Notes

1   Spreadsheets and guidance can be downloaded from www.homesandcommunities.co.uk/economic-appraisal-tool.
2   www.healthyurbandevelopment.nhs.uk/pages/hudu_model/hudu_model.html.
3   www.sportengland.org/facilities__planning/planning_tools_and_guidance/planning_kitbag.aspx.
4   www.southwark.gov.uk/YourServices/planningandbuildingcontrol/S106/S106dev.html.
5   www.securedbydesign.com/index.aspx.

# References

BRE (Building Research Establishment), 2006. *EcoHomes: The environmental rating for homes: The Guidance – 2006, Issue 1.2.* Watford: BRE. Available at www.breeam.org/filelibrary/EcoHomes_2006_Guidance_v1.2_-_April_2006.pdf (accessed 19 February 2010)

CLG (Department for Communities and Local Government), 2006. *Planning Policy Statement 3, PPS3, Housing.* London: TSO. Available at www.communities.gov.uk/publications/planningandbuilding/pps3housing (accessed 26 January 2010)

CLG, 2007. *The Indices of Deprivation.* London: HMSO

Colantonio, A. and Dixon, T., 2009. *Measuring Socially Sustainable Urban Regeneration in Europe.* Oxford Institute for Sustainable Development, Oxford Brookes University

ODPM (Office of the Deputy Prime Minister), 2004. *Safer Places: The Planning System and Crime Prevention.* London: TSO. Available at www.communities.gov.uk/documents/planningandbuilding/pdf/147627.pdf (accessed 12 August 2010)

ODPM, 2005. *ODPM Circular 05/2005: Planning Obligations.* London: ODPM. Available at www.communities.gov.uk/documents/planningandbuilding/pdf/147537.pdf (accessed 19 February 2010)

# 2.5
# Economic Sustainability

## Introduction

It could be argued of course that the objectives of sustainability and economic growth are incompatible and that development of any sort that creates new businesses, jobs and buildings is potentially increasing global carbon emissions and resource depletion. This is related to the discussion in Chapter 2.2 Land Use and Density on refurbishment/re-use of existing buildings versus demolition and new build, although there are national and regional policy drivers that impact on the volume and rate of economic growth which may not be compatible with international agreements to reduce carbon emissions. The outcome of the UN Framework Convention on Climate Change (COP 15) in Copenhagen in 2009 and the refusal of China to compromise its economic growth plans in committing to legally binding carbon reduction targets is a prime example of this at the global geo-political level.

Economic sustainability is closely linked to social sustainability since the economic vitality of an area has significant impact on social factors such as crime, activity and quality of services, amenities and facilities.

The potential for a development to drive economic growth in an area is dependent to a large extent on its size and diversity. For example multi-million pound mixed-use developments such as the Battersea Power Station redevelopment (see the case history in Chapter 2.1) and the urban extensions of Malmo and Hammarby in Sweden are being used to provide a catalyst for growth in their respective areas. In all these examples economic growth is being driven by a combination of improved transport links, new businesses and jobs created to serve the area and draw in visitors and shoppers from the region.

## Job creation

Every construction project, no matter how small, will have an economic impact through a combination of bringing work to contractors involved in the construction phase, maintenance personnel during the life of the building and local retail and service businesses supplying residents and/or workers' shopping needs. If the development caters for businesses, some of which may be completely new to the area, then it may have a significant long-term impact on employment in the area. The assessment of any impact on local businesses and employment may not be that straightforward since there may be a negative impact on existing local businesses such as can occur when a new supermarket diverts business away from existing corner shops or high street retail premises.

A developer will need to undertake a series of assessments in liaison with local planning authorities and specialist agents in order to determine the best mix of retail, commercial and leisure businesses to attract to the development, the impact they will have on local businesses, and the net number of jobs created.

# 2.6
# Energy Strategy and Infrastructure

## Introduction

The decisions made during the masterplanning phase of development establish the foundations for the detailed energy strategy and eventual carbon emissions over the lifetime of buildings. It is at this stage that fundamental decisions are made such as target carbon emissions and to what extent on-site renewable energy generation and district energy systems can be employed.

One of the first decisions to be made by the masterplanning team is the 'carbon target' for the project. This may be driven by local or regional planning requirements, the Building Regulations which will apply to the development, a funding condition or a commercial or moral imperative identified by the developer. The approach to so-called 'zero carbon' development will be very different to one that is designed to the England and Wales Building Regulations Part L 2010, (CLG, 2009) for example.

A preliminary energy strategy should be developed based on estimates of carbon emissions associated with operational energy consumption for each of the building uses. At the masterplanning stage there may not have been sufficient design development to carry out calculations or computer modelling, however, it is important to establish targets for energy consumption expressed in terms of carbon emissions and the extent to which those emissions are offset by on-site generation of renewable energy, district energy, combined heat and power, etc. For example, if the target is to achieve zero carbon emissions averaged over the year, including 'unregulated' emissions – that is carbon emissions associated with cooking, white goods and household appliances in dwellings; and lifts, office and/or process equipment in commercial buildings – then the reduction in 'regulated' emissions and how this is to be achieved should be stated.

## Existing utilities assessment

Local authorities in the UK are responsible for producing Development Plan Documents (DPD) in accordance with PPS12 (CLG, 2008), paragraph 4.8 of which states that:

The core strategy should be supported by evidence of what physical, social and green infrastructure is needed to enable the amount of development proposed for the area, taking account of its type and distribution. This evidence should cover who will provide the infrastructure and when it will be provided. The core strategy should draw on and in parallel influence any strategies and investment plans of the local authority and other organisations.

Hence a project that is recognized within a Core Strategy should have been assessed by the local authority and an infrastructure plan established. Some local authorities have commissioned utility capacity studies, a good example of which can be drawn from Exeter and East Devon Local Authority.[1]

During the development of a masterplan it is important to gain an understanding of the current utilities capacities in the vicinity of the site compared with the likely demand, along with any reinforcement proposed in the area. This will include the potential for district energy systems serving other sites in the area.

In the UK this will normally mean approaching the energy and water companies who own and operate the local utilities infrastructure, contacting the local planning authority to establish what other developments are planned locally that might benefit from connection to a district energy scheme and estimating demand for gas, electricity, water, sewage and surface water drainage. In London and other large cities in the UK there is ongoing work on coordinating energy generation schemes across multiple developments whilst exploring the potential to serve existing buildings and using low and zero carbon fuels. Similar projects have already been undertaken elsewhere in the world, including schemes in Austria and the Isle of Man, for example, that recover energy from sewage and household waste (see the Case Histories below).

The demand analysis referred to above would normally be carried out by a services consultant based on a preliminary accommodation schedule and using rules of thumb derived from typical loads on gas, electricity, water, sewage and surface water infrastructure for the different types of accommodation, taking into account simultaneous demand for the worst case scenario and allowing for potential expansion. For a development built to modern sustainable design principles these are likely to be excessive; for example if it is anticipated that the development programme will result in buildings having to meet UK Building Regulations requirements that come into force in 2016 for dwellings or 2019 for commercial buildings, then it will be necessary to design to achieve net zero carbon emissions averaged over a 12 month period. This does not mean of course that there will be no instantaneous demand for mains electricity or gas, but the demand is likely to be significantly lower than would be required to meet current (2010) Building Regulations requirements. Similarly, if dwellings are to be designed to a Level 5 or 6 Code for Sustainable Homes (CSH) rating then water consumption of 80 litres/person/day must be achieved, which is around 50 per cent of what a typical person uses; based on an average of 152 litres/person/day for unmetered dwellings in 2005.[2] This will have a commensurate impact on sewage capacities. The increase in the use of sustainable urban drainage (SUDS) through permeable landscaping and green roofs, etc. to achieve CSH and BREEAM criteria for surface water run-off will also reduce the demand on the stormwater drainage infrastructure.

# Energy efficient design at concept

A golden rule for low carbon design of buildings is that no matter where or how the energy is generated the aim should be to use good design to reduce energy demand as much as possible, commensurate with occupant comfort and amenity (see Chapter 3.2 Operational Energy and Carbon for a more detailed analysis of the options). This doesn't mean having walls that have 10m thick insulation and slits for windows, but requires an optimization process the results of which will depend partly on end-use and partly on the climate. The building is in effect a 'coarse climate modifier', with the fine tuning carried out by the environmental services. The less fine tuning the services have to provide, the less energy will be required to create thermal and visual comfort conditions.

The rules for low carbon design at the masterplanning stage have to be adjusted for the building type and end-use, but the following key points apply across all building types:

- Maximize the use of natural forces from the wind, convection, sunlight and daylight, taking into account the potential for ingress of external noise and pollutants (Chapters 2.3 Massing and Microclimate, 3.4 Design for Natural Ventilation and 3.6 Light and Lighting).
- Utilise the thermal inertia in building materials.
- Consider the relationship between buildings and the landscape, both existing and future.

During the masterplanning of a development, it is necessary to establish dimensions for the individual buildings and the gaps between, as well as a preliminary landscaping strategy. Although it is not necessary to establish insulation properties and exact window sizes at this stage, floor depths, building heights, sun path, overshadowing, exposure to wind and orientations of longest walls all have significant impacts on energy performance and strategy.

During the detailed design, the thermal performance of the buildings will need to be modelled and it may be worth considering carrying out preliminary computer simulation at the masterplanning stage in order to verify some of the more critical decisions that impact on massing (Chapter 3.7 Computer Simulation of Building Environments). These need to be coordinated with assessing the impact of massing on the neighbourhood through modifications to the wind environment and availability of sunlight and daylight, as well as noise and air pollution impacts on the future occupants of the proposed development (Chapter 2.3 Massing and Microclimate).

For a large development a decision will have to be made on the number of buildings required to meet accommodation and circulation requirements and the area to be allocated to landscaping. This may depend on context, the value of the site and the limits on development density. For example a site in a tall building cluster in the centre of a large city will almost certainly require most of the land to be occupied by the building footprint and for the building to be tall. A very large site might be able to cater for multiple tall buildings or some mix of building heights, in which case the wind environment and overshadowing within the site becomes a major design issue as well as the impact to and from surrounding buildings and open areas. City centre sites always present complex issues of building interaction and the challenge is to maximize opportunities for using natural energy from the wind and the sun without allowing noise and pollution from surrounding streets to enter the building. This is why, in a couple of recent examples of tall buildings in London, the lower floors have been designed with sealed windows and full air conditioning whilst upper floors have employed mixed mode natural ventilation via double skin facades and light wells (See the case history for 30 St Mary Axe below).

The proportions of energy/carbon attributed to each load, such as heating, cooling, lighting, etc. will vary significantly between different building types, end usages and locations. The pie chart in Figure 2.7 gives proportions of average $CO_2$ emissions associated with energy consumption for typical air-conditioned UK office buildings (based on 1999 data).

Figure 2.8 shows energy use for a typical UK dwelling, based on 2008 data.

Comparing the two pie charts in Figure 2.7 and 2.8 indicates how much more important artificial lighting is for an office building than in the home. For low energy buildings – in which heating loads will be a significantly lower proportion than indicated here – the importance of reducing lighting energy becomes even more important. Hence the design of facades and spacing between buildings to maximize daylighting is critical.

In order to minimize operational carbon emissions for an office building located in the UK or a similar climate the following issues should be addressed:

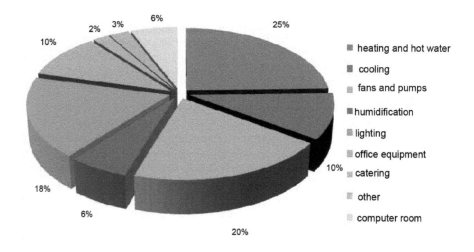

**Figure 2.7** *$CO_2$ emissions associated with energy use in a typical UK air-conditioned office building*

*Source*: Energy Consumption Guide 19 (Action Energy, 2000).

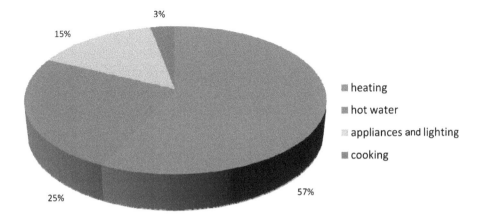

3%

15%

- heating
- hot water
- appliances and lighting
- cooking

25%

57%

**Figure 2.8** *Typical dwelling energy use*

*Source*: UK Low Carbon Transition Plan (DECC, 2009).

- A different strategy may be necessary for bespoke headquarters type buildings compared with speculative office buildings. An understanding of market expectations is required, which will be different for a business park on the outskirts of a provincial English town to downtown Manhattan or the City of London. Financial dealing rooms have particularly high equipment loads and spatial requirements. In the UK the market might be expecting a particular British Council of Offices (BCO) specification.[3] Historically, this has tended to drive the demand for full air conditioning, however the 2009 version has increased the recommended summer temperatures from 22°C to 24°C, which opens the door for mixed-mode solutions.
- A decision is required on whether the strategy is to be based around atria/ light wells, with or without natural ventilation or air conditioning (or mixed-mode, a mixture of both) and the corresponding impact on floor to ceiling heights, building footprint, relationship to noise and pollution sources and plant room sizes and locations.
- The extent to which daylight and views out are to be maximized needs to be established.
- If the natural ventilation route is to be followed, the maximum number of storeys possible with openable windows and an atrium needs to be determined, typically 10–12, in order to avoid excessive velocities at openings. An innovative design can be seen in the Deichtor building in Hamburg the cross-section of which is shown in Figure 2.9.[4]
- Tall buildings require a special approach to natural ventilation, for example by dividing a full height atrium into 10–12 storey sections, as was done at the Commerzbank tower in Frankfurt am Main (see Chapter 3.4 Design for Natural Ventilation).
- For cellular offices without an atrium the maximum depth to capitalize on daylight and single-sided natural ventilation needs to be determined. This is typically around 13m including a 1m wide corridor.

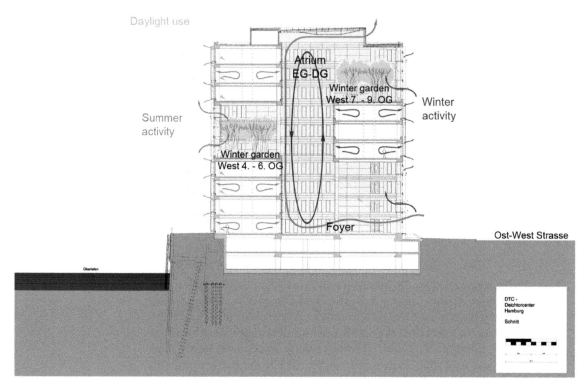

**Figure 2.9** *Deichtor building, Hamburg showing natural ventilation strategy*

*Source*: Image courtesy of BRT Architekten, Hamburg.

- Note that greater floor depths are possible with an atrium, but this depends on whether an open plan is possible, the depth of daylight penetration and whether adequate daylight from the atrium is feasible, all of which is likely to require verification by computer simulation.
- Orientation of the longest window walls should be as close to north/south facing as possible.
- Consideration should be given to a solar shading strategy, which should relate to current and future overshadowing by neighbouring buildings and allow for a sufficiently deep zone for external shading/fins for southerly orientated windows.

For a residential development it is important to know what proportion of the total development will be social units and whether they will be separated from the market units. The CSH Level to be targeted will impact on the energy strategy (a minimum will be specified for Homes and Communities Agency funding). A low energy design will impact on wall thickness but will have little impact on the overall massing.

Retail development is even more market driven than speculative office development, although the anchor tenants for a shopping centre may have strong corporate sustainability drivers. Because of the requirements for display lighting, refrigeration, ventilation, etc. the energy demand profile is dominated by electricity and opportunities for carbon reduction are limited. At the masterplanning stage any innovations such as maximizing daylighting needs to be addressed. For example Sainsburys' landmark store in Greenwich was the first supermarket to be naturally lit, using an extensive array of rooflights (Figure 2.10).

**Figure 2.10** *Sainsburys supermarket, Greenwich, showing rooflights*

*Source*: Author

## On-site energy generation

In order to achieve a low or zero carbon (LZC) outcome some on-site low or zero carbon energy generation will be required, unless the scheme can be connected to an LZC neighbourhood scheme (see below). In the UK many regional and local planning authorities require a specified percentage of carbon emissions or energy consumption to be derived from on-site renewable energy sources. This specification was developed by the London Borough of Merton in 2003 and subsequently known as the Merton Rule.[5] The percentages adopted vary from the original 10 per cent adopted by London Borough of Merton to the 20 per cent which was incorporated into the consolidated London Plan in 2008. It is interesting to note that this has been dropped in the 2009 consultation for the London Plan in favour of an overall percentage reduction to be achieved through a hierarchy of energy efficiency and on-site LZC generation measures (see below), along with overall targets for renewable technologies to be installed in Greater London by 2015, 2020 and 2025.

The 2008 version of the London Plan also encourages developers to incorporate combined heat and power (CHP or cogeneration) or combined cooling, heating and power (CCHP or trigeneration) into developments. This usually involves the generation of electricity from a gas-fired generator or turbine, the surplus heat from which is used to provide heating, and cooling via an absorption refrigeration plant in the case of CCHP/trigeneration. This can only be considered as a renewable energy source if the primary fuel is defined as renewable. For example gas derived from gasification of biomass or solid waste, gas from biodigestion of compost, methane from sewage, recycled vegetable oil or biomass (such as wood chip or pellets).

Below we look at the key characteristics of the various renewable energy sources, including LZC fuels, and their potential impact on masterplanning and meeting the objectives of an energy strategy. The technologies covered fall into various categories:

- renewable technologies that convert energy from the environment directly into usable energy, such as solar photovoltaic (sunlight into electricity), solar hot water (sunlight into hot water), wind turbines (wind into electricity), hydro power (flow of water into electricity), geothermal (heat or 'coolth' into hot or cooling water);
- low carbon technologies that convert energy from the environment through the addition of primary energy, such as ground source heat pumps (using a refrigeration cycle to upgrade heat from earth);
- low carbon technologies that use fuels which have absorbed $CO_2$ during the growing process or are a waste product, such as biomass, biogas (methane), bio-oil and household waste (incinerated or gasified).

The technologies which are described as 'low carbon' herein are those which use carbon in the process of converting the source into its final form, whether it be the electricity used to drive a heat pump or the carbon used to harvest, process and deliver biomass. However, in terms of their whole life carbon footprint, none of these technologies deliver completely renewable energy. All LZC products use carbon in their manufacture, which in some cases, such as a wind turbine mounted in a sheltered location, results in more carbon being used in manufacture than in its useful life (see below).

## Solar photovoltaics

Photovoltaic (PV) panels convert light to direct current electricity through photon-to-electron energy transfer within the dielectric materials from which they are manufactured. PV panels can generate electricity even on overcast days, although peak output occurs when the sky is clear and the angle of the sun's rays is closest to normal to the panel surface. Hence the amount of energy available from the sun for a site on the equator is around twice that available in London, taking account of typical cloud cover. Research comparing the performance of PV panels in different climate zones (Gottschalg et al, undated) found that PV panels installed in Mallorca, at a latitude of 39°N, generated around 70 per cent more electrical energy in a year than a site in Oxford (52°N).

Direct current generated from the PV panels is usually converted to alternating current via an inverter and two-way metering system which provides an interface with the incoming mains and electrical distribution system for the building served. Hence any surplus electricity generated is transmitted back to the grid through arrangement with the electricity company. Arrangements vary from country to country, but normally the District Network Operator (DNO) will offer a feed-in tariff for the electricity returned to the grid. In most European countries the feed-in tariff is between 0.3 and 0.65 with terms varying from 10 to 25 years[6] and with premiums being offered in many countries for building integrated systems (BIPV), the economics of on-site generation are much enhanced. From April 2010 a UK feed-in tariff was introduced with generation tariffs in the first year varying from 36.1p per kWh for installations smaller than 4kW rated power, to 29.3p/kWh for installations between 100kW and 5MW rated power (DECC, 2009). The tariffs are index linked and last for 25 years. There is an additional export tariff which is fixed at 3p/kWh across the board.

Photovoltaic materials fall into four categories: polycrystalline, monocrystalline, thin film and hybrid. Polycrystalline and monocrystalline cells differ only in the way in which the silicon used in their manufacture is processed. The monocrystalline panel uses a purified thin wafer of silicon which is more costly but results in higher efficiencies compared with the material used for a polycrystalline panel. In both cases the material is laminated into a glass or plastic sandwich. The monocrystalline panel has a uniform black or dark blue finish, whereas polycrystalline panels have a textured appearance and can be manufactured in different colours. Some manufacturers have devised products that offer a degree of transparency and can be integrated into window design, although with decreased efficiency, limited daylight transmission and increased cost.

Thin film PV is manufactured by depositing a thin layer of the dielectric material in vapour form within a vacuum onto a substrate such as glass, plastic or metal. There are a number of substances available that have suitable

**Figure 2.11** *Large scale photovoltaic array at Nellis USAF base, Nevada*

*Source:* In the public domain (US Air Force)

characteristics including amorphous silicon (a-Si), cadmium telluride (CaTe) and copper indium gallium sellenide (CIGS). They have lower efficiencies than the mono- or polycrystalline products, but they tend to be lower cost, although the cost per unit of electricity generated is similar.

The term 'hybrid PV cells' covers a number of recent developments in PV technology, including, for example, photosynthetic hybrid PV cells which use genetically engineered proteins that generate electricity through photosynthesis. Although these technologies have not yet reached the marketplace, it has been claimed that the unit costs will be very much lower than existing.

The major barrier to the uptake of photovoltaics has been their cost. In the UK there has been some degree of subsidy in the form of grant programmes, such as the Large Scale Building Integrated PV programme introduced in 2000, which has since evolved to become the Low Carbon Buildings Programme,[7] whilst in Germany the Solar Roof Programme resulted in some 140,000 photovoltaic installations receiving government funding.

According to a paper comparing the cost and carbon effectiveness of thin film a-Si PV and solar panels (Croxford and Scott, 2006), it would typically take around 6 years to recover the carbon used in the manufacture of photovoltaic panels, whilst it would take more than a thousand years to recover the initial investment based on a simple payback calculation, not taking into account any subsidy from a feed-in tariff and no cost increase in electricity. However, this reduces to around 30 years if electricity prices increase by 10 per cent annually and the cost of the installation is reduced by a 60 per cent Government grant. It has been claimed that a typical payback period of 2 years can be expected on a residential PV installation, taking into account the feed-in tariff.

For the future, however, it appears that capital costs will fall dramatically as the volume of demand increases, lower cost products are developed and new factories come on-line, in China in particular.

## Solar hot water

Solar hot water systems use the heat from the sun to heat water via heat exchangers exposed to solar radiation. The most common use for this heat is for domestic hot water, although it can be used for heating swimming pool water, pre-heating central heating water and in combination with underground storage to provide long-term storage of solar energy.

The two most common types of heat exchanger are the evacuated tube collector and the flat plate collector. The former comprises an array of annealed glass tubes within which a vacuum is created and either a heat pipe or pipe loop installed. The heat pipe contains a low boiling point fluid such as alcohol which becomes vapour and migrates to the header end (Figure 2.12) where it heats circulating fluid containing an antifreeze, or directly heats water in a cylinder (integrated tank solar collector).

The heat pipe evacuated tube collectors are very efficient and work better at low outdoor temperatures and overcast conditions than the flat plate type. However, because temperatures in the heat pipe can reach 200°C they require continual flow and good safety controls.

Flat plate collectors comprise an array of tubes embedded in a dark coloured absorber plate covered with a glass or transparent plastic sheet. Where there is

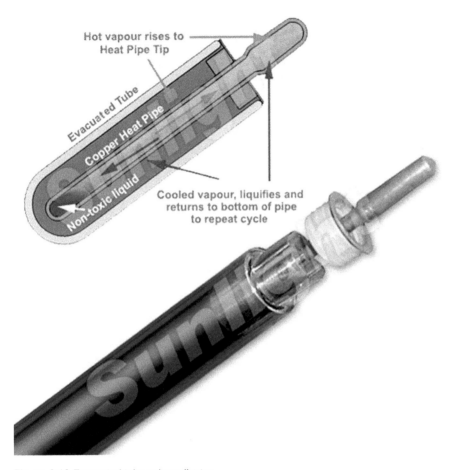

**Figure 2.12** *Evacuated tube solar collector*

a risk of freezing, a water and glycol mix is circulated through the tubes and through a heat exchanger in a hot water cylinder.

Generally there will be times when there is insufficient heat available from the sun so the cylinder will be fitted with either an electric or water fed immersion heater. The most common arrangement, which has become standard practice across the Mediterranean, is for each dwelling to have its own solar panels and hot water cylinder, frequently of the integrated or 'batch' type. However, an option for hotels and apartment buildings is for a roof-mounted array of collectors to be connected to central cylinders which in turn serve a hot water distribution system, which in the case of apartments can be metered for hot water use.

According to Croxford and Scott (2006) solar collectors are more carbon- and cost efficient than photovoltaics, with a carbon payback of around 2 years and a cost payback of hundreds of years, rather than more than a thousand, but payback is possible within the life of a system with either fuel price inflation at 7 per cent or a 50 per cent initial grant.

## Wind turbines

Wind turbines are available in horizontal (HAWT) and vertical axis (VAWT) configurations. The horizontal type (Figure 2.13) varies in size from 500mm diameter, 60W rated output typically used for charging batteries on boats and caravans to a 126m diameter rotor blade delivering a maximum of around 7MW, such as the one recently completed in Emden, Germany.

A HAWT has the turbine in the upper housing, connected in-line with the blade shaft, whilst a VAWT has the turbine at its base and is therefore inherently less top-heavy and easier to work on. There are a number of different blade designs based on either the Darrieus or 'eggbeater' blade configuration or the Savonius design, which uses sails or cups to catch the wind. A sub-group of the

**Figure 2.13**
*Horizontal axis wind turbines near Winterton, Norfolk*

*Source*: Author

**Figure 2.14** *Helical vertical axis wind turbine*

*Source*: Image courtesy of quietrevolution

**Figure 2.15** *Savonius vertical axis wind turbine at Tesco store, Dudley*

*Source*: Image by Matt Buck

Darrieus VAWT is the helical bladed type (Figure 2.14) which has gained popularity amongst architects in particular for its aesthetic qualities. These are usually pole mounted, with taller poles used for ground mounting and shorter poles for roof mounting, the aim being in all cases to avoid turbulent conditions and capture the highest and most reliable winds. There are some large Darrieus VAWT installations that are mounted close to the ground, usually in exposed locations and stabilized with guy ropes (Figure 2.16).

The cost effectiveness of large-scale wind turbines is well established and the business model for wind farms is sufficiently persuasive for a competitive marketplace to have been established globally for specialist wind farm developers and operators. The economic case for integrating wind turbines into

**Figure 2.16** *Large Darrieus VAWT at Cap-Chat, Quebec Canada*

*Source*: Spiritrock4u at en.wikipedia

building developments is more difficult however, primarily because of the negative impact of buildings on the performance of the turbines and potential noise and vibration disturbance from the turbines to building occupants.

For example the three 9m diameter HAWTs built into the roof of the Strata residential tower in the Elephant and Castle, London (Figure 2.17) are in the ideal location to capture maximum wind speeds, but the rotational speed of the wind turbines has had to be limited to reduce noise and vibration transmission to the apartments below.

**Figure 2.17** *Wind turbines integrated into the roof of the Strata tower, London*

*Source:* Author

As with photovoltaic panels, wind turbines generate direct current and require an inverter to produce alternating current at the desired voltage and two-way metering to provide an interface with the electricity grid.

The feed-in tariff associated with wind turbines is similar to that for photovoltaics, although the rates for generation for the UK range from 34.5p/kWh for turbines rated at lower than 1.5kW to 18.8p/kWh for those rated at between 100 and 500kW, but dropping to 4.5p/kWh for larger sizes up to 5MW. As for PV installations there is an additional 3p/kWh for export to the grid and the rates are index linked, although at 20 years the duration is shorter.

The life cycle carbon emissions associated with wind turbines are highly dependent upon location. One assessment of a 1.5kW roof mounted horizontal axis turbine showed that a carbon payback of between 10 and 39 months was achieved provided an output of between 1000–4000kWh was generated annually (Rankie et al, 2006). However, BRE have determined that actual outputs vary enormously with location such that a domestic scale turbine located on a 2 storey house in Manchester was found to deliver only 150kWh per year, whilst the same turbine in an exposed location in Scotland (Lerwick) generated around 3000kWh (Phillips et al, 2007). The embodied carbon in manufacture also varied widely – from 180kgCO$_2$ to 1444kgCO$_2$, whilst that associated with delivery, installation and maintenance varied from 18 to 147kgCO$_2$, depending on the product and location. Hence the rapid payback predicted by Rankie et al (2006) was only achieved in exposed locations, whilst for sheltered locations the embodied carbon is unlikely to be recovered during the life of the turbine.

## Geothermal energy

The temperature of rock at the centre of the Earth has been estimated at around 6000°C. The Earth's centre is approximately 6400km from the surface, whilst the crust is between 8 and 40km deep. Below this lies the 2900km deep mantle which encases the 2200km deep liquid outer core and the solid inner core which has a radius of around 1300km. Although historically it was thought that volcanic eruptions were caused by molten rock from the earth's core finding its way to the surface, in fact they occur when the movement of neighbouring tectonic plates creates high pressures that cause the rock to melt. This molten rock (or magma) can find its way to the surface through vent holes due to its buoyancy. In granite rock heat is generated through radioactive decay. In locations such as Iceland, California, Italy, New Zealand and Japan sources of geothermal energy are located quite close to the surface. For example, Iceland derives more than 50 per cent of its energy from geothermal power stations that inject water typically 1–5km into high temperature rocks that turn the water into steam that is drawn through a separate borehole to the surface where it is used to drive a turbine and provide district heating.

There are potential sources of geothermal energy across the planet and atlases have been produced that show their location and potential. Examples include *The Renewable Energy Atlas of the West*, which can be downloaded from the internet and covers the eleven western states of the US,[8] and the *Atlas of geothermal resources in Europe* (Hurther and Schellschmidt, 2003). There are a number of locations in the UK which are already being exploited for their geothermal energy, including Southampton where the geothermal heating district energy system has been operating since 1986 and been extended to cover a large area of waterfront development. There are also sites in Cleethorpes, Scotland and Cornwall which are yet to be fully exploited.

Geothermal energy can also be extracted from cool water in aquifers using similar technology to the above, that is by drilling boreholes into the water-bearing rocks. The aquifer can form part of a circulation system via heat exchangers to provide cooling through a building, without recourse to refrigeration. If the yield is adequate and the geology suitable, with no or very low horizontal flows within the aquifer, the ground around the aquifer can

**Figure 2.18** *Blue Lagoon, near Reykjavik, Iceland, with geothermal power station in background*

*Source*: SketchUp

provide long-term storage of energy, known as aquifer thermal energy storage (ATES). Dedicated wells can be drilled for storing excess heat in the summer (hot wells) and 'coolth' in the winter (cold wells), reversing the flows in winter from the hot well and in the summer from the cold well (see Figure 2.19).[9]

Systems that tap into the aquifer are known as 'open loop' and benefit from the low temperatures normally available in an aquifer. 'Closed loop' systems do not use aquifer water directly and therefore suffer from higher water temperatures entering the building, hence they tend to be combined with a refrigeration plant – a reversible heat pump – to generate chilled water suitable for comfort cooling (refer to the section discussing ground source heat pumps below).

None of these systems are entirely 'zero carbon' because of the electrical energy required to pump water through the wells/boreholes. Unless the geology under a development site has previously been surveyed using test boreholes to determine the location of rock or aquifers of suitable temperatures or yield, the pre-development costs and risks are high because of the geological uncertainties involved. The condition of the upper layers is also important because of the risk of contaminating aquifers via a borehole. With some five projects in the

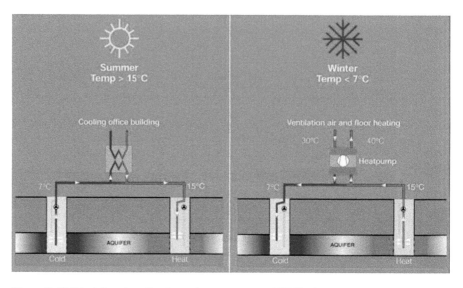

**Figure 2.19** *Principles of aquifer thermal energy storage (ATES) with heat pump*

*Source*: Courtesy of IfTech Ltd

pipeline, but only a couple completed at the time of writing, there is very little experience with ATES in the UK. There are, however, hundreds of installations in The Netherlands, Belgium and Germany where the propensity of low flow aquifers close to the surface provides optimum ground conditions. Drilling boreholes in the UK requires a permit from the Environment Agency, necessitating the submission of an EIA predicting the impacts of the proposed installation on aquifer levels, quality and temperatures.

There are very few data available on the cost of geothermal plants as an on-site renewable energy source for individual developments. Studies carried out in the US in 2007 found that life cycle costs per MWh for geothermal power stations were similar to wind farms and gas-fired power stations.[10] However, an ATES plant is only likely to be cost effective for large developments where there is a reasonable balance between heating and cooling demands, such as that proposed for the South Kensington Cultural and Academic Estate.[11]

## Ground source heat pumps (GSHP)

A ground source heat pump is a refrigeration machine that has its evaporator connected to a water circuit which runs through the ground and its condenser either used as a heating coil as part of an air handling unit or connected to a water circuit supplying a space heating system. Because condensing temperatures are usually lower than is required for the efficient sizing of radiator heating, GSHPs are usually linked to underfloor heating or convectors, using flow temperatures of around 40°C. However, for buildings with very low heating requirements the lower temperatures may not result in excessively large radiators. The market is dominated by vapour compression heat pumps that use an electrically driven compressor. These have an average coefficient of

**Figure 2.20** *Ground source heat pump 'slinky' ground coil being installed*

*Source:* Image Copyright 2009 Central Heating New Zealand Ltd (www.centralheating.co.nz)

performance (CoP – heat output divided by electrical power required by the compressor) of between 3.0 and 4.0, which means that their carbon performance compared with gas-fired heating is marginal, depending on the carbon intensity of electricity, which varies from country to country. For example, the ratio of carbon content in electricity to that in natural gas in the UK is currently around 2.72. This figure is above 3 in the US, whilst in France, with its high proportion of nuclear generation, the ratio is around 0.45, making GSHPs extremely carbon efficient in that country.

Ground coupling can either be via a closed loop coil buried between 600mm and 1m below an external space (Figure 2.20), a series of closed circuit boreholes, open loop boreholes (which may be part of an ATES – see Figure 2.19) or bespoke piles (also known as 'energy piles' or thermo-active foundations, Figure 2.21).[12]

Simple GSHP systems are relatively cost effective, the payback period depending on the differential between gas and electricity pricing. Historically, they have been used for larger dwellings that do not have access to a gas supply but do have a garden of a sufficient size for the coils to be buried underground. Closed loop systems rely on a good thermal conductivity through the ground, hence a ground conductivity test is required as part of the viability assessment. Installations that require the extensive use of bore hole drilling or integration with foundation laying result in a front-heavy construction programme, requiring particularly good coordination between trades.

**Figure 2.21** *Pipework and reinforcement bars for energy pile prior to installation*

*Source*: Cementation Skanska Energypile (RTM)

## Small-scale hydroelectricity

Large-scale hydroelectric schemes have been used to generate electricity since the first installation was built in the US in 1882, followed shortly by the turbine constructed alongside Niagara Falls, which is still operating. By the early 1900s hydro schemes were generating some 40 per cent of the US's electricity. Since then some of the largest civil engineering projects in the world – such as the Hoover dam in the US in the 1930s and more recently the Three Gorges hydroelectric scheme in China – have involved constructing dams and associated hydroelectric schemes, in many cases creating irreversible damage to displaced communities and ecosystems. According to a report from the International Commission on Large Dams (ICOLD), in 2003 there were some 45,000 large dams globally, of which 22,000 were in China.[13] China continues with its hydroelectric programme, growing from 125GW in 2010 to 150GW in 2015. In 2000 ICOLD collaborated with the International Energy Agency (IEA) to produce a long-term global strategy for hydroelectricity, entitled 'Hydropower and the World's Energy Strategy'.[14]

Part of this strategy includes the promotion of small-scale hydroelectricity and the IEA has set up a 'Small-scale hydro Annex' to its 'Implementing Agreement for Hydropower Technology and Programs'.[15]

Small-scale hydroelectric installations associated with a commercial masterplan are rare and depend upon there being the right combination of resource and topography on the site. In the US there is a small but well-developed marketplace for do-it-yourself 'micro-hydro' systems, mostly catering for rural home-owners with streams running through their land.

A number of well established turbine designs are available including the Pelton wheel, which is best applied to situations where there is a head of water of 50m or more, whilst for lower heads there is a choice between the Francis propeller, the Banki pressurized self-cleaning crossflow waterwheel and the Archimedes screw (Figures 2.22 and 2.23).

Archimedes screw schemes that divert water through a penstock have proven to be the least harmful to fish and can be quite cost effective, particularly where access is reasonable and the terrain is not compatible with wind or solar alternatives, for example in a sheltered river valley.

In the UK hydroelectricity generation represents around 2 per cent of total electrical demand, most of which is produced at large-scale hydroelectric dams in Scotland and Wales. Recently a number of small-scale schemes have been developed as community projects, including Archimedes screw based installations on the River Ribble at Settle in Yorkshire and the Torrs gorge at New Mills at the confluence of the Rivers Sett and Goyt in the Peak District.[16]

Hydroelectric schemes are also subject to feed-in tariffs, which in the UK vary from 19.9p/kWh for those rated at less than 15kW to 4.5p/kWh for schemes capable of generating 2–5MW, plus 3p/kWh for export, and lasting 20 years index linked.

## Biomass, bioliquids and energy from waste

The EU Renewable Energy Directive has the following definitions:

- Biomass is 'the biodegradable fraction of products, waste and residues from agriculture (including vegetal and animal substances), forestry and related industry as well as biodegradable fraction of industrial and municipal waste'.
- Bioliquids refers to the liquid fuel for energy purposes produced from biomass.
- Biofuels are the liquid or gaseous fuels for transport produced from biomass.

In the 2007 UK Biomass Action Plan, potential sources of biomass are given as industrial and municipal wastes, sewage sludge, food waste, forest wood fuel, agricultural residues, wood waste and energy crops, from which energy can be derived through either combustion, gasification, pyrolysis, anaerobic digestion or conversion to a liquid (biofuel) for use in vehicles.

The primary difference between the sustainability of biomass and that of coal or petroleum products is that biomass should be derived from short

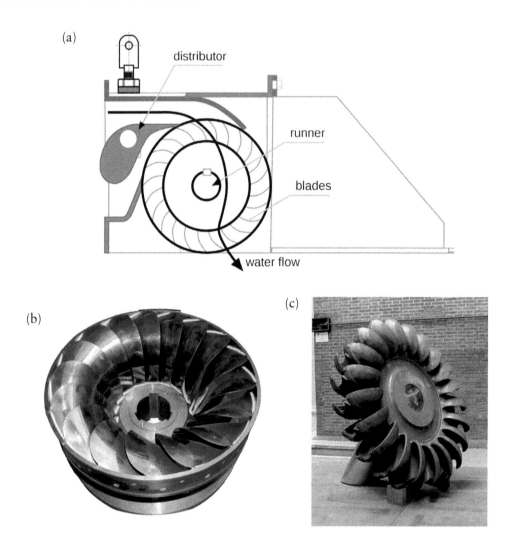

**Figure 2.22** *(a) Banki, (b) Francis and (c) Pelton hydro turbines*

*Sources*: Banki – European Small Hydrower Association; Pelton – Photograph by Andy Dingley

rotation crops or waste products. However, biomass cannot be considered a zero carbon fuel because of the carbon emissions associated with harvesting, processing and transportation. For wood based material produced and consumed in the UK this corresponds to typically around 3 per cent of the total carbon content for wood chip and 6 per cent for wood pellets. However, for biomass imported from overseas, the transportation carbon content increases dramatically. In a report for an IEA conference on biomass, it was stated that importing wood products from the Baltics to The Netherlands added 5 per cent to the carbon content of the resultant fuel, whereas importing from Latin America added 10 per cent. According to a report by independent consultants Verdantix published in November 2009, the UK Government target for 15 per cent of renewable energy to be obtained from biomass by 2020 through the construction of seven biomass-fired power stations generating some 2100MW between them, along with local authorities investing in biomass technologies through the Carbon Reduction Commitment (see Box 1.7), can only be met through importing millions of tonnes of biomass fuel.[17] This is likely to create a significant bottleneck in the marketplace for those wishing to obtain biomass for use in building integrated or district energy plants.

Biomass installations require significant space for fuel storage: typically 100m³ for a 500kW boiler based on 100 hours of operation. Space is also required for a hopper and mechanized feed system, cleaning equipment, ash storage and access for deliveries and maintenance. Dust from the fuel and ash can also present a significant maintenance burden.

A study carried out in 2007 by AEA Energy & Environment for the London councils reviewed the potential impact of various scenarios of biomass uptake in Greater London on meeting the Air Quality Objectives (AQO) required of local authorities under the 2007 Air Quality Standards Regulations. For example the 24h AQO for particulate material smaller than 10 microns ($PM_{10}$) for 2020 of 32 micrograms/m³ will be exceeded by an assumed growth model for biomass by some 5 micrograms, whilst the objective for $PM_{2.5}$ will be exceeded by a similar amount and for nitrogen dioxide ($NO_2$) by 3–10 micrograms/m³.

The burning of biomass for heat and cooking is not new of course, having started soon after humans discovered fire. However, the widespread burning of solid fuel in congested communities has for centuries been associated with air quality problems and was seriously restricted in the UK with the introduction of smoke free zones in 1956, currently referred to as smoke control areas and controlled by the Clean Air Act 1993.[18] However, a new generation of biomass boilers and combined heat and power plant is being developed so that the emissions from combustion can be brought down to acceptable levels. With current technologies, the clean combustion of biomass is problematic for a small-scale plant, although wood pellet burning stoves and residential central heating boilers remain popular in rural communities and in countries such as Canada, the northern states of the US, Sweden, Germany and Austria, mostly using quite sophisticated hopper feeds and thermostatic control (Chapter 3.2 Operational Energy and Carbon).

The keys to reducing emissions of particulate matter and oxides of nitrogen ($NO_x$) from boilers are efficient combustion and removing pollutants from the combustion products. Alternatively, biomass can be converted to gas through 'advanced thermal treatment', that is by gasification and/or pyrolysis.

## Efficient combustion

Larger solid fuel boilers are usually either of the grate or fluidized bed type. The key to achieving high combustion efficiencies is to maximize combustion temperatures by forcing air through the combustion chamber. The fluidized bed achieves this by burning the fuel on a bed of sand or other mineral which is violently agitated by the combustion air, resulting in temperatures of 800–900°C. This also allows a mix of fuels such as combustible waste, wood and coal to be burnt, which is not possible for the grate type of boiler. However,

**Figure 2.23** *Archimedes screw water turbines*

the minimum thermal duty currently available is around 40MW. Fluidized bed boilers have been used in coal-fired power stations for many years and are currently having a renaissance in large-scale biomass combined heat and power plants serving district energy systems across mainland Europe. The same technology but using a pressurized bed is applied to gasification (see below).

As with all solid fuels, biomass does not respond well to rapidly changing demand and seasonal efficiencies are reduced considerably compared to gas- or oil-firing if instantaneous loads vary rapidly. An economic analysis carried out by the DTI for the UK Biomass Strategy (DTI, 2007) indicates that neither wood chips nor pellets compete economically with gas-firing under any of the scenarios studied, although, if utilization is high, they do compete with oil-firing for middle-sized applications.

## Cleaning of combustion products

To achieve the level of emissions of particulates and $NO_x$ associated with gas-fired boilers it is necessary to clean the combustion products from a biomass boiler. Particulates can be removed using some combination of cyclone dust collectors, electrostatic precipitators and fabric filters. Cyclone collectors work through centrifugal force precipitating particles onto the external wall of a cylinder (similar to a bagless vacuum cleaner) and are efficient for larger particle sizes. Fine particles have to be removed by either electrostatic precipitators or fibre filters. This technology is costly and tends only to be available for larger plant. The $NO_x$ emissions from biomass-fired boilers can be up to 10 times that from gas-firing, but the technologies required to reduce $NO_x$ emissions, such as selective catalytic reduction (SCR) using ammonia and urea, tend only to be available for large-scale plant (>50MW), although research is ongoing on their application to smaller plants.

## Gasification and pyrolysis

Again there is nothing new about gasification, which is the manufacture of a synthesis gas (syngas) by the application of heat to a feedstock. In the case of pyrolysis, this is carried out in the absence of oxygen, but the dominant process in modern gasification is partial oxidation, involving small quantities of oxygen, although the process usually does include pyrolysis as a brief intermediate step. Pyrolysis was used in the manufacture of town gas in the UK towards the end of the 18th century and continued in 'gas works' until the discovery of natural gas under the North Sea in the 1960s and the conversion process carried out for all gas-fired equipment between 1967 and 1977. In the early days, town gas was also manufactured locally in the US by gasification and natural gas was mainly flared off from oil wells, although the first natural gas pipeline was built in 1925 and its distribution has been regulated since the 1930s.

In recent times gasification has primarily been used as part of an industrial process, for example in the manufacture of paper using waste pulp as feedstock and using the syngas in kilns. Gasification of biomass occurs at temperatures between 800°C and 1000°C, with a number of different processes having been developed operating at atmospheric pressure or under pressurized conditions. As well as syngas, gasification produces tar and char, whilst emissions to atmosphere are extremely low, thus largely overcoming the air quality problems associated with the combustion of biomass or the incineration of waste. Char is a solid residue which can either be burnt to dry the feedstock or recovered for use as activated carbon. The biggest challenge to designers is preventing the internal surfaces being coated with the tar. This has resulted in some high profile failures of biomass gasification, such as the small-scale plant installed at the BedZed development in Beddington, UK, which was never satisfactorily commissioned.

Larger-scale installations have overcome this problem, however, and the pilot fast internal circulating fluidized bed (FICFB) wood chip gasification plant at Güssing in Austria, rated at 2MW electrical output and 4.5MW heat output (see Figures 2.24 and 2.25), has been successfully providing community heat and power since 2002.

A number of waste gasification plants have been installed around the world, including the Thermoselect plant in Chiba, Japan which, although operating since 2000, has not yet recorded a net carbon saving. An Energos waste gasifier entered its commissioning phase in the Isle of Wight in the autumn of 2008, designed to process some 30,000 tonnes of waste per annum and generate up to 6MW of electricity from a steam turbine.

## Bioliquids

Bioliquids used as a fuel for heating and power compete with biofuels in the form of biodiesel and bio-ethanol as fuels for road vehicles. Biodiesel is a bioliquid having similar properties to diesel derived from fossil fuels and, as sold on the forecourt, is frequently derived from a mixture of biofuel and fossil-fuel based diesel. It is processed from a reaction with an alcohol in the presence of a catalyst to yield mono-alkyl esters and glycerine, which is subsequently removed.

Biodiesel is either derived from oily nuts and seeds such as rape seed, palm nut, sunflower, ricinus (castor oil), thistle and jatropha, or through recycling of cooking fats, both from recycled vegetable oil (RVO) and from animal fats, collectively called used cooking oil (UCO) or bio-oils. Biodiesel processing is defined by EU Standard EN 14214 or US Standards ASTM D6751, both of which refer to the percentage of biodiesel within the mix $n\%$ as B$n$. For example, pure biodiesel is B100. A major use for biodiesel mixes is for road vehicles. In the UK all biodiesel sold at the pump is blended, typically at 5 per cent by volume (B5).

Bio-ethanol is an alcohol derived from fermentation of plant starches derived typically from sugar cane, grains, switchgrass and agricultural waste. In the UK, to be exempt from excise duty, it must be 'a petrol quality liquid fuel consisting of ethanol produced from biomass' (HMRC, 2004).

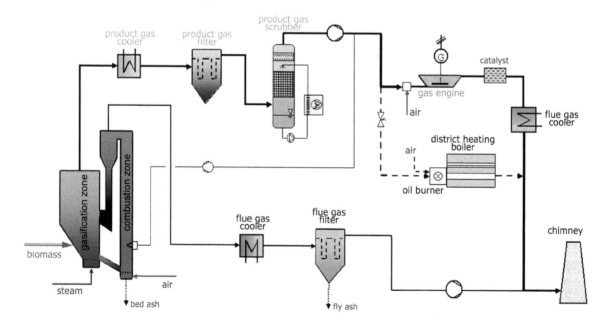

**Figure 2.24** *Biomass CHP demonstration plant in Güssing, Austria – schematic*

**Figure 2.25** *Biomass CHP demonstration plant in Güssing, Austria – elevation*

*Source:* Reinhard Rauch

According to a study by Cornell University and the University of California, Berkeley, turning plants such as corn, soybeans and sunflowers into fuel uses much more energy than the resulting ethanol or biodiesel generates.[19] There is concern that the current growing demand for vegetable oil is causing deforestation, with centuries old forests being replaced by oil palm plantations. As far back as 2000 Friends of the Earth were reporting that 87 per cent of deforestation in Malaysia was due to oil palm plantations. When land is cleared it is often burned, which releases large quantities of $CO_2$. Vegetable oil production would have to increase substantially to meet current EU targets for displacing fossil fuel petrol and diesel consumption – the RTFO (Renewable Transport Fuels Obligation) requires 5 per cent of transport fuels to be derived from biofuels by 2010/11. With current technology, such an increase in production would have a substantial environmental impact.

Raw vegetable oil has a high viscosity and is not suitable for many diesel engines. The process of transesterification involves adding alcohol and a catalyst – usually potassium hydroxide – which drives off glycerine and produces a lower viscosity biodiesel which has similar characteristics to fossil fuel diesel.

The schematic in Figure 2.26 shows the process with inputs of vegetable oils, recycled greases and sulphuric acid, and outputs of biodiesel and glycerine. Glycerine can be sold on for use in the pharmaceutical industry. This is not a carbon neutral process since it requires temperatures of around 65°C and additional chemicals that have their own carbon footprint.

Raw vegetable oils, including recycled oils from cooking and food processing can be used directly in boilers and CHP engines, but because of their high viscosity specialist equipment is required.

Tall oil is a by-product of the Kraft wood pulp manufacturing process for the paper industry and some 1.2 million tonnes is produced annually, some 60 per cent of which originates from the US. A process has been developed in Canada that converts this to biodiesel, involving de-pitching and catalytic hydrogenation.

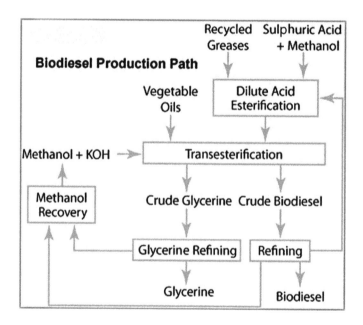

**Figure 2.26** *Typical biodiesel production diagram*

*Source*: Intercontinental Fuels, US

The potential for obtaining oil from algae has been known for many years and the US National Renewable Energy Laboratory (NREL) worked on this as part of their Aquatic Species Programme between 1978 and 1996.[20] This led to claims that the entire oil consumption of the US could be supplied from 'algae farms' located in deserts.

The theoretical advantages of algae as a biofuel feedstock are well-established. Micro-algae can be composed of up to 50 per cent oil and can grow at a far faster rate than any other plant life, sometimes doubling or trebling their mass in a single day. It is this that explains their potentially huge yields compared with other feedstocks. Micro-algae therefore represent the highest yield biofuel crop, with some manufacturers claiming that it is possible to produce up to 184,000 litres of biodiesel per hectare per annum, although this is in locations with high incident solar radiation. This compares with palm oil that can produce 5950 litres per hectare.

It also requires few resources: that is, prime agricultural land is not displaced, and only sunlight, water and carbon dioxide are required for optimum growth conditions. It can be grown on either marginal land that is unsuitable for food crops, or in photo-bioreactors, which can be located almost anywhere so long as sunlight, water and carbon dioxide are in plentiful supply. $CO_2$ from chimneys can be captured to use as a feedstock for the manufacture of algae.

One such system has been developed by Eco-Synthesis Limited (ESL) using a bioreactor to produce biomass and biodiesel from micro-algae using captured $CO_2$ as a feedstock. This bioreactor provides the capability to use $CO_2$ emissions from power plants or industry as a feedstock to cultivate micro-algae converting $CO_2$ wastes into a high yield biomass fuel. The biomass has a high oil content making it ideal for processing for biodiesel or alternatively used for co-firing back into the power plant. There are also anaerobic digestion processing options.

Cultivating biodiesel from micro-algae using $CO_2$ emissions as a feedstock is not a new concept, and organizations such as GreenFuel Technologies in the US have been successful with small-scale laboratory units. Although they have had some difficulty in scaling up their technology, they have recently announced a $92m contract to build a plant in Europe (country not specified). ESL's bioreactor incorporates techniques that are proven in other industries to accurately control the rate of growth of the micro-algae, preventing the catastrophic failure experienced by GreenFuel. The design also has a very high yield having the capability to convert 275 tonnes of $CO_2$ each day into biomass on a process footprint of just 1.35ha. The resultant biomass is potentially available at a lower cost to the consumer than mains gas.

PetroSun Algae Biofuels have developed 1100 acres of saltwater ponds in Rio Hondos, Texas to cultivate algae for biodiesel. The company projects that this algae farm will produce a minimum of 4.4 million gallons (17,800 tonnes) of algal oil and 50,000 tonnes of biomass on an annual basis. Numerous other projects are in the planning stage, but the demand for biodiesel from the road vehicle sector is likely to dominate the marketplace.

The combustion characteristics of B100 biodiesel are very similar to 35 sec fuel oil and hence most oil-fired boilers would require only a minor adjustment to their burner nozzles to run off biodiesel. Numerous boiler manufacturers now promote their products as being suitable for biodiesel. A number of manufacturers have developed waste oil boilers which can run off anything from sump oil to recycled vegetable and animal fats from cooking and food processing.

Diesel combined heat and power generators are available that can run on untransesterified oil using special injector nozzles, increased injector pressure, stronger glow plugs and fuel pre-heating. A number of German engine manufacturers, such as Elsbett Technologie and MAN Diesel, have been manufacturing engines with these features for over 30 years. There are a number of installations across Europe that incorporate a waste oil fuelled CHP plant – see the case history below for the vegetable oil and sewage gas CHP facility in Fritzens, Austria.

## Energy from waste

Historically, energy-from-waste installations have been developed and operated by municipal authorities or their contractors and use either moving grate or fluidized bed incinerators that incorporate heat exchangers for creating steam for use in a turbine to generate electricity and hot water for district heating schemes if there is a demand nearby. Because of the mixture of materials used as a feedstock, the efficiency of flue gas treatment and disposal of solid wastes, which can be hazardous, are critical. Flue gas treatment will be more complex than for biomass boilers, requiring cyclones and filtration to remove particles and usually some form of scrubbing and activated carbon filtration to remove gaseous pollutants.[21]

Alternatively, waste can be gasified and the syngas used for generating steam for a turbine and district heating scheme but with much lower airborne emissions, as referred to in the previous section. More detail of the issues surrounding waste management and processing is given in Chapter 3.11 Waste Management and Recycling.

## Biogas and anaerobic digestion

Biogas is a mixture of methane and $CO_2$ produced through bacterial degradation of organic matter. It occurs naturally through the rotting of vegetation, sewage or organic waste in landfill.

Anaerobic digestion, also known as biodigestion, occurs when microorganisms breakdown biodegradable materials in the absence of oxygen to produce methane/biogas. It occurs naturally in swamps, sediments and manure, which led to its first reported discovery in the 17th century. It has been widely used in wastewater treatment and the biogas formed has been harnessed since the mid-19th century. The same process occurs within landfill sites, although other gases may be present, creating an unpredictable mix dependent upon the types of waste found.

Modern anaerobic digestion plants are designed to create optimum conditions for biogas production, whilst the resultant digestate makes a nutrient-rich fertilizer. In the UK, the Environment Agency (EA) and Waste and

Resources Action Programme (WRAP) have developed a quality protocol for anaerobic digestate (PAS 110) so that in law it doesn't have to be treated as waste.[22]

Many biodigestion plants are associated with sewage treatment works which use the biogas to provide electrical power for the operation of the works and waste heat for accelerating the bio-digestion process, however, an increasing number of plants are being built in rural communities to make use of biodegradable agricultural wastes. Kitchen waste generally includes an ideal combination of biodegradable material. The South Shropshire biodigester described in the Case Histories section below is a good example of this application.

In Europe, funding for anaerobic digestion facilities is available through the EC Intelligent Energy for Europe Programme that has identified a number of 'Biogas Regions' across Europe which are learning from the experience gained from the operation of plants in Styria, Austria and Baden Wurttemberg, Germany.[23] In the UK, the Severn Wye Energy Agency (SWEA) is managing the project covering Wiltshire, Gloucestershire, Monmouthshire and Powys.

Seepage of landfill gas has been responsible for a number of disasters resulting in damage to buildings and loss of life, such as when a property was destroyed in Lescoe, England in 1986 and two people were killed in Skellingsted, Denmark in 1991. Whilst requiring Member States to reduce the volume of biodegradable material going to landfill, the 1999 EC Landfill Directive (see Box 1.4) also requires suitable monitoring and controls whilst the UNFCCC Clean Development Mechanism is sponsoring landfill methane recovery projects globally, with a particular focus on the developing world. With around 100,000 tonnes of waste going to landfill annually in the UK alone and 1 tonne of biodegradable waste potentially emitting between 200 and 400m$^3$ of gas, the potential for the recovery of methane is clearly substantial and many of the current operating companies are already generating electricity using gas from the sites they manage.

## Combined heat and power (CHP) and fuel cells

Combined heat and power (CHP or cogeneration) and combined cooling, heat and power (CCHP or trigeneration) involve the generation of thermal and electrical energy concurrently. Unless an LZC fuel is used, they cannot be classified as renewable energy sources, however, they do represent efficient methods of generating energy on site and hence reducing carbon emissions compared with offsetting all of the energy demand from boilers and grid electricity.

Generating electricity always involves the concurrent generation of heat and it is relatively straightforward to tap into that heat. However, CHP viability is highly dependent upon there being a demand for heat coincident with the demand for electricity for as many hours as possible. In the UK, CHP installations can qualify for exemption from the Climate Change Levy if the developer or owner can prove that they meet the Government's definition of 'good quality' CHP.[24] The definition is provided in Guidance Note 10 and requires a minimum electrical efficiency of 20 per cent and a minimum Quality

Index (QI) of between 95 for initial operation and 105 for the 'Design MaxHeat' conditions, based on minimum periods during which the plant operates at maximum heat output. These periods vary from 500 hours for residential community heating to 1000 hours for mixed use schemes having less than 10 per cent qualifying heat output supplied to dwellings. The QI is based on a formula that is calculated from electrical and thermal efficiencies and factors that depend on fuel type and rated electrical capacities.

In the US the EPA has introduced the Combined Heat and Power Partnership, which is a 'voluntary program seeking to reduce the environmental impact of power generation by promoting the use of CHP'.[25]

In the UK the Combined Heat and Power Association provides invaluable information on technical aspects of CHP along with links to their member's websites.[26]

For most CHP plant the amount of heat available is 1.5–3 times that for electricity, although this can be lower for fuel cells and much higher for steam turbines. For most buildings there is a fairly consistent demand for electricity, but in temperate climates the only year-round heating requirement is typically for hot water services. CCHP on the other hand uses some of the surplus heat to provide chilled water via an absorption chiller which uses a closed circuit refrigeration cycle based on refrigerants such as ammonia or lithium bromide to convert low grade heat to 'coolth' (see Chapter 3.2 Operational Energy and Carbon).

The choice of technologies and range of products in the CHP/CCHP marketplace is growing all the time. Most of the technologies have been around for many years and have been adapted from other applications, such as power stations, jet engines, road vehicle engines and fuel cells. CHP plant falls broadly into the categories of internal combustion engines, gas and steam turbines and fuel cells, with electrical outputs varying from 1kWe to hundreds of MWe. The terms micro- and mini-CHP are used for the smaller end of the market, although confusingly the term has been used for outputs as high as 350kWe. Units are available for individual dwellings, which will be covered in more detail in Chapter 3.2 Operational Energy and Carbon.

As with photovoltaics and wind turbines, CHP installations require two-way metering for connection to the grid and in European countries such as Germany and The Netherlands benefit economically from an increased feed-in tariff (FiT). In the UK the primary fuel will have to be biogas from anaerobic digestion for a FiT to apply, although 'micro-CHP' generating less than 2kW also qualifies, but only for 10 years.

## Internal combustion engine generators

Internal combustion engine generators are usually reciprocating engines similar to a road vehicle but designed for running on gas, diesel or fuel oil. The former normally uses spark ignition and the last two use compression ignition. The range of sizes available run from 1kWe (Honda ECOWILL micro CHP, currently only available in Japan[27] up to around 4.5MWe (CIBSE, 1999).

These are the most common type of CHP plant used in buildings. They are reliable and easy to service and for the smaller duties required for most building applications, they tend to be more efficient than gas turbines. Part load operation is possible down to 25 per cent of electrical demand, although efficiencies suffer. If thermal demands do not track electrical demands, then thermal storage can be used to cater for asymmetric conditions. Electrical efficiencies for a typical gas spark ignition machine lie between 28 and 35 per cent.

## Gas turbines

Gas turbines use as the prime mover a jet similar to that used to power an aircraft. The exhaust gases from this are channelled into a turbine which in turn drives an electricity generator, with electrical efficiencies typically in the range 25–33 per cent. Waste heat is recovered via a heat exchanger in the exhaust and used for heating purposes. Historically, this technology has been applied to large-scale power station plant, frequently in combination with steam turbines (see the discussion of combined cycle turbines below), whilst individual CHP plants are available capable of generating up to 200MWe. However, recent developments have seen the introduction of packaged plant rated as low as 30kWe into the small-scale CHP marketplace.

For the smaller plant in the 30–350kWe range, sometimes referred to as microturbines, electricity is normally generated as direct current and hence inverters are required to convert to an alternating current for connecting with the main distribution circuit and interfacing with the grid.

Part load operation may be controlled using inlet guide vanes and/or by cycling multiple turbines. Thermal storage is essential to cater for varying thermal loads.

## Steam turbines

A steam turbine uses the same principle as a gas turbine but the turbine's blades are rotated using the pressure from steam or a vapourized organic fluid such as toluene (Organic Rankine Cycle – see below). Despite their low electrical efficiency – a basic plant generates up to 10 times more heat than electricity (Figure 2.28) – the steam turbine has been used as the primary means of electricity generation for over 100 years. The technology is both well developed and reliable. Most plants are very large, 10MWe or greater, however, smaller packaged plants have been developed generating as little as 100kW electricity.

Steam is generated in a boiler, which can be fired from any fuel, including biomass and biogas, and condenser water is normally returned to the boiler as feed water. For CHP installations either a back pressure or extraction condensing turbine is used. Because of their low electrical output, small-scale plants have found limited application, although one such plant marketed in the US by Carrier as the Microstream turbine generator has been designed to tap steam from existing steam installation pressure reducing sets, generating up to 275kWe through a Euler steam turbine when steam is not required for other purposes.

Because of the volatility of the heat exchange medium the Organic Rankine Cycle (ORC) allows a steam turbine to operate at much lower temperatures and is therefore compatible with biomass and even geothermal energy sources. ORC-based products are available for a wide range of duties and products are in various stages of development globally for single dwelling use (Chapter 3.2 Operational Energy and Carbon).

## Combined cycle turbines

The steam generated by a gas turbine is used to power a steam turbine and both generate electricity resulting in a much higher electrical efficiency than either gas or steam turbines in isolation (see Figure 2.29). Historically, the main application for combined cycle plant has been for power stations, hence the installations tend to be 20MWe and upwards.[28]

**Figure 2.27** *Steam turbine internal view*

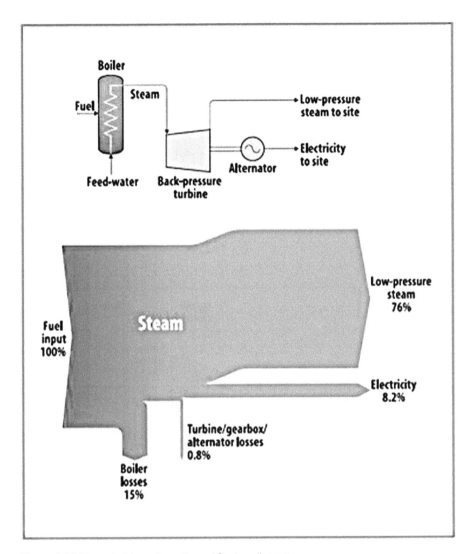

**Figure 2.28** *Steam turbine schematic and Sankey diagram*

## Fuel cell CHP

A fuel cell is essentially a battery that uses an external fuel source and can be used to generate electricity and heat simultaneously, the main discharge being water. It was originally developed as a method for generating power and producing pure water for the US space programme.

The process uses a catalyst to convert hydrogen and oxygen to electricity, heat and water. There is an extremely active research programme globally, both in the application of fuel cells to CHP and to powering motor vehicles (see Chapter 3.10 Design to Reduce Vehicle Impacts). The main types of fuel cell currently available are:

- proton exchange membrane/polymer electrolyte membrane (PEFC/PEM);
- phosphoric acid (PAFC);
- molten carbonate (MCFC);
- solid oxide (SOFC);
- alkaline (AFC);
- direct methanol (DMFC).

The electro-chemical reactions for the PEM fuel cell are shown in the schematic in Figure 2.30. All fuel cells contain an electrolyte sandwiched between an anode and a cathode. In the case of a PEM fuel cell, hydrogen is supplied to the anode on which a platinum catalyst has been seeded. This separates the hydrogen into positive and negative ions driving the positive ions (protons) through the solid polymer catalyst towards the cathode which is fed with oxygen, whilst the negative ions generate electricity in the process, and heat and

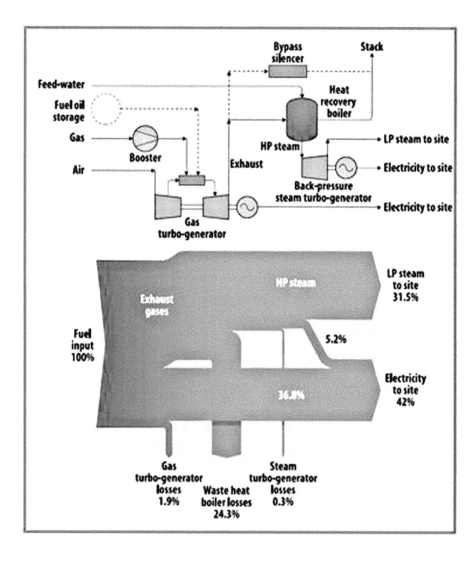

**Figure 2.29** *Combined cycle turbine Sankey diagram and schematic*

water emanate from the combining of the hydrogen and oxygen at the cathode.

PEM fuel cells operate at around 80°C and are the lightest of the currently available technologies, which is why they are favoured for road vehicle applications. The platinum catalyst is expensive, however, and they are prone to carbon monoxide poisoning of the platinum.

The PAFC uses liquid phosphoric acid as an electrolyte, the acid is contained in a Teflon-bonded silicon carbide matrix and porous carbon electrodes containing a platinum catalyst. Figure 2.31 shows a cutaway image of a 400kWe UTC packaged PAFC plant that has gained widespread application in the US. The process is similar to PEM fuel cells but the package is heavier and more robust. PAFC are commonly called 'first generation' fuel cells since they were the first to be incorporated into commercial packages on a large scale. They typically operate at around 200°C at electrical efficiencies of between 40 and 45 per cent. Costs, again, are very high, primarily because of the platinum.

Molten carbonate fuel cells operate at around 650°C and electrical efficiencies between 50 and 60 per cent; they are not as durable as PAFC, but they are more resistant to poisoning than PEM fuel cells.

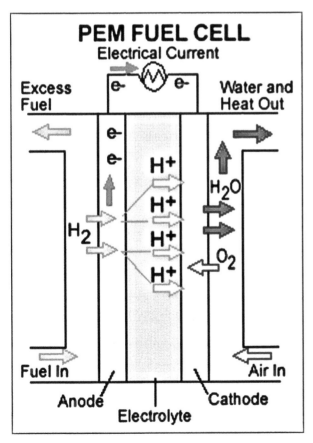

**Figure 2.30** *Proton Exchange Membrane (PEM) process*

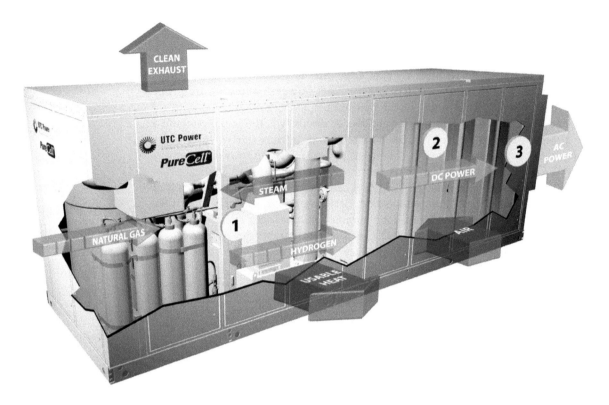

**Figure 2.31** *Packaged phosphoric acid fuel cell CHP unit*

*Source*: Image courtesy of UTC Power

Solid oxide fuel cells have a ceramic electrolyte that operates at 1000°C with similar efficiencies to MCFC. Although they have mainly been used for large plants because of the need for shielding against the high temperatures, they have recently been adapted for use in micro CHP plants (Chapter 3.2 Operational Energy and Carbon).

Alkaline fuel cells were the first type used in the space programme with an electrolyte of potassium hydroxide operating at temperatures of up to 250°C with a 60 per cent electrical efficiency, although they are susceptible to $CO_2$ poisoning.

Recent developments include fuel cells which work directly from methanol, rather than hydrogen (DMFC) and a cell that uses electricity from photovoltaics or wind turbines (regenerative fuel cells), although products are still under development using these technologies.

Hydrogen can be manufactured separately and stored locally (in a moving vehicle for example), whilst oxygen is obtained from a supply of ambient air. Alternatively, hydrogen can be produced on-site from any energy source, such as natural gas, biogas, biomass, wind turbines or photovoltaics using such technologies as the electrochemical hydrogen separator. The packaged plant shown in Figure 2.31 comprises a PAFC fed with hydrogen derived from gas by steam reforming within the package. A 200kWe version of this plant was used in the first fuel cell CHP installation in the UK at Woking, fuelled by a combination of natural gas and gas recovered from an anaerobic digester.

## Connection to neighbourhood energy schemes

There are a number of key issues that drive the decision whether to develop a district or community energy scheme, either within the boundaries of a development or a neighbourhood scheme in collaboration with others:

1 carbon efficiency;
2 economies of scale and diversity;
3 potential for external funding, design, construction and operation;
4 energy strategy of local authority.

For a development to achieve zero carbon emissions, it is necessary to open the door for zero carbon energy generation. Although this is possible for single buildings such as the Kingspan Lighthouse and Cub House pilots at the UK Building Research Establishment (BRE) Innovation Park near Watford and elsewhere (Chapter 3.2 Operational Energy and Carbon), the choices of technology increase significantly if heating, cooling and electricity can be generated from large scale plants serving mixed uses.[29] Economies of scale have an impact at a number of levels, not just the financial, but also in terms of demand profile and hence carbon efficiency. A mix of uses results in a demand profile that is much steadier than for an individual building, whilst energy generation can be achieved much more efficiently if plant doesn't have to cycle rapidly in response to load changes. In particular, a mix of residential and commercial uses will guarantee operation for at least 18 hours a day and throughout the weekend. Continuous industrial demand, swimming pools and uses such as airport terminals, newspaper offices and tele-support facilities can potentially provide a useful 24 hours a day, 365 days a year demand.

District energy schemes require large central plant rooms and usually extensive lengths of buried pipework. The cost of providing these and the transmission losses associated with them will depend primarily on the density of the development. A low rise development will need a separate building to house energy plant and perhaps many kilometres of pipework buried externally, whilst a high density development with, say, one or two high rise buildings on a small site is likely to require plant and pipework to be contained within the shell of the buildings. In both cases costs and energy losses will need to be compared with a decentralized approach in testing the viability of the scheme. Historically, large-scale district energy has been used for entire cities, towns or districts. It has been used extensively across mainland Europe, including the Scandinavian countries, Iceland and Russia, the US and Canada; with the earliest commercial system constructed in Lockport, New York in 1877. Iceland derives approximately 95 per cent of its heating from geothermal district heating, whilst Denmark offsets some 60 per cent of its total heat demand from district systems, 80 per cent of which uses CHP plant. Although Russia obtains some 50 per cent of its electricity from district schemes, many of these are very old and inefficient. Uptake in the UK has been very slow in comparison;

although the earliest system was installed in Pimlico, London in 1950, penetration by 2000 was only 1 per cent of heating, including large schemes in Nottingham, Sheffield and Southampton. District heating gained a bad reputation in the UK and elsewhere because many of the early steam and high temperature hot water systems suffered from leaks and low thermal efficiency, with basic temperature controls leading to discomfort, whilst the lack of heat metering in individual dwellings meant there was no incentive to save energy. With pre-insulated pipework, modern jointing techniques, thermostatic control and local heat metering, these problems are a thing of the past.

Whereas currently there are some 2500 district energy schemes in the US generating some 9 per cent of electricity consumed, the Thermal Energy Efficiency Act (US Congress, 2009), which was referred to the Committee on Energy and Natural Resources in August 2009, would, if enacted, require that this value be increased to 20 per cent through CHP by 2030, suggesting that this would result in some 800 million tonnes of carbon being saved per annum (equivalent to taking half the vehicles off the road in the US).

There are numerous permutations of energy source that can be used in matching generation profile to demand. These do not have to rely totally on central plant; for example building integrated photovoltaic cells or wind turbines could be linked to an electrical distribution network also fed from a combined heat and power plant, provided that the demand profile maximizes the utilization of all sources (see Figure 2.32, which shows an integrated utility approach at Hammarby, Sweden). The economic equation will depend in part on the availability of a feed-in tariff for electricity and the differential between tariffs for different sources.

If there is a good rate for selling renewable electricity to the grid, then there is a benefit in generating more electricity than is required to meet demand. In the UK, for example, there is a premium for generating electricity from PV compared with biomass CHP, which could provide an incentive to maximize the area of PV installed on available surfaces of a suitable orientation and solar exposure.

Any of the LZC energy sources referred to above can be considered for inclusion in a district heating system, but during the development of an energy strategy at the masterplanning stage of a project the following key questions should be considered:

- Is the local or regional authority, or development agency planning a district energy system for an area that could include the site under consideration, or is there one existing that has spare capacity or could be extended to serve the site?
- What renewable fuels or energy sources exist in the locality that might be tapped into; for example geothermal boreholes, waste transfer station, sewage treatment works?
- Does the site lend itself to regular deliveries of fuels such as biomass; for example having a jetty nearby suitable for deliveries by barge?
- Are there suitable locations for central plant, chimney, fuel storage and delivery access?
- Would the site lend itself to industrial type plant, such as gasification of biomass or waste?

- Would an Energy Services Company (ESCo) or Multi-Utility Services Company (MUSCo) be interested in taking over the design, finance, installation and operation of some or all of the installation?
- What sources of funding and subsidy could be available?

It is necessary to assess space requirements for major plant items, such as an energy centre, during the masterplanning stage. However, the rules of thumb available are quite limited when it comes to the 'alternative' technologies such as gasification and fuel cell CHP. For example, the UK-based BSRIA produces rules of thumb (Pennycock, 2003) that only cover boilers, calorifiers, air handling plant, chillers, cold water tanks, hot water storage, cooling towers, electrical sub-stations and standby generation. Most district energy schemes will incorporate some element of CHP and, although a gas-fired CHP plant should be no larger than a boiler of the same duty, consideration should be given to the potential additional space requirements for thermal storage, absorption chillers, ice storage, fuel storage and/or fuel cells. In many cases

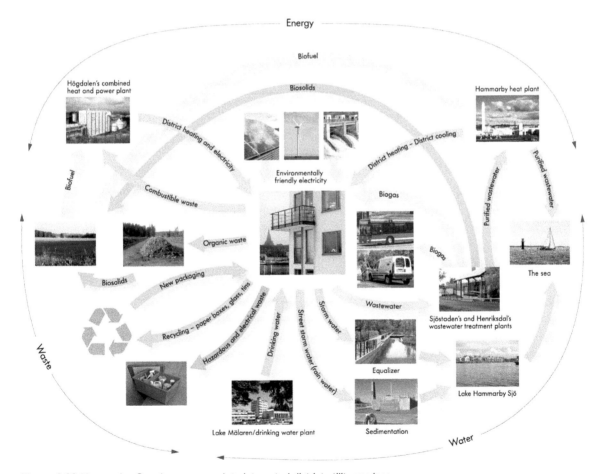

**Figure 2.32** *Hammarby, Sweden – approach to integrated district utility services*

*Source*: Lena Wettren, Bumlings AB

sufficient design development will be required at the concept stage to determine approximate plant and equipment space and access requirements.

The 'eco-cycle' shown in Figure 2.32 is a representation of this multi-source integrated approach to energy and utility supply applied to the town of Hammarby in Sweden, also known as the Hammarby Model.

There are numerous bodies that have been established to support district heating and energy services, such as the Danish District Heating Association, the French District Heating and Cooling Association and in the US the International District Energy Association.[30] There is currently no equivalent UK organization apart from the Combined Heat and Power Association.

## Funding and energy services

It is worth investigating options for funding and contracting of technologies and energy services for the proposed development at the masterplanning stage because it can impact on spatial planning and economic viability.

Government subsidies in the form of grants and awards such as the UK Government's Low Carbon Building Programme[31] are available in many countries globally and in certain cases there may be funding available from other sources such as federal funding programs in the US or the European Local Energy Assistance scheme (ELENA)[32] and the Intelligent Energy – Europe Programme[33] in the EU, although many of these are designed to support local authorities, rather than individual projects.

The practice of using an ESCo to finance, design, install and operate district or community energy systems has been applied in the US, many European countries and elsewhere. There is a long history of district energy being the responsibility of municipal authorities, particularly in mainland Europe, Russia and the former Eastern European and USSR countries. This was the case for the

**Figure 2.33** *Hammarby Sjostad apartments*

*Source*: Hans Kylburg

early schemes in Britain also, such as those in Glasgow and Manchester which were installed in the early 1900s.

The modern trend has been towards the privatization of energy services, for example in the UK companies have been established either as extensions of existing utility companies, energy consultants, CHP installers or facility management companies, or as stand-alone ESCos. There have been examples of ESCos being formed for specific projects, such as the joint venture between Woking Borough Council and ESCO International A/S, a Danish green energy company that created Thameswey Energy Ltd as a vehicle for developing a district energy system for the town of Woking in England. Other partnerships have involved local communities, such as those mentioned above for the Settle and Torrs hydro schemes. Communal energy projects involving local stakeholders are common across mainland Europe and Scandinavia in particular.

In the UK there are a number of issues that have to be addressed when using an ESCo for a specific project:

- Private wire network (PWN) versus separate meter connections to the licensed distribution system: the former avoids utility company distribution and connection charges within the development, but removes the protection that these offer the consumer.
- Consumer choice of electricity supplier: a PWN can make it more difficult for individuals to purchase their electricity from a competing utility company, which in the UK, for example, requires agreement with a licensed supplier.
- Metering and rates: modern pulsed metering means that it is possible to meter not only electricity consumption and cold water, but centrally provided heating, cooling and hot water services, with continuous monitoring of consumption and automatic billing. To minimize the risk of consumers opting to purchase their electricity from a competing utility company most ESCos will offer a discounted rate compared with that offered by utilities; this is possible because of the savings in utility service charges and income from Renewable Obiligation Certificates (ROCs) or their equivalent.
- Risk of the ESCo defaulting on the agreement: it is important that each project is protected from the ESCo going bankrupt or otherwise defaulting. In the UK this is normally achieved through the establishment of a 'Special Purpose Vehicle' (SPV) based on a carefully worded agreement between the developer and ESCo.

Similar issues will need to be addressed for all energy services contracts globally.

## Case histories

### Integrated multi-utility CHP plants, Fritzens, Austria

This installation is an integrated multi-utility waste to energy facility that comprises a CHP plant that uses unprocessed UCOs as a fuel and another that uses biogas (methane) from a sewage treatment plant. The plant processes

270kg/h of cooking fat to generate 1127kW electricity and 1500kW heat, producing some 9000MWh electricity and 12,000MWh heat annually. The UCO is obtained locally from restaurants, factories and some 10,000 households. The facility also includes a sewage treatment plant, the solid waste from which combines with other organic waste in a fermentation process producing methane which fuels the separate CHP plant. The heat from both these plants is used for district heating and to dry sewage, which in turn is available as a solid fuel.

## South Shropshire biodigester, Ludlow

The South Shropshire biodigester in Ludlow is an anaerobic digestion plant that has been operating since 2006 and that now handles around 5000 tonnes per annum of household kitchen waste producing 190kW of electricity and fertilizer that is returned to local agricultural land.[34] It has two stages of digestion with a residence time of 25–30 days, twin 900m³ digestion tanks and a 150m³ gas storage tank. Anaerobic digestion is controlled at a temperature of 42°C.

## 30 St Mary Axe

30 St Mary Axe, commonly known as 'The Gherkin', is located in the City of London at the heart of the financial sector.[35] It was developed by Swiss Re and designed by Foster & Partners architects with Hilson Moran Partnership building services engineers and Arup structural engineers and opened in 2003.

**Figure 2.34** *Anaerobic digestion tanks at South Shropshire facility, Ludlow*

*Source*: Biogen Greenfinch

**Figure 2.35** *30 St Mary Axe in City of London*

*Source*: John Armagh

At 180m the building was the second tallest in London at the time it was opened. It has space for some 4000 office workers and has a fully glazed double skin with blinds in the cavity. Each office floor has access to the spiralling light wells which allow for a degree of natural ventilation, if occupants choose the mixed-mode operation which also allows for full air conditioning with a 4-pipe fan coil system to meet the British Council of Offices (BCO) Class A specification. The fully glazed walls and exposed location allow for significant daylighting provided the blinds are not closed, and the intelligent lighting controls allow for zoned control of lights from daylight and occupancy detectors.

## Notes

1　www.exeterandeastdevon.gov.uk/Utilities-Capacity-Study/.
2　www.defra.gov.uk/sustainable/government/progress/regional/summaries/16.htm.
3　www.bco.org.uk/research/researchreports/detail.cfm?rid=135.
4　www.centerforlys.dk/lysetsdag/pdf/DTC_en.pdf.
5　www.merton.gov.uk/mertonrule.html.
6　A table for feed-in tariffs is available at www.energy.eu/#feedin.
7　www.lowcarbonbuildings.org.uk/.
8　www.energyatlas.org/contents/default.asp.
9　www.iftech.co.uk/index.php?option=com_content&task=view&id=21&Itemid=33.
10　www.geo-energy.org/aboutGE/powerPlantCost.asp.
11　intraweb.stockton.edu/eyos/energy_studies/content/docs/FINAL_PAPERS/5A-2.pdf.
12　See www.energypiles.com/page/index2.html for an example of open loop boreholes.
13　www.icold-cigb.net/.
14　www.ieahydro.org/reports/Hydrofut.pdf.
15　www.small-hydro.com/index.cfm?fuseaction=taskForce.home.
16　For more information on the Archimedes screw installations on the River Ribble and the Rivers Sett and Goyt, see www.communityshares.org.uk/case_studies/torrs/.
17　www.sourcewire.com/releases/rel_display.php?relid=51740.
18　A map of smoke control areas in the UK can be found at www.uksmokecontrolareas.co.uk/locations.php.
19　www.news.cornell.edu/stories/july05/ethanol.toocostly.ssl.html.
20　www.unh.edu/p2/biodiesel/article_alge.html.
21　For details of the new waste from energy 'incinerator' on the Isle of Man see www.iomguide.com/government/isle-of-man-incinerator.php.
22　For details on PAS 110, see the DECC biogas portal at www.biogas-info.co.uk/.
23　Information on the EC Intelligent Energy for Europe Programme can be found at www.swea.co.uk/downloads/Biogas_Brochure.pdf.
24　www.chpqa.com/html/notes.htm. Guidance Note 10 can be found at www.chpqa.com/html/notes.htm.
25　www.epa.gov/chp/.
26　www.chpa.co.uk.
27　For details of the Honda ECOWILL micro CHP, see www.micropower.co.uk/about/mchp.html.
28　chp.decc.gov.uk/cms/combined-cycle.
29　www.kingspanlighthouse.com/ and www.cubhousingsolutions.com/about.htm.
30　www.districtenergy.org/.
31　www.lowcarbonbuildings.org.uk/.
32　www.eib.org/products/technical_assistance/elena/index.htm.
33　ec.europa.eu/energy/intelligent/.
34　www.greenfinch.co.uk.
35　www.30stmaryaxe.com/.

## References

Action Energy, 2000. *Energy consumption guide 19: Energy uses in offices*. London: Action Energy. Available at http://217.10.129.104/Energy_Benchmarking/Offices/ECON19reprintMarch03.pdf (accessed 12 August 2010)

CIBSE, 1999. *Small-scale combined heat and power for buildings*. Application Manual AM12. London: CIBSE

CLG (Department for Communities and Local Government), 2008. *Planning Policy Statement 12: Local Spatial Planning.* London: CLG

CLG, 2009. *Proposed Changes to Part L and Part F of the Building Regulations: A Consultation Paper. Vol. 2: Proposed Technical Guidance for Part L.* London: CLG. Available at www.communities.gov.uk/documents/planningandbuilding/pdf/partlf2010consultation (accessed 21 February 2010)

Croxford, B. and Scott, K., 2006. *Can PV or Solar Thermal Systems be cost effective ways of reducing CO$_2$ emissions for residential buildings?* London: University College London. Available at http://eprints.ucl.ac.uk/2642/1/2642.pdf (accessed 3 March 2010)

DECC (Department of Energy and Climate Change), 2009. *Closed Consultation: Consultation on Renewable Electricity Financial Incentives.* Available at www.decc.gov.uk/en/content/cms/consultation/elec_financial/elec_financial.aspx (accessed: 3 March 2010)

DTI (Department of Trade and Industry), 2007. *UK Biomass Strategy 2007.* Working paper 1 Economic analysis of biomass energy. London: DTI. Available at www.berr.gov.uk/files/file39040.pdf (accessed 3 March 2010)

Gottschalg, R., Jardine, C.N., Rüther, R., Betts, T.R., Coribeer, G.J., Close, J., Infield, D. G., Kearney, M. J., Lan, K. H., Lane, K., Pang, H. and Tscharner, R. undated. *Performance of amorphous silicon double junction photovoltaic systems in different climatic zones.* Available at www.eci.ox.ac.uk/research/energy/downloads/pvcomp-cj-02.pdf (accessed 5 March 2010)

HMRC, 2004. Hydrocarbon oils: introduction of a reduced rate of duty on bioethanol. HMRC ref BN CE27/04

Hurther, S. and Schellschmidt, R., 2003. *Atlas of geothermal resources in Europe.* Amsterdam: Elsevier

Pennycock, K. (ed.), 2003. *Rules and Thumb. Guidelines for building services.* (BSRIA Guide: UK 4th edition) Bracknell: Building Services Research and Information Association

Rankie, R., Chick, J., and Harrison, G., 2006. Energy and carbon audit of rooftop wind turbine. *Proc. Institution of Mechanical Engineers Part A: Journal of Power & Energy.* vol 20, no 7, pp643–654

Phillips, R. Blackmore, P., Anderson, J., Clift, M., Aguilo-Rullon, A. and Pester, S. 2007. *Micro wind turbines in urban environments* (BRE Report FB17) London: BRE

US Congress, 2009. *S.1621: Thermal Efficiency Act of 2009.* Washington: US Congress. Available at www.govtrack.us/congress/bill.xpd?bill=s111-1621 (accessed 5 March 2010)

# 2.7

# Integrated Sustainable Transportation Planning

## Introduction

The development of an integrated sustainable transport strategy is a fundamental component in the masterplanning of a sustainable community. This is recognized by Government policy in most countries as highlighted in previous chapters on Policy, Legislation and Planning, Sustainable Communities and Social Sustainability.

In 2007 carbon emissions associated with transport globally represented some 23 per cent of the total, including some 16.7 per cent associated with road transport (IEA, 2009). In the US the contribution was 31 per cent, with 26.4 per cent from road transport, whilst in the UK 25 per cent of carbon emissions arose from transport, with 22.8 per cent from road transport, compared with 24.5 per cent and 19.1 per cent for the European Union as a whole. These high proportions from road transport for the US and UK are indicative of the reliance placed on the motor car and the historical lack of investment in public transport in the two countries.

Governments worldwide are having to address the carbon associated with transport as part of their strategy for addressing the contribution of transport to climate change and how to achieve long-term strategies for achieving domestic targets. In China, for example, the percentage of carbon emissions from transport in 2007 was only 6.8 per cent, including 4.6 per cent from road transport, however car ownership, although starting from an admittedly low base, has been rising exponentially in China in recent years and is expected to rise to around 4 per cent of the population in 2010 (compared to 76.5 per cent in the US and 30 per cent in Europe). The current production rate of road vehicles in China is approximately 9 million vehicles per year.

The EU Greening Transport Package (EU, 2008) focuses on introducing measures to further reduce emissions from road vehicles, expanding congestion charging schemes and funding rail, inland waterways and maritime transport projects under the Trans-European Networks and Marco Polo programmes to encourage a shift from road transport.

Through both existing transport planning policy (CLG, 2001) and publication of their latest consultation document on planning and transportation: *Delivering a Sustainable Transport System* (DfT, 2008) the UK Government has recognized that it is necessary to encourage people to find alternatives to using motor vehicles, whilst also increasing utilization of public transport so that it can run more efficiently.

The Department for Transport (DfT) has also published its strategy for meeting the UK Government's commitments under the 2009 Climate Change Act. *Low Carbon Transport: A Greener Future* (DfT, 2009) refers to a shift to new technologies and fuels, including sustainable biofuels, and commits to offering a significant subsidy towards low emission cars. Other key commitments cover making public transport accessible, attractive, low carbon and easy to use, integrated with other modes such as buses and cycles. It refers to a number of ongoing initiatives, such as Sustainable Travel Towns and Cities,[1] the National Cycle Plan and the requirement for local authorities to produce Local Transport Plans addressing sustainability issues. The Sustainable Travel Towns initiative was launched in England in 2004 with schemes for Darlington, Peterborough and Worcester, whilst other towns are being encouraged to follow suit following significant reductions in car use during the first year of what were essentially marketing/education exercises. A more ambitious exercise was launched in Cardiff in 2009 funded by the Welsh Assembly including free bike hire and bus travel in the city, a new park and ride service and improved cycling and walking routes. The National Cycle Plan will attempt to increase cycle use in the UK from the current 2 per cent of journeys, which compares with 27 per cent in The Netherlands.

Clearly the US has a much greater dependence on the motor car and a public transport network which, although well utilized in city centres, suffers from low demand elsewhere. The acronym-ready *Safe, Accountable, Efficient Transportation Equity Act: A Legacy for Users* (SAFETEA-LU) (USDoT, 2005) includes a number of measures that address sustainability issues including funding for:

- a pilot programme for 'non-motorized transportation' and 'Safe Routes to School';
- public transport and a series of congestion mitigation measures, including road pricing and high occupancy vehicle lanes;
- recreation trails
- a Transportation, Community and System Preservation Program (TCSP) involving the private sector in integrating plans for local communities with transport infrastructure and its long-term maintenance.[2]

The Federal Highways Administration (FHWA) has set out its 'Vital Few Objectives' on Environmental Stewardship and Streamlining to federal transportation projects including (USDoT, 2009):

- an integrated approach to multimodal planning, the environmental process and project development at a systems level;
- Context Sensitive Solutions (CSS) at a project level, including the integration of environmental and community values into transportation decisions from inception through to delivery;

- accelerated Environmental Assessment and Environmental Impact Statement processes;
- increased ecosystem and habitat conservation.[3]

# Sustainable transport strategy

There should be no further housing development in sites which are poorly located with respect to public transport; and no more road building to unlock the (car-dependent) 'potential' of development sites. New building should be at densities and in a form which supports a wide range of local facilities and makes it easy to reach them on foot or by bike. (Taylor and Sloman, 2008).

This is one of the principal recommendations from a report by a British organization called 'Transport for Quality of Life' (TQL) and commissioned by the Campaign for Better Transport, an influential non-governmental organization (NGO).

Taylor and Sloman's (2008) checklist reflects many of the issues covered in Chapter 2.4 Social Sustainability, including the transport credits that are used in Ecohomes, the predecessor to the CSH. LEED for Homes 2008 (see Chapter 1.3 Assessment Methodologies) has a similar requirement under LL5: Community Resources/Transit which sets targets for maximum distances from a range of amenities and transit nodes to 'encourage the building of ... homes in development patterns that allow for walking, biking or public transit (thereby minimizing dependency on personal automobiles and their associated environmental impact).'

The TQL Masterplanning Checklist sets out pointers towards achieving a sustainable transport strategy under the headings of:

- location of new developments;
- density of development;
- local facilities and jobs;
- street layout and design;
- public transport;
- parking;
- restraint on car movement;
- smart travel behaviour change programmes.

## Location of new developments

In its introduction PPG13 (CLG, 2001) states that 'Land use planning has a key role in delivering the Government's integrated transport strategy. By shaping the pattern of development and influencing the location, scale, density, design and mix of land uses, planning can help to reduce the need to travel, reduce the length of journeys and make it safer and easier for people to access jobs, shopping, leisure facilities and services by public transport, walking, and cycling.'[4]

TQL goes further by suggesting that new developments should not be located close to a motorway or high-speed dual carriageway, but be within safe walking distance of major public transport links and be adjacent to or within urban centres. This is based on research carried out on housing developments in the Oxford area in the late 1990s which demonstrated, not surprisingly, that people who lived close to main roads but suffered from poor public transport used their private cars for the majority of journeys (96 per cent) whilst those in close proximity to frequent and rapid public transport services used their cars less (65 per cent).

## Density of development

TQL suggests that higher densities of residential development (>200 dwellings/hectare) can be accommodated in locations with excellent public transport links.[5] Excellent public transport would typically be defined as a 'turn up and go' service, with waiting times of a couple of minutes or less during peak periods, but clearly there needs to be commensurate availability of local amenities.

## Local facilities and jobs

In the UK food alone is currently transported around 30 billion kilometres per year to arrive on our plates.[6] This corresponds to approximately 19 million tonnes $CO_2$, or about 17 per cent of the total UK carbon footprint, of which some 2 million tonnes is associated with trips to shops.

It follows from the findings reported by TQL on location of residential development referred to above, that ideally all dwellings should be within walking distance of local centres and that access can therefore be dedicated to pedestrians and cyclists only, with car access engineered to become more expensive, less convenient and less rapid. This is the principle behind 'filtered permeability' practised in Germany and elsewhere (see the discussion of street layout and design below). Convenient walking distance depends on the amenity, for example BRE's Ecohomes stipulates that post boxes and convenience stores should be within 500m to be truly convenient, whereas other amenities, such as schools, banks, medical centres, community facilities, etc. can be up to 1km away. Points awarded for access to a transport node double if it is within 500m of the dwelling compared to between 500m and 1km, and there are additional points for a high frequency of service during peak periods. LEED for Homes has a similar gradation, but with higher scores awarded the more amenities there are located within a quarter of a mile (400m), whilst transit services within 0.5 mile (800m) are awarded points based on the number of 'rides per weekday'.

Surveys in the UK dating from 1993 (ECOTEC Research and Consultancy, 1993) looked at the travel mode split for journeys to 'local centres'. For journeys of up to 1km, walking was the dominant mode (63 per cent of trips), followed by cycling (19 per cent). Between 1km and 1.6km cycling dominated (27 per cent) followed by walking (20 per cent). For longer trips (1.6–5km) cycling and walking were replaced by public transport and the car, in roughly equal proportions. The amenities at these local centres that were most visited were food shops, followed by newsagents, banks, post offices and medical facilities.

Where a development creates significant employment then consideration should be given not only to the commuting of residents outwards but also the journeys of those employed within the development. A 'win–win' outcome would be for most residents to be employed on-site, either working from home, in 'live/work' units or employed by the businesses located within or adjacent to the development, and for the balance to be drawn from the local community. For this reason Ecohomes rewards designs that encourage home working by ensuring occupants have sufficient space and connectivity to enable setting up an office (see Chapter 3.10 Design to Reduce Vehicle Impacts). A new planning statement for England and Wales (PPS4, 2009, CLG, 2009) refers under Policy EC2: Planning for Sustainable Economic Growth to the requirement for regional planning bodies and local planning authorities to 'ensure that their development plan ... facilitates new working practices such as live/work'. To this end a number of networks have sprung up in recent years to support and promote live/work; for example the live/work network in the UK and live/work world in the US.[7] However, according to liveworkhomes.co.uk, accessed in mid-2010 most UK mortgage lenders will not lend to a purchaser on a unit that has more than one third of its floor area designated as a workplace, which is compatible with a 3-storey unit having its ground floor as workplace and top two floors as living space.

## Street layout and design

The provision of safe and adequate routes for pedestrians and cyclists is fundamental to a modal shift in transportation planning. A safe pedestrian route is defined in Ecohomes (Chapter 2.4 Social Sustainability) in terms of pavement width, safe crossing points, visibility, speed limits, etc. and is factored into the 'Secured by Design' key points set out in Box 2.1. TQL reports on a study from 1997 that found that 'more than half of the residential areas of Seattle do not have sidewalks' (Taylor and Sloman, 2008). In contrast, in towns such as Freiberg and Groningen in Germany the concept of 'filtered permeability' has been introduced, in which through-traffic is channelled into purpose-designed routes and 'vehicular access points to residential areas are limited, while bridges, tunnels, bus gates and a panoply of short cuts assist the more sustainable modes.'

Access and distribution roads to residential areas should not present 'rat runs' and speed limits should be no more than 20mph, achieved through suitable design measures. Access around the dwellings should follow the principle of 'home zone' residential street design, physically restricting vehicles to approximately walking speed, with pedestrians given priority over vehicles.[8] This opens the whole street area to pedestrian activity, incorporating trees, planters, play equipment and seating to create a pleasant environment for pedestrians, whilst obstructing driver sight lines and obliging drivers to move slowly. Home Zones or *woonerf* were pioneered in The Netherlands in the 1970s but have only recently been piloted in the UK, with nine schemes receiving funding from the DETR from 2000.

**Figure 2.36** *Home zone at Rowling Gate development, Bristol, England*

*Source:* Photograph by Linda Bailey

Subsequently DfT has published the Quiet Lanes and Home Zones (England) Regulations (DfT, 2006) which set out the legal framework for establishing a home zone, including requiring consultation with those affected. They also provide guidance, indicating that home zones should only be considered when traffic volumes will be less than 100 vehicles in the peak hour during the afternoon. Speed limits should be considerably less than 20mph, but to make this enforceable requires the approval of the Secretary of State, which has historically not been given due to speedometer inaccuracy at low speed.

Local centres should be pedestrianized and designed with good access for cycles and a 'people-centred' street approach. They should include plants and trees, seats and play equipment and comply with the principles of 'Secured by Design' with clear sight lines and good lighting; interesting interactions between the street and shops, cafes and open spaces; generous pavement widths with space for street cafes and stalls. The design should be 'legible' so that pedestrians can easily establish their route, with clearly signposted access routes and parking/set-down zones for public transport, taxis and deliveries, where possible, segregated from the pedestrianized zone. Large areas of tarmac that have unrestricted access to vehicles should be avoided. Clearly signposted, accessible and sheltered parking for cycles should be available close to key destinations, such as transport nodes and retail centres.

## Public transport

For new developments in Vienna, a tram service must be provided with a maximum interval of 6 minutes even before the first occupants move in. TQL quotes a city official as saying:

It does look a bit wasteful. But the city council has a policy that when new housing is built there should never be a time when anybody has to be dependent on a car, because once they get into the habit it is hard to change. So before anybody can move in, the public transport has to be in place, working to the same standard we expect everywhere else in the city. (Taylor and Sloman, 2008)

There are numerous examples of successful tram systems across Europe, such as the one in Freiberg, Germany that was extended to serve the new suburb of Vauban during the early stage of development, resulting in only 16 per cent of journeys being in private vehicles (see Figure 2.37).

Some important lessons from these experiences should be considered by the masterplanner:

- Residential development should fall within 800m to 1km of a public transport node having good connectivity to regional hubs.
- Public transport provision should be planned so that it is running at maximum intensity in time for the first phase of a development.
- Wherever possible public transport should be provided with dedicated routeways.
- Priority should be given to access to public transport hubs by pedestrians and cycles, with sheltered cycle storage and only limited car parking.

In London, planning applications are expected to have assessed the level of accessibility to public transport using the Public Transport Accessibility Level (PTAL – DfT, 2007) which is determined from:

**Figure 2.37** *Tram station in Vauban, Freiberg, Germany*

*Source:* Photograph by Kaffeeeinsten

- the walking time from the development to the public transport access points;
- the reliability of the service modes available;
- the number of services available within the catchment;
- the level of service at the public transport access points – i.e. average waiting time.

The PTAL is determined within a scale from 1–6, where 1 is very poor and 6 is excellent.

For assessing access to amenities the 'Accession' software package was developed for the DfT by MVA Consulting. It is an integrated Geographic Information System (GIS) and analytical tool that imports information from a range of data sets then calculates and maps accessibility to key services such as hospitals and schools using public transport timetable data and geo-demographic data to assess the status of the transport network.

## Car parking

In the UK Planning policy for residential development (PPS3: Housing, CLG, 2000) it states that local planning authorities should take account of 'the extent to which the proposed development ... takes a design-led approach to the provision of car-parking space, that is well-integrated with a high quality public realm and streets that are pedestrian, cycle and vehicle friendly', whilst leaving the detail of car parking standards to individual local authorities.

On the other hand, Annex D to PPS13: Transport (CLG, 2001) sets out maximum numbers for non-residential facilities, whilst stating that there should be 'no minimum standards for (car parking in) development, other than for disabled people.'

The UK standards vary enormously between rural and urban areas; for example Annex 4 of the London Plan (2008) states that 'all developments in areas of good public transport accessibility and/or town centres should aim for less than 1 space per unit'. This compares with the requirement set out in the 'Supplementary Planning Document for provision of car parking space in residential development for the New Forest District' in Hampshire which allows 1.4 spaces per one bed unit sharing communal parking and space for three cars on-plot for a four bedroomed house.

The TQL report draws from case histories across mainland Europe that have adopted significantly lower residential car parking numbers combined with very good public transport links to local centres, whilst catering for cyclists and pedestrians, in proposing a maximum provision of 0.5 spaces per unit, including a proportion of units contracted to be 'car-free' but with access to a local car club (see below). The report also suggests that residents should be charged the full cost of parking provisions and that where possible car parking should be segregated from homes, using such devices as edge of development car parks, which can host car club vehicles and charging points for electric vehicles.

### Restraint to car movement

The key task for a masterplanner is to design developments so that other modes are faster and more convenient than the car. Local trips to public transport nodes, shops and other amenities should be easier and faster on foot or cycle than by car. Routes by road should be labyrinthine through 'home zone' environments that necessitate ultra low-speed driving and giving way to pedestrians. Furthermore, the lack of available space and the high cost of parking should also create significant deterrents to gratuitous driving for local trips.

## Smart travel behaviour

Travel behaviour is dependent, in part, on the options presented by the physical environment provided by the masterplan and its relationship to surrounding destinations and, in part, on the long-term management of the development. In general people will chose the option which gives the shortest travel time, the lowest cost and greatest convenience. The problem is that there is strong resistance to change that makes it very difficult to persuade people who are used to the independence and convenience of car ownership to change their habits.

However, there are a number of initiatives that are having some success worldwide, including car clubs, car sharing and travel plans for developments, schools, hospitals and workplaces:

- Car clubs provide members with access to a pre-booked car parked in a designated bay in their neighbourhood using a swipe card. In the UK there is a mix of commercial and community car clubs. In the US they are available in many major cities and on university campuses.[9]
- Travel planning can either form part of the development process or be used as a management tool.[10] Development transport plans will normally be required by local planning authorities to accompany a planning application (PPG13, CLG, 2001, paras 89–90). PPG13 also refers to local authorities setting targets for businesses, schools, hospitals, etc. to develop travel plans and para. 32 states that 'Businesses should make every effort – for instance by adopting travel plans – to encourage car sharing, and use of non-car modes of transport.'

In the UK the sustainable transport organization, Sustrans, provides a useful resource for schools to develop strategies for encouraging alternatives to the 'mum dropping kids off in her four wheel drive' mentality through developing safe alternatives.[11] In the US federal funding has been used to establish the National Center for Safe Routes to School which has been established to 'enable community leaders, schools and parents across the United States to improve safety and encourage more children to safely walk and bicycle to school'.[12] Funding is available for improving pedestrian and cycle routes, providing information, cycle racks and so forth.

## Prediction of transport-related emissions

Significant development in the UK will require a Transport Impact Assessment.[13] In order to predict any increase in $CO_2$, pollutant and noise emissions associated with a development, it is necessary to establish the baseline condition and then predict the change in emissions, both with the new development in place and with other proposed developments that will also impact conditions locally.

Baseline conditions may be determined from either a database, using information obtained from local monitoring stations or monitoring carried out specifically for the planning application. The Design Manual for Roads and Bridges provides guidance on assessing air quality, noise and vibration impacts for road projects, including reference to a number of computer models.[14] However, an air quality or noise impact assessment must take into account other significant sources (see Chapters 3.5 Air Quality, Hygiene and Ventilation and 3.8 Noise and Vibration).

In the US guidance for transportation impact analysis is provided by the Institute of Transportation Engineers,[15] whilst the US EPA have produced a vehicle emission model known as the Motor Vehicle Emission Simulator (MOVES) (USA EPA, 2010) that can be used for development projects. The FHWA's Traffic Noise Model (US FHWA DoT, 2008) allows the simulation of traffic noise.

## Notes

1    www.dft.gov.uk/pgr/sustainable/demonstrationtowns/.
2    www.fhwa.dot.gov/tcsp/.
3    For more detail refer to the US Department of Transportation FHWA website on Planning and Environment Linkages at: www.environment.fhwa.dot.gov/integ/index.asp.
4    It is recommended that this subsection is read with Chapter 2.2 Land Use and Density in mind.
5    This subsection should be read with Chapters 2.2 Land Use and Density and 2.4 Social Sustainability borne in mind.
6    Read this subsection in conjunction with Chapters 2.4 Social Sustainability and 2.5 Economic Sustainability.
7    Not surprisingly, both of these networks have websites: www.liveworknet.com/ in the UK and www.liveworkworld.com/ in the US.
8    www.homezones.org/.
9    www.carclubs.org.uk/.
10   Guidance can be downloaded from www.dft.gov.uk/pgr/sustainable/travelplans/work/.
11   www.sustrans.org.uk/what-we-do/safe-routes-to-schools.
12   www.saferoutesinfo.org/.
13   See the Planning Advisory Service guidance at www.pas.gov.uk/pas/core/page.do?pageId=119232.
14   www.standardsforhighways.co.uk/dmrb/vol11/index.htm.
15   www.allbusiness.com/transportation-warehousing/1085310-1.html.

## References

CLG (Department for Communities and Local Government), 2000. *Planning Policy Statement 3: Housing (PPS3)*. Available at www.communities.gov.uk/planningandbuilding/planning/planningpolicyguidance/planningpolicystatements/planningpolicystatements/pps3/ (accessed 5 August 2010)

CLG, 2001. *Planning Policy Guidance 13: Transport (PPG13)*. Available at www.communities.gov.uk/publications/planningandbuilding/ppg13 (accessed 6 March 2010)

CLG, 2009. *Planning Policy Statement 4: Planning for Sustainable Economic Growth (PPS4)*. Available at www.communities.gov.uk/planningandbuilding/planning/

planningpolicyguidance/planningpolicystatements/planningpolicystatements/pps4/ (accessed 5 August 2010)

DfT (Department for Transport), 2006. *Circular 02/2006: Quiet Lanes and Home Zones (England) Regulations.* Available at www.dft.gov.uk/pgr/sustainable/homezones/ular22006thequietlanesan5740.pdf (accessed 6 March 2010)

DfT, 2007. *Guidance on Transport Assessment.* London: TSO. Available at http:// webarchive.nationalarchives.gov.uk/+/http://www.dft.gov.uk/ adobepdf/165237/202657/guidanceontapdf (accessed 4 August 2010)

DfT, 2008. *Delivering a Sustainable Transport System: Consultation on planning for 2014 and beyond.* Available at www.dft.gov.uk/consultations/archive/2009/planning/ dastsconsultation.pdf (accessed 6 March 2010)

DfT, 2009. *Low Carbon Transport: A Greener Future.* Available at www.dft.gov.uk/pgr/ sustainable/carbonreduction/low-carbon.pdf (accessed 6 March 2010)

ECOTEC Research and Consultancy, 1993. *Reducing Transport Emissions through Planning.* London: HMSO

EU, 2008. *The Greening Transport Package.* Available at http://ec.europa.eu/transport/ strategies/2008_greening_transport_en.htm (accessed 6 March 2010)

IEA, 2009. *World Energy Outlook 2009.* Paris: IEA. Available at www. worldenergyoutlook.org/docs/weo2009_es_english.pdf (accessed 6 March 2009)

Taylor, I. and Sloman, L., 2008. *Masterplanning Checklist for Sustainable Transport in New Developments. Transport for Quality of Life.* Available at www.bettertransport. org.uk/media/press_releases/ocotober_2008/checklist (accessed 12 January 2010)

USDoT, US Department of Transport, 2005. *Safe, Accountable, Flexible, Efficient Transportation Equity Act.* Available at www.fhwa.dot.gov/safetealu/legis.htm, Washington: Federal Highway Administration (accessed 6 March 2010)

USDoT, 2009. *Planning & Environment Linkages: A Guide to Measuring Progress in Linking Transportation Planning and Environmental Analysis.* Washington: FHWA (Federal Highway Administration). Available at www.environment.fhwa.dot.gov/ integ/meas_progress.asp#s2 (accessed 6 March 2010)

US EPA, 2010. *Policy Guidance on the Use of MOVES 2010 for State Implementation Plan Development, Transportation, Conformity, and Other Purposes.* Available at www.epa.gov/OMS/models/moves/420b09046.pdf (accessed 6 February 2010)

US FHWA DoT, 2008. *Traffic Noise Model Version 3.* Available at www.fhwa.dot.gov/ environment/noise/tnm/index.htm (accessed 6 February 2010)

# Part 3

# Sustainability and Building Design

# 3.1
# Sustainability Strategy

## Introduction

The following chapters deal with the issues that need to be considered in the sustainable design of individual buildings. In this chapter we establish what is involved in developing a strategy for a planning application and how that relates to environmental impact assessment and strategic issues such as climate change, sick building syndrome (SBS), life cycle assessment, cost and commercial factors.

## Developing a sustainability framework for design

Sustainability involves every member of a design team and must be considered from the earliest conceptual stages and monitored through to commissioning and delivery of the completed building. It impacts on almost every decision the design team has to make and hence a framework is required within which decisions can be checked. This framework should include all relevant targets that the development must incorporate in order to meet legislative, planning, performance and funding requirements, including any targets that the developer wishes to meet for commercial or moral reasons.

An appropriate assessment scheme, such as BREEAM, the CSH or LEED, makes a very useful framework for sustainable design. In many instances a specified performance under one of these schemes will be required either by the planning authority, funder or developer. Most of the schemes are primarily based on environmental criteria, so it will be necessary to add socio-economic sustainability criteria based on either local/regional planning requirements or commercial factors. Planning for future changes in building regulation, such as zero carbon emissions, may impact on targets for buildings that are not going to be commenced until after these regulations come into force.

Other environmental criteria that are not found in the chosen assessment scheme should be imported from the EIA into this framework. For tall buildings this might include criteria for wind environmental impact and overshadowing, for example (Chapter 2.3 Massing and Microclimate).

A good example of a regional planning framework that covers most of the above issues is provided by the London Plan (Mayor of London, 2008), which is currently under review, and the corresponding supplementary planning guidance (SPG) on Sustainable Design and Construction (Mayor of London, 2006). Although planning applications are made to the relevant London

Borough, the London Mayor is a consultee on specified major projects. These documents are used by the Mayor's officers at the Greater London Authority to review applications, whilst the Boroughs are expected to apply the principles in the London Plan and SPG in considering all planning applications. The SPG includes 'Essential Standards' and 'Mayor's Preferred Standards' for applying to new projects in London. Many of these are similar to criteria which are included in BREEAM or CSH. Box 3.1 summarizes the Essential Standards and their relationship to BREEAM/CSH. Although intended for London, they could be applied to most building projects worldwide with minor adjustments. Also included are the BREEAM and CSH credits that do not overlap with Essential Standards and a column that references key legislation or policy documents along with the section within this book that covers the area concerned.

---

## Box 3.1 London Plan sustainable design and construction essential standards

| SPG Sustainable Design and Construction Essential Standard | BREEAM Offices 2008/CSH 2009 | Key Legislation or Policy (England & Wales)/Cross reference |
|---|---|---|
| 100% of development on previously developed land, unless very special circumstances can be demonstrated | BREEAM Offices LE1 – At least 75% of the proposed development's footprint is on an area of land which has previously been developed for use by industrial, commercial or domestic purposes in the last 50 years. | PPS3: Housing / Ch. 2.2 Land Use and Density |
| | BREEAM Offices LE2 – To encourage positive action to use contaminated land that otherwise would not have been remediated and developed. | PPS3: Housing / Ch. 2.2 Land Use and Density |
| Development density should be maximized based on local context, design principles, open space provision and public transport capacity – based on Densities Matrix in London Plan which gives a range from 150 habitable rooms per hectare (hr/ha) for suburban location having PTAL of 0–1 to 1100hr/ha for central location with PTAL of 4–6. | CSH Eco 5 – To promote the most efficient use of a building's footprint by ensuring that land and material use is optimized across the development. | PPS1: Sustainable Development. PPS13: Transport, Annex D / Ch. 2.2 Land Use and Density, Ch. 2.7 Integrated Sustainable Transportation Planning |
| Follow principles of good design:<br>• maximize the potential of sites<br>• promote high quality inclusive design and create or enhance the public realm<br>• contribute to adaptation to, and mitigation of the effects of climate change<br>• respect local context, history, built heritage, character and communities | BREEAM Offices Mat 7 – To recognize and encourage adequate protection of exposed parts of the building and landscape, therefore minimizing the frequency of use of replacement materials. | Climate Change Act 2009. By Design – Urban Design in the Planning System: Towards Better Design, CABE, 2000. Dept of Health (2003) Tackling Health Inequalities: A programme for action. / Ch. 2.4 Social Sustainability. See also Sustainability and Climate Change below |

- provide for or enhance a mix of uses
- be accessible, usable and pe rmeable for all users
- be sustainable, durable and adaptable in terms of design, construction and use
- address security issues and provide safe, secure and sustainable environments
- be practical and legible, be attractive to look at and, where appropriate, inspire, excite and delight
- respect the natural environment and biodiversity, and enhance green networks and the Blue Ribbon Network (rivers)
- address health inequalities.

| | | |
|---|---|---|
| Existing buildings are reused where practicable, where the density of development and residential amenity are optimized and where the building conforms or has the potential to meet the standards for energy, materials, biodiversity and water conservation below | BREEAM Offices Mat 3 – To recognize and encourage the in-situ reuse of existing building facades. Mat 4 – To recognize and encourage the reuse of existing structures that previously occupied the site. | 3.12 Materials Specification |
| Minimize need for and use of mechanical ventilation, heating and cooling systems | BREEAM Offices – inherent in energy credits which reward low carbon design. Hea7– To recognize and encourage adequate cross flow of air in naturally ventilated buildings and flexibility in air-conditioned/ mechanically ventilated buildings for future conversion to a natural ventilation strategy. | Part L: 2010 Building Regulations / Ch. 3.2 Operational Energy and Carbon, Ch. 3.4 Design for Natural Ventilation |
| Buildings provide for flexibility of uses during their projected operational lives | | |
| Buildings adapt to and mitigate for the effects of the urban heat island and the expected increases in hot dry summers and wet mild winters | | Climate Change Act 2009 / Ch. 2.2 Land Use and Density. See also Sustainability and Climate Change below |
| Design in facilities for bicycles and electric vehicles | BREEAM Offices Tra 3 – To encourage building users to cycle by ensuring adequate provision of cyclist facilities. CSH Ene 8 – Cycle Storage to encourage the wider use of bicycles as transport by providing adequate and secure cycle storage facilities, thus reducing the need for short car journeys | PPS13: Transport / Ch. 2.7 Integrated Sustainable Transportation Planning |

| | BREEAM Offices Tra 4 – To recognize and encourage the provision of safe and secure pedestrian and cycle access routes on the development. | PPS13: Transport / Ch. 2.7 Integrated Sustainable Transportation Planning |
|---|---|---|
| Carry out an energy demand assessment<br>Maximize energy efficiency<br>Major commercial and residential developments to demonstrate that consideration has been given to the following ranking method for heating and where necessary cooling systems:<br>• passive design<br>• solar water heating; then<br>• combined heat and power for heating and cooling (i.e.trigeneration), preferably fuelled by renewables; then<br>• community heating and cooling; then<br>• heat pumps; and then<br>• gas condensing boilers | BREEAM Offices Ene 1 – To recognize and encourage buildings that are designed to minimize the $CO_2$ emissions associated with their operational energy consumption.<br>CSH Ene 1 – To limit emissions of carbon dioxide ($CO_2$) to the atmosphere arising from the operation of a dwelling and its services. Ene 2 – To future proof the energy efficiency of dwellings over their whole life by limiting heat losses across the building envelope. | Part L: 2010 Building Regulations / Ch. 3.2 Operational Energy and Carbon<br>Part L: 2010 Building Regulations / Ch. 3.2 Operational Energy and Carbon<br>Part L: 2010 Building Regulations. UK Energy Efficiency Action Plan, 2007. UK Renewable Energy Strategy, 2009, Energy White Paper, 2007 / Ch. 2.6 Energy Strategy and Infrastructure, Ch. 3.2 Operational Energy and Carbon |
| | CSH Ene 3 – To encourage the provision of energy efficient internal lighting, thus reducing the $CO_2$ emissions from the dwelling. | Part L1: 2010 Building Regulations / Ch. 3.2 Operational Energy and Carbon, Ch. 3.6 Light and Lighting |
| | BREEAM Offices Ene 2 – To recognize and encourage the installation of energy sub-metering that facilitates the monitoring of in use energy consumption. Ene 3 – To recognize and encourage the installation of energy sub-metering that facilitates the monitoring of in use energy consumption by tenant or end user. | Ch. 3.2 Operational Energy and Carbon |
| | BREEAM Offices Ene 8 – To recognize and encourage the specification of energy efficient lifts. Ene 9 – To recognize and encourage the specification of energy efficient escalators and travelling walkways | Ch. 3.2 Operational Energy and Carbon |
| | CSH Ene 4 – To provide a reduced energy means of drying clothes. | Ch. 3.2 Operational Energy and Carbon |

|  |  |  |
|---|---|---|
|  | CSH Ene 5 – To encourage the provision or purchase of energy efficient white goods, thus reducing the $CO_2$ emissions from appliance use in the dwelling. | EC Regulation No 1980/2000 on revised Community ecolabel award scheme / Ch. 3.2 Operational Energy and Carbon |
| Wherever on-site outdoor lighting is proposed as part of a development it should be energy efficient, minimizing light lost to sky | BREEAM Offices Ene 4 – To recognize and encourage the specification of energy efficient light fittings for external areas of the development. Pol 7 – To ensure that external lighting is concentrated in the appropriate areas and that upward lighting is minimized, reducing unnecessary light pollution, energy consumption and nuisance to neighbouring properties. CSH Ene 6 – To encourage the provision of energy efficient external lighting, thus reducing $CO_2$ emissions associated with the dwelling. | Part L: 2010 Building Regulations / Ch. 3.2 Operational Energy and Carbon, Ch. 3.6 Light and Lighting |
| Carbon emissions from the total energy needs (heat, cooling and power) of the development should be reduced by at least 10% by the onsite generation of renewable energy. (London Plan 2008 increased this to 20%, but 2009 consultation has different approach) | BREEAM Offices Ene 5 – To reduce carbon emissions and atmospheric pollution by encouraging local energy generation from renewable sources to supply a significant proportion of the energy demand. CSH Ene 7 – To reduce carbon emissions and atmospheric pollution by encouraging local energy generation from renewable sources to supply a significant proportion of the energy demand. | Part L: 2010 Building Regulations / Ch. 2.6 Energy Strategy and Infrastructure, Ch. 3.2 Operational Energy and Carbon |
|  | BREEAM Offices Mat 2 – To recognize and encourage the specification of materials for boundary protection and external hard surfaces that have a low environmental impact, taking account of the full life cycle of materials used. | Strategy for Sustainable Construction 2008 Ch13 / Ch. 3.12 Materials Specification |
| 50% timber and timber products from Forest Stewardship Council (FSC) source and balance from a known temperate source | BREEAM Offices Mat 5 – To recognize and encourage the specification of responsibly sourced materials for key building elements. CSH Mats 2/3 – To recognize and encourage the specification of responsibly sourced materials for the basic/finishing building elements. | Strategy for Sustainable Construction 2008 Ch13 / Ch. 3.12 Materials Specification |

| | | |
|---|---|---|
| Insulation materials containing substances known to contribute to stratospheric ozone depletion or with the potential to contribute to global warming must not be used | BREEAM Offices Mat 6 – To recognize and encourage the use of thermal insulation which has a low embodied environmental impact relative to its thermal properties and has been responsibly sourced.<br><br>CSH Pol 1 – To reduce global warming from blowing agent emissions that arise from the manufacture, installation, use and disposal of foamed thermal and acoustic insulating materials. | Strategy for Sustainable Construction 2008 Ch13 / Ch. 3.12 Materials Specification |
| | BREEAM Offices Pol 1 – To reduce the contribution to climate change from refrigerants with a high global warming potential. Pol 2 – To reduce the emissions of refrigerants to the atmosphere arising from leakages in cooling plant. | Climate Change Act 2008 / Ch. 3.13 Pollution |
| Minimize use of new aggregates | BREEAM Offices Wst 2 – To recognize and encourage the use of recycled and secondary aggregates in construction, thereby reducing the demand for virgin material. | Aggregates Levy (Registration and Miscellaneous Provisions) (Amendment) Regulations 2007. Strategy for Sustainable Construction 2008 Ch12 / Ch. 3.12 Materials Specification, Ch. 4.3 Construction Waste Management |
| Residential developments to achieve average water use in new dwellings of less than 40m³ per bedspace per year (approximately 110litres/head/day) | CSH Wat 1 – To reduce the consumption of potable water in the home from all sources, including borehole well water, through the use of water efficient fittings, appliances and water recycling systems. Wat 2 – To encourage the recycling of rainwater and reduce the amount of mains potable water used for external water uses. | Part G Building Regulations. / Ch. 3.9 Water Conservation |
| | BREEAM Offices Wat 1 – To minimize the consumption of potable water in sanitary applications by encouraging the use of low water use fittings. | Ch. 3.9 Water Conservation |
| 100% metering of (water supplies to) all newly built property | BREEAM Offices Wat 2 – To ensure water consumption can be monitored and managed and therefore encourage reductions in water consumption. | Ch. 3.9 Water Conservation |

| | BREEAM Offices Wat 3 – Major leak detection to reduce the impact of major water leaks that may otherwise go undetected. Wat 4 – Sanitary supply shut-off to reduce the risk of minor leaks in toilet facilities. | Ch. 3.9 Water Conservation |
| --- | --- | --- |
| Demonstrate that adverse impacts of noise have been minimized, using measures at source or between source and receptor (including choice and location of plant or method, layout, screening and sound absorption) in preference to sound insulation at the receptor, wherever practicable | BREEAM Offices Pol 8 – To reduce the likelihood of noise from the new development affecting nearby noise-sensitive buildings. CSH Hea 2 – To ensure the provision of improved sound insulation to reduce the likelihood of noise complaints from neighbours. | Part E Building Regulations. PPG24 Planning and Noise / Ch. 3.8 Noise and Vibration |
| All new gas boilers should produce low levels of $NO_x$ | BREEAM Offices Pol 4 – To encourage the supply of heat from a system that minimizes $NO_x$ emissions, and therefore reduces pollution of the local environment. CSH Pol 2 – To reduce the emission of nitrogen oxides ($NO_x$) into the atmosphere. | Ch. 3.13 Pollution |
| Use Sustainable Drainage Systems (SDS) measures, wherever practical | BREEAM Offices Pol 5 – To encourage development in low flood risk areas or to take measures to reduce the impact of flooding on buildings in areas with a medium or high risk of flooding. CSH Sur 1 – To design housing developments which avoid, reduce and delay the discharge of rainfall to public sewers and watercourses. This will protect watercourses and reduce the risk of localized flooding, pollution and other environmental damage. Sur 2 – To encourage housing development in low flood risk areas, or to take measures to reduce the impact of flooding on houses built in areas with a medium or high risk of flooding | PPG25 Development and Flood Risk / Ch. 2.2 Land Use and Density, Ch. 3.14 Landscaping, Ecology and Flood Risk |

| | | |
|---|---|---|
| Achieve 50% attenuation of the undeveloped site's surface water run-off at peak times | BREEAM Offices Pol 5 – To ensure effective operation of the water run-off attenuation measures, the facilities must discharge half their volume within 24–48 hours (unless advised otherwise by a statutory body) of the storm event in readiness for any subsequent storm inflow. CSH Sur 1 – Ensure that the peak rate of run-off into watercourses ... is no greater for the developed site than it was for the pre-development site (Mandatory) | PPG25 Development and Flood Risk / Ch. 2.2 Land Use and Density, Ch. 3.14 Landscaping, Ecology and Flood Risk |
| | BREEAM Offices Pol 6 – Minimize watercourse pollution to reduce the potential for silt, heavy metals, chemicals or oil pollution to natural watercourses from surface water run-off from buildings and hard surfaces. | Ch. 3.14 Landscaping, Ecology and Flood Risk |
| Mitigate any negative impact on the microclimate of existing surrounding public realm and buildings to meet the Lawson criteria for wind comfort and safety | | CABE Guidance on Tall Buildings 2003 / Ch. 2.3 Massing and Microclimate |
| Inert and low emission finishes, construction materials, carpets and furnishings should be used wherever practical. | BREEAM Hea 9 – To recognize and encourage a healthy internal environment through the specification of internal finishes and fittings with low emissions of volatile organic compounds (VOCs). Mat 1 – To recognize and encourage the use of construction materials with a low environmental impact over the full life cycle of the building. CSH Mat 1 – To encourage the use of materials with lower environmental impacts over their life cycle. | Strategy for Sustainable Construction 2008 Ch 13 / Ch. 3.12 Materials Specification |
| All plant and machinery should be accessible for easy maintenance | | |
| All developments should meet the principles of inclusive design, adopting the principles of SPG 'Accessible London: Achieving an Inclusive Environment', including requirement for Access Statement. (Mayor of London, 2004) | | |

All residential development should meet Lifetime Home standards and 10% should meet wheelchair accessibility standards as set out in Delivering Housing Adaptations for Disabled People (CLG, 2006)
Developments should incorporate principles of 'Secured by design'

CSH Hea 4 – To encourage the construction of homes that are accessible and easily adaptable to meet the changing needs of current and future occupants.

BREEAM Offices Man 8 – To recognize and encourage the implementation of effective design measures that will reduce the opportunity for and fear of crime on the new development.
CSH Man 4 – To encourage the design of developments where people feel safe and secure; where crime and disorder, or the fear of crime, does not undermine quality of life or community cohesion.

Lifetime Homes, Lifetime Neighbourhoods: A National Strategy for Housing in an Ageing Society, CLG, 2008 / Ch. 3.15 Security and Flexibility

Safer Places: The Planning System and Crime Prevention, ODPM, 2004 / Ch. 2.4 Social Sustainability, Ch. 3.15 Security and Flexibility

No net loss of publicly accessible open space

Create appropriate new open, green, publicly accessible spaces where these can redress identified areas of deficiency of public open space

CABE Open Place Strategies, 2009 / Ch. 2.2 Land Use and Density

CABE Open Place Strategies, 2009 / Ch. 2.2 Land Use and Density

CSH Eco 2 – To encourage development on land that already has a limited value to wildlife, and discourage the development of ecologically valuable sites.

PPS9: Biodiversity and Geological Conservation / Ch. 2.2 Land Use and Density, Ch. 3.14 Landscaping, Ecology and Flood Risk

No net loss of biodiversity and access to nature on the development site

BREEAM Offices LE4 – To minimize the impact of a building development on existing site ecology. LE6 – To minimize the long-term impact of the development on the site's, and surrounding area's, biodiversity.
CSH Eco 4 – To reward steps taken to minimize reductions and to encourage an improvement in ecological value.

PPS9: Biodiversity and Geological Conservation / Ch. 2.2 Land Use and Density, Ch. 3.14 Landscaping, Ecology and Flood Risk

BREEAM Offices LE5 – To recognize and encourage actions taken to maintain and enhance the ecological value of the site as a result of development.

PPS9: Biodiversity and Geological Conservation / Ch. 2.2 Land Use and Density, Ch. 3.14 Landscaping, Ecology and Flood Risk

| | CSH Eco 2 – To enhance the ecological value of a site. Eco 4 – To reward steps taken to minimize reductions and to encourage an improvement in ecological value. | |
| --- | --- | --- |
| Reduction in areas of deficiency in access to nature | | |
| Minimize, reuse and recycle demolition waste on site where practical | BREEAM Offices Wst 2 – To recognize and encourage the use of recycled and secondary aggregates in construction, thereby reducing the demand for virgin material. | Aggregates Levy (Registration and Miscellaneous Provisions) (Amendment) Regulations 2007. Strategy for Sustainable Construction 2008 Ch12 / Ch. 3.12 Materials Specification, Ch. 4.3 Construction Waste Management |
| Specify use of reused or recycled construction materials | BREEAM Offices Mat 1 – To recognize and encourage the use of construction materials with a low environmental impact over the full life cycle of the building. CSH Mat 1 – To encourage the use of materials with lower environmental impacts over their life cycle. | Strategy for Sustainable Construction 2008 Ch13 / Ch. 3.12 Materials Specification |
| Provide facilities to recycle or compost at least 35% of household waste by means of separated dedicated storage space. | CSH Was 1 – To recognize and reward the provision of adequate internal and external storage space for non-recyclable waste and recyclable household waste. CSH Was 3 – To encourage developers to provide the facilities to compost household waste, reducing the amount of household waste sent to landfill. | Waste Strategy for England 2007 / Ch. 3.11 Waste Management and Recycling |
| Recycling facilities should be as easy to access as waste facilities | BREEAM Offices Wst 3 – To recognize the provision of dedicated storage facilities for a building's operational-related recyclable waste streams, so that such waste is diverted from landfill or incineration. CSH Was 1 – To recognize and reward the provision of adequate internal and external storage space for non-recyclable waste and recyclable household waste. CSH Was 3 – To encourage developers to provide the facilities to compost household waste, reducing the amount of household waste sent to landfill. | Waste Strategy for England 2007 / Ch. 3.11 Waste Management and Recycling |

| | BREEAM Offices Wst 6 – To encourage the specification and fitting of floor finishes selected by the building occupant and therefore avoid unnecessary waste of materials. | |
| --- | --- | --- |
| Reduce waste during construction and demolition phases and sort waste stream on site where practical | BREEAM Offices Wst 1 – To promote resource efficiency via the effective and appropriate management of construction site waste.<br>CSH Was 2 – To promote reduction and effective management of construction related waste through the use of a Site Waste Management Plan (SWMP). | Site Waste Management Plans Regulations 2008, Waste Strategy for England 2007/ Ch. 4.3 Construction Waste Management |
| | BREEAM Offices Man 1 – To recognize and encourage an appropriate level of building services commissioning that is carried out in a coordinated and comprehensive manner, thus ensuring optimum performance under actual occupancy conditions. | |
| Reduce the risk of statutory nuisance to neighbouring properties as much as possible through site management | BREEAM Offices Man 3 – To recognize and encourage construction sites managed in an environmentally sound manner in terms of resource use, energy consumption and pollution.<br>CSH Man 3 – To recognize and encourage construction sites managed in a manner that mitigates environmental impacts. | Town & Country Planning (Environmental Impact Assessment) (England) (Amendment) Regulations 2007 / Ch. 4.4 Considerate Contracting and Construction Impacts |
| All developers should consider and comply with the Mayor / London Councils Best Practice Guidance 'Control of dust and emissions during construction and demolition' (2006) | BREEAM Offices Man 3 – To recognize and encourage construction sites managed in an environmentally sound manner in terms of resource use, energy consumption and pollution.<br>CSH Man 3 – To recognize and encourage construction sites managed in a manner that mitigates environmental impacts. | Town & Country Planning (Environmental Impact Assessment) Regulations 1999 / Ch. 4.4 Considerate Contracting and Construction Impacts |
| Comply with protected species legislation | BREEAM Offices LE3 – To encourage development on land that already has limited value to wildlife and to protect existing ecological features from | PPS9: Biodiversity and Geological Conservation / Ch. 4.4 Considerate Contracting and Construction Impacts |

|  |  |  |
| --- | --- | --- |
|  | substantial damage during site preparation and completion of construction works.<br>CSH Eco 3 – To protect existing ecological features from substantial damage during the clearing of the site and the completion of construction works. |  |
| All developers should sign up to the relevant Considerate Constructors Scheme or in the City of London to the Considerate Contractor Scheme | BREEAM Offices Man 2 – To recognize and encourage construction sites which are managed in an environmentally and socially considerate and accountable manner.<br>CSH Man 2 – To recognize and encourage construction sites managed in an environmentally and socially considerate and accountable manner. | Ch. 4.4 Considerate Contracting and Construction Impacts |
|  | BREEAM Offices Man 4 – To recognize and encourage the provision of guidance for the non-technical building user so they can understand and operate the building efficiently.<br>CSH Man 1 – To encourage and reward provision of guidance enabling occupants to understand and operate their home efficiently and make the best use of local facilities | Part L: 2010 Building Regulations / Ch. 4.6 Commissioning and Handover |
|  | BREEAM Offices Hea 1 – To give building users sufficient access to daylight.<br>CSH Hea 1 – To improve the quality of life in homes through good daylighting and to reduce the need for energy to light the home. | Ch. 2.3 Massing and Microclimate, Ch. 3.6 Light and Lighting |
|  | BREEAM Offices Hea 2– To allow occupants to refocus their eyes from close work and enjoy an external view, thus reducing the risk of eyestrain and breaking the monotony of the indoor environment. | Ch. 3.6 Light and Lighting |
|  | BREEAM Offices Hea 3 – To reduce problems associated with glare in occupied areas through the provision of adequate controls. | Ch. 3.6 Light and Lighting |

| | |
|---|---|
| BREEAM Offices Hea 4 – High frequency lighting. To reduce the risk of health problems related to the flicker of fluorescent lighting. | Ch. 3.6 Light and Lighting |
| BREEAM Offices Hea 5 – To ensure lighting has been designed in line with best practice for visual performance and comfort. | Ch. 3.6 Light and Lighting |
| BREEAM Offices Hea 6 – To ensure occupants have easy and accessible control over lighting within each relevant building area. | Ch. 3.6 Light and Lighting |
| BREEAM Offices Hea 8 – To reduce the risk to health associated with poor indoor air quality. | Ch. 3.5 Air Quality, Hygiene and Ventilation |
| BREEAM Offices Hea 10 – To ensure, with the use of design tools, that appropriate thermal comfort levels are achieved. | Ch. 3.3 Thermal Comfort |
| BREEAM Offices Hea 11 – Thermal zoning to recognize and encourage the provision of user controls which allow independent adjustment of heating/cooling systems within the building. | Ch. 3.3 Thermal Comfort |
| BREEAM Offices Hea 12 – To ensure the building services are designed to reduce the risk of legionellosis in operation. | ACoP L8 Legionnaires' disease The control of legionella bacteria in water systems, HSE, 2001 / Ch. 3.5 Air Quality, Hygiene and Ventilation |
| BREEAM Offices Hea 13 – To ensure the acoustic performance of the building meets the appropriate standards for its purpose | Ch. 3.8 Noise and Vibration |
| CSH Hea 3 – To improve the occupiers' quality of life by providing an outdoor space for their use, which is at least partially private. | PPG13: Transport / Ch. 2.7 Integrated Sustainable Transportation Planning |
| BREEAM Tra 1 – To recognize and encourage development in proximity to good public transport networks, thereby helping to reduce transport-related emissions and traffic congestion. | PPG13: Transport / Ch. 2.7 Integrated Sustainable Transportation Planning |

| | |
|---|---|
| BREEAM Tra 2 – To encourage and reward a building that is located in proximity to local amenities, thereby reducing the need for extended travel or multiple trips. | PPG13: Transport / Ch. 2.7 Integrated Sustainable Transportation Planning |
| BREEAM Tra 5 – Travel plans to recognize the consideration given to accommodating a range of travel options for building users, thereby encouraging the reduction of user reliance on forms of travel that have the highest environmental impact. | PPG13: Transport / Ch. 3.10 Design to Reduce Vehicle Impacts |
| BREEAM Tra 6 – Limiting car parking to encourage the use of alternative means of transport to the building other than the private car, thereby helping to reduce transport related emissions and traffic congestion. | PPG13: Transport / Ch. 2.7 Integrated Sustainable Transportation Planning |
| CSH Ene 8 – To reduce the need to commute to work by providing residents with the necessary space and services to be able to work from home. | Ch. 3.10 Design to Reduce Vehicle Impacts |

## Sustainability strategy and climate change

Long-term climate change, which includes the potential for global warming, is an issue in which politics and science mingle in sometimes unhelpful ways. At the time of writing there is a great deal of controversy about the reliability of the science, undermined as it has been by a small number of over enthusiastic advocates, whilst an ever louder community of 'climate change sceptics' is still questioning the link between global warming and human activity, or even the very existence of global warming. The other facet of this issue that must not be overlooked is the sustainability of the natural resources that are used to fuel human activity. Coal, oil and gas are all depleting resources, although as with global warming there is considerable disagreement about when global oil, gas and coal reserves will peak. Some would argue that in the case of oil this happened in July 2008, when production peaked at around 75 million barrels/day, whilst an authoritative paper from the UK Energy Research Centre (Sorrell and Spiers, 2009) recognizes the difficulty in predicting production from the data on reserves currently available by concluding that 'peak oil' could be reached between 2009 and 2031. Although only about 15 per cent of gas reserves are associated with oil production, the Association for the Study of

Peak Oil and Gas report that a Vice President from Shell[1] has stated that peak gas 'maybe not too different from peak oil', peaking between 2019 and 2030. Similarly, according to German research from 2007 reported by the Energy Watch Group, 'peak coal' could occur around 2025.[2] (However much credence these reports can be given, the overriding message is that it is becoming steadily more difficult and expensive to extract these resources, hence, when added to the growing evidence for global warming, there is an irrefutable argument for reducing the energy consumption and greenhouse gas emissions associated with human activities.

It is self-evident that climate change is a cross-cutting theme that has a bearing on a number of aspects of sustainable building design. Bearing in mind it has been estimated that buildings are responsible for 48 per cent of global carbon dioxide emissions, it is clear that new buildings will both impact on future climate change and need to be future-proofed against the weather changes predicted. These issues have to inform everything from global policy – through the activities of the UN Framework Convention on Climate Change (UNFCCC), for example – through Government policy and legislation, such as the British and American climate change acts – to local government strategy, masterplanning and the design of individual buildings. We have addressed all but the last point in Parts 1 and 2 and we will be covering actions that the building design team can take to reduce carbon emissions and other greenhouse gases in later sections on Operational Energy and Carbon (3.2), Design for Natural Ventilation (3.4), Light and Lighting (3.6), Water Conservation (3.9), Design to Reduce Vehicle Impacts (3.10), Materials Specifications (3.12), Pollution (3.13) and Considerate Contracting and Construction Impacts (4.4). We will also be looking at what the design team has to address in mitigating against the effects of climate change, such as flood risk (3.14) and the impact of higher temperatures on the potential for natural ventilation (3.4).

## Strategy for avoiding sick buildings

The idea that buildings can make people ill is one that undermines deep-seated beliefs in the function of buildings, which are supposed to provide protection, not make you ill. Sick building syndrome (SBS) has no readily identifiable cause, nor is it easily definable, but it can be described only by a group of symptoms including the following which have been reported by numerous researchers:

- irritated, itching, dry or watering eyes;
- irritated, itching, runny, dry or blocked nose;
- sore or constricted throat;
- dry mouth;
- headache, lethargy, irritability, difficulty in concentrating;
- dry, itching or irritated skin and rashes.

SBS may be diagnosed when a significant proportion of a building's occupants report a large number of these symptoms during the time they spend in the building, but that they disappear or steadily reduce when people are away from the building.

There may be a single and identifiable cause of these symptoms, such as the failure of the building's heating, ventilation and air conditioning (HVAC) system, which may be easily corrected, in which case it would not be defined as SBS. In the US during the 1970s there were numerous reports of clusters of symptoms, usually associated with discomfort, that were tracked down to outdoor air dampers in air-conditioning plant having been closed to save energy following the oil crisis earlier that decade. This resulted in an elevation in the concentration of indoor pollutants and poor indoor air quality (IAQ) and came to be known as 'Tight Building Syndrome'.

The causes of SBS have been the subject of numerous studies and guidance from the Health and Safety Executive (HSE, 1995). More recently the HSE sponsored a study of 4000 civil servant workers from 44 office buildings in London that concluded that there was a significant correlation between SBS symptoms and work-related stress. The HSE has reported elsewhere that:

- 1 in 5 workers report feeling extremely stressed at work.
- Half a million people in the UK suffer work-related stress at a level they believe is making them ill.
- A total of 12.8 million days a year are lost through injury or ill health as a result of stress.[3]

However, work-related stress alone cannot explain all of the sick buildings and the numerous epidemiological studies and investigations of reported symptoms have identified a range of workplace characteristics and environmental factors (Box 3.2).

It can be seen from the list of risk factors that, although there are a number of operational issues outside the remit of the design team, decisions that are made during design development can have a significant influence on the long-term risk of SBS developing. These issues are dealt with later under the headings of Thermal Comfort (3.3), Design for Natural Ventilation (3.4), Air Quality, Hygiene and Ventilation (3.5), Light and Lighting (3.6) and Noise and Vibration (3.8). Other building-related illnesses such as Legionnaires' Disease and Humidifier Fever are also discussed under air quality and hygiene (3.5).

## Sustainability and commercial viability

It has frequently been argued that sustainable design and commercial viability pull in opposite directions; that the capital cost of incorporating sustainability into a new development increases unit costs to such a level that it cannot compete in the marketplace. For example, in the UK there has been a great deal of concern about the impact of the Government's drive towards zero carbon on capital costs and hence the saleability of property in a depressed marketplace (CLG, 2009).

However, the predicted 8–18 per cent increase in capital cost to achieve zero carbon dwellings built in 2017 represents a level playing field for developers to which the market must adjust. In other words, if sustainability is supported by legislation then the market should be able to stand the increased capital cost provided such that the burden is felt evenly by all, until a point is reached where property becomes unaffordable to a significant proportion of the market.

---

## Box 3.2 Risk factors for sick building syndrome

Characteristics of work and building:

- Sedentary occupation, clerical work
- More than half of occupants using display screen equipment for more than 5 hours a day
- Maintenance problems identified
- Low ceilings – typically lower than 2.4m (between floor finish and underside of ceiling)
- Many changes or movements of furniture and equipment (high churn rate)
- Public sector tenant or occupants
- Large areas of open shelving and exposed paper
- Sealed building and city centre location
- Large size – typically an occupied floor area greater than 2000m²
- Building more than 15 years old
- Scruffy appearance
- Large areas of soft furnishings, carpets and fabrics

Environmental factors:

- Low room humidities
- Low supply rate of outdoor air
- Smoking permitted in work areas
- Damp areas and mould growth
- Dust, solvents and ozone emissions from printers and photocopiers
- Low frequency fluorescent lamps creating subliminal flicker
- High room temperatures
- Dusty atmosphere
- Gaseous emissions (VOCs) from building materials and cleaning materials
- Low frequency noise

Source: Appleby, 1997.

---

The extent to which a developer is willing to invest in sustainability measures that go beyond legislation is more complex. Some measures, such as on-site renewable energy sources and the provision of social housing, may be driven by negotiations with the local planning authority and vary from place to place. Achieving exemplary scores under an assessment scheme such as BREEAM or LEED on the other hand may improve the likelihood of obtaining planning permission in marginal circumstances or help leverage funding, but in many cases the benefits to the developer will be more nebulous. A 2005 report from the Building Research Establishment and Cyril Sweett cost consultants suggested that a BREEAM Excellent rating could be achieved with an increase in cost of between 2.5 and 3.4 per cent compared with a base case of borderline Pass/Good (a score of around 40%) for a naturally ventilated office building, depending on proximity to amenities and transport nodes (BRE and Cyril Sweett, 2005). An Excellent rated air-conditioned office building however, would be 3.3–7 per cent more costly compared with a base case achieving a Pass rating. All these percentage costs must be treated with caution since it could be argued that a fully integrated approach to sustainable design will achieve much lower capital costs than 'green washing' an existing design. Also, some sustainability measures such as natural ventilation can result in a reduction in both capital and running costs. Decisions about the economics of sustainability should ideally be based on whole life costing (WLC) or net present value (NPV) calculations – a policy requirement for all publicly funded projects under the UK Government Sustainable Procurement Action Plan (DEFRA, 2009). However, for speculative developments the developer is rarely going to occupy the building and hence will not benefit from any savings in running costs arising from investment in sustainability measures. Hence, if a developer is basing its decision on economic or commercial criteria alone, the following factors will have to be considered:

- Will a 'green label' such as BREEAM Excellent or LEED Platinum give a commercial edge that can justify a premium on sales value or make the units or floor space more saleable in a competitive marketplace?
- Reputation, credibility and corporate sustainability – will association with a premium sustainable product have a positive impact on corporate reputation and possibly enhance share value and public profile?

Many of the larger and niche developers are cultivating a reputation in sustainability through presenting at conferences and sitting on influential committees because of their involvement with high profile exemplar sustainable projects.

## Notes

1    www.peakoil.net/headline-news/shell-vice-president-peak-gas-could-come-earlier-than-we-think.
2    www.energybulletin.net/node/28287.
3    www.focuseap.co.uk/Portals/0/newsletters/Summer%202006.pdf.

## References

Appleby, P., 1997. 'Building Related Illnesses'. In D. Snashall (ed.) *ABC of Work Related Disorders*. London: British Medical Journal Publishing Group

BRE and Cyril Sweett, 2005. *Putting a Price on Sustainability*. Watford: BRE

CABE (Commission for Architecture and the Built Environment), 2000. *Urban Design is the Planning System: Towards Better Design*. Available at: www.cabe.org.uk/files/by-design-urban-design-in-the-planning-system.pdf. Accessed on 4 October, 2010

CABE, 2009. *Open Space Strategies: Best practice guidance*. Available at www.cabe.org.uk/publications/open-space-strategies (accessed 5 August 2010)

CLG (Department of Communities and Local Government), 2006. *Delivering Housing Adaptations for Disabled People*. London: TSO

CLG, 2008. *Lifetime Homes, Lifetime Neighbourhoods: A National Strategy for Housing in an Ageing Society*. Available at www.communities.gov.uk/publications/housing/lifetimehomesneighbourhoods (accessed 5 August 2010)

CLG, 2009. *Costs and benefits of alternative definitions of zero carbon homes*. Available at www.communities.gov.uk/publications/planningandbuilding/definitionszero carbonhomes (accessed 20 January 2010)

DEFRA (Department for Environment, Food and Rural Affairs), 2009. *UK Government Sustainable Procurement Action Plan: Securing the Future*. Available at www.defra.gov.uk/sustainability/government/documents/SustainableProcurementActionPlan.pdf (accessed 7 March 2010)

HSE (Health and Safety Executive), 1995. *How to deal with SBS – Guidance for employers, building owners and building managers*. London: HMSO

HSE, 2002. HSL-L8. *Approved Code of Practice (ACOP) and Guidance – Legionella Bacteria in Water Systems*. Sadbury, Suffolk: HSE

Mayor of London, 2004. *London Plan Supplementary Planning Guidance: Accessible London: Achieving an Inclusive London Plan Environment*. London: Greater London Authority (GLA)

Mayor of London, 2006. *Supplementary Planning Guidance: Sustainable Design and Construction*. Available at: http://legacy.london.gov.uk/mayor/strategies/sds/docs/spg-sustainable-design.pdf (accessed 6 March 2010)

Mayor of London, 2008. *The London Plan: Spatial Development Strategy for Greater London – Consolidated with alterations since 2004*. Available at www.london.gov.uk/thelondonplan/docs/londonplan08.pdf (accessed 6 March 2010)

ODPM (Office of the Deputy Prime Minister), 2004. *Safer Places: The Planning System and Crime Prevention*. London: TSO. Available at www.communities.gov.uk/documents/planningandbuilding/pdf/147627.pdf (accessed 12 August 2010)

Sorrell, S. and Spiers, J., 2009. *Global Oil Depletion: An Assessment of the Evidence for Near-Term Physical Constraints on Global Oil Supply*. UK Energy Research Centre (UKERC) Report. London: UKERC

# 3.2
# Operational Energy and Carbon

## Introduction

This chapter aims to help the reader implement the energy strategy developed during the conceptual stages of design or masterplan as outlined in Chapter 2.6 Energy Strategy and Infrastructure. Operational energy is defined here as that consumed by the environmental services, heating of domestic hot water, circulating pumps, lighting, consumer goods and any processes involved in the operation of the building and any activities therein. Operational carbon is the carbon dioxide emissions associated with the generation of operational energy. The operational energy is dependent upon the energy demand, the efficiency of generation within the building or development and the transmission losses between the point of generation and demand. The associated carbon dioxide emissions depend on the fuel used to generate heat and electricity and the 'carbon intensity' of that fuel in $kgCO_2/kWh$. The carbon intensity of electricity generated at power stations depends on the mix of fuels used and varies from country to country and even from hour to hour in any given country. For example, because of its high proportion of nuclear power, France generates electricity with an average carbon intensity which is 15–20 per cent of that in the UK, whilst its dependence on coal gives Poland electricity generation an intensity around 80–90 per cent higher than the UK; China emits 50–60 per cent more carbon per kWh than the UK. The carbon intensity of natural gas on the other hand is around 35 per cent of that for electricity generated in the UK, based on the most recent $CO_2$ emission factors (carbon intensity) proposed by BRE for the revised Standard Assessment Procedure (SAP) for dwellings (BRE, 2009) for use in the 2010 revision to Part L1A of the Building Regulations. The carbon intensities proposed in Table 12 of SAP 2009 are $0.591 kgCO_2/kWh$ for grid electricity and $0.206 kgCO_2/kWh$ for natural gas.[1] These figures are higher than in previous manifestations of SAP and Part L and do not differentiate between electricity obtained from the grid or displaced from the grid.

## Detailed energy strategy

Some planning authorities provide detailed guidance on what they expect to see in a detailed energy strategy accompanying a planning application for a new development. This will usually need an assessment of carbon dioxide emissions using a methodology that will demonstrate compliance with National Building Regulation targets. The planning authorities will normally require evidence that a specified strategy for integrating renewable energy technologies into a development meets their specified percentage of total operational carbon dioxide emissions. Although sometimes not clear in planning guidance, it should be assumed that the percentage carbon reduction from renewable uses a legally compliant $CO_2$ emission rate as a baseline.

In the UK this will require preliminary calculations based on SAP for dwellings and Simplified Building Energy Model (SBEM) for non-residential buildings (CLG, 2009). It will be necessary to develop the design in sufficient detail to be confident that it can be achieved in the final design within the cost parameters set for the project. The strategy should be based on the following structure:

1   Baseline operational $CO_2$ emissions, giving assumptions and design features required to achieve or better target emissions that comply with legal requirements.
2   Design features should be set out in the following hierarchy of measures:
    • passive design features;
    • energy efficiency features of building services;
    • on-site energy generation measures, such as combined heat and power (CHP).
3   Renewable energy technologies proposed to achieve or exceed the specified reduction in the baseline $CO_2$ emissions.

When setting out the reasons for choosing a particular solution it is important to develop economic comparisons for the various carbon reduction strategies based on whole life costing or net present value (WLC/NPV) techniques and preferably using costs based on a preliminary design and current quotations (see the section on the Cost of carbon saving below).

The strategy should be summarized by setting out graphically the predicted month by month energy demand compared with energy generated on-site and from the grid presented on the same axis in terms of kilogrammes of $CO_2$.

## Passive design[2]

As the name implies, passive design uses free energy from the building's microclimate along with the thermodynamic properties of the building to reduce its carbon footprint. The main examples of passive design are natural ventilation and daylighting to mitigate the need for artificial cooling and lighting, respectively. These will be examined in more detail in Chapters 3.4 Design for Natural Ventilation and 3.6 Light and Lighting. However, good passive design will also reduce heating demand and cooling load where air conditioning cannot be avoided through:

• suitable orientation, size, shape, and shading of windows (facade engineering);
• optimizing glazing characteristics;
• use of thermal mass to attenuate fluctuations in temperature and facilitate the use of 'night cooling' to reduce the temperature rise the following day;

- use of shading from trees and evaporation from water bodies to reduce incident solar radiation and external temperatures;
- use of light-coloured (high emissivity) external surfaces to reflect solar radiation;
- optimizing thermal insulation and air leakage characteristics of structure;
- examining the feasibility of green roofs (see Chapter 3.14 Landscaping, Ecology and Flood Risk).

The carbon impact of options will normally need to be compared using computer simulation software (see Chapter 3.7 Computer Simulation of Building Environments). Ideally, options should be compared in terms of both kilogrammes of $CO_2$ saved and whole life cost:net present value or payback period.

## Shading

Shading design must be integrated into the engineering of the facade and tailored for the orientation and daylighting strategy. Shading both reduces solar gain through glazing and the amount of daylight that enters the building. The aim is to reduce solar gain and hence either the risk of overheating or the cooling energy required to offset solar heat gain. Direct solar gain can also create discomfort from glare and by directly heating the surface of the body. Unshaded glass reduces the intensity of solar radiation and the amount of daylight that enters the space by amounts that depend on the emissivity of the outer surface, the amount of body tinting of the glass and the number of glass layers and their respective thicknesses. Glass manufacturers have developed surface coating and body tinting techniques that optimize the balance between maximizing daylight transmission and minimizing solar transmission (solar gain factor). In some instances an optimized glazing solution will provide enough solar control and daylighting to provide the lowest summertime temperatures or summer heat gain because the photoelectric lighting controls allow the electric lighting to remain off for longer (see Chapter 3.6 Light and Lighting). This could be true for easterly orientated glazing that only receives high solar gains in the morning, when outdoor temperatures are relatively low, for example. Northerly glazing, on the other hand, is unlikely to require much solar protection at all. In these cases the primary function of shading will be to provide glare control, hence a simple internal or between-pane roller or venetian blind could suffice. However, there is evidence that these tend to be left drawn unnecessarily by occupants, thus leading to lighting being left on, with associated electricity consumption and heat gain. Perforated between-pane venetian blinds allow more daylight to penetrate and are less likely to be damaged, particularly where windows are openable (see Chapter 3.4 Design for Natural Ventilation).

External shading provides the most effective barrier to solar radiation; however, to be effective during periods of low solar altitude angle, they tend to reduce daylighting and interfere with views. For this reason the optimum orientations for the long elevations of a building are north/south. This allows the use of simple 'brises soleil' above glazing to block the high solar altitude angles, which occur in the middle of the day, whilst lower angles in summer occur on the east and west elevations (Figure 3.1).

As can be seen from Figure 3.1 the ideal design for a brises soleil incorporates a louvred construction that reflects sunlight and increases the daylight penetration into the space. External 'fins' can be extended into the space to create light shelves that reflect sunlight onto the ceiling and improve daylighting further (Chapter 3.6 Light and Lighting). Other options, such as external awnings, have been used, however, they have a dramatic impact on the 'view of

**Figure 3.1** *Brises soleil at Yorkshire sculpture park underground gallery*

*Source*: Photograph by Mcginnly

the sky' from inside and hence daylight levels and use of electric lighting. Adjustable awnings can be withdrawn when there is no direct sunlight, however, they are prone to wind damage.

The dimensions for external shading will depend on solar geometry and it is recommended that options be modelled using dynamic simulation to establish the optimum design (Chapter 3.7 Computer Simulation of Building Environments).

## Insulation

For most residential applications the approach to passive design will be different to that for an office building. The usage patterns of a home are very different to a workplace; occupancy periods are less predictable and the degree of user intervention, such as light switching and window opening, much greater. The application and utilization of air conditioning is also very different to non-residential applications. Although the installation of residential air conditioning is increasing in cities such as London, experience indicates that utilization may be restricted to short periods in mid-summer in many instances.

Passive design of dwellings should focus on reducing the heating demand whilst minimizing the risk of overheating; many highly insulated dwellings with virtually no heat loss in deepest winter experience excessive summer time temperatures because of inadequate window openings and the inability of the fabric to dissipate heat in summer. The greatest problems occur with dwellings that overlook sources of noise and pollution (Chapter 3.8 Noise and Vibration) and hence experience unacceptable conditions with the windows open.

Appendix P of the UK SAP (BRE, 2009) includes a procedure for the 'assessment of internal temperatures in summer' which is used to determine the likelihood of a high internal temperature during hot weather, using default values of internal gain and a calculation of solar gain, accounting for any benefit from thermal mass and ventilation from opening windows. Internal heat gains include lighting, appliances, cooking, water heating and occupants, but do not take into account unusually high gains from appliances such as home cinema equipment or equipment used for home working.

For dwellings in the UK there are currently no prescribed minimum U-values, only 'limiting fabric parameters', which for a wall is given as 0.35W/m²K. This represents a 'worst acceptable value', area weighted for all external walls; hence the actual U-values will be determined from an optimization process to achieve the TER for the whole dwelling. As carbon targets tighten towards zero carbon in 2016, it is anticipated that U-values for all elements will need to fall accordingly. A number of standards have been developed across Europe in response to the Energy Performance of Buildings Directive (see Chapter 1.2 Policy, Legislation and Planning) including the German PassivHaus standard, adapted for use in the UK and the US, the Swiss MINERGIE Standard, and the Association of Energy Conscious Builders' (AECB) CarbonLite Programme. All of these standards are similar and the requirements for the UK version of PassivHaus are set out in Box 3.3.[3]

Optimal thermal insulation thicknesses depend on minimum outdoor air temperatures experienced. However, the maximum thickness that can be incorporated into a wall will depend on available construction methods. The

---

## Box 3.3 Outline specification for PassivHaus in the UK

**PassivHaus standard**

| | |
|---|---|
| Compact form and good insulation | All components of the exterior shell of a PassivHaus are insulated to achieve a U-Value that does not exceed 0.15W/m²/K |
| Southern orientation and shade considerations | Passive use of solar energy is a significant factor in PassivHaus design |
| Energy efficient window glazing and frames | Windows (glazing and frames, combined) should have U-Values not exceeding 0.80W/m²/K, with solar heat-gain coefficients around 50% |
| Building envelope air-tightness | Air leakage through unsealed joints must be less than 0.6 times the house volume per hour (this is the equivalent of an air permeability value of less than 1m³/hr/m² @ 50Pa) |
| Passive preheating of fresh air | Fresh air may be brought into the house through underground ducts that exchange heat with the soil. This preheats fresh air to a temperature above 5°C (41°F), even on cold winter days |
| Highly efficient heat recovery from exhaust air using an air-to-air heat exchanger | Most of the perceptible heat in the exhaust air is transferred to the incoming fresh air (heat recovery rate over 80%) |
| Energy-saving household appliances | Low energy refrigerators, stoves, freezers, lamps, washers, dryers, etc. are indispensable in a PassivHaus |
| Total energy demand for space heating and cooling | Less than 15kWh/m²/yr |

---

*Source*: www.passivhaus.org.uk/index.jsp?id=669.

challenge increases as the building height increases; for example single storey houses have been built from rendered straw bales with an overall thickness of 480mm, achieving a thermal transmittance coefficient (U-value) of between 0.17 and 0.20W/m²K. The conductivity ($\lambda$) of straw has been reported variously at between 0.048 and 0.06W/mK, which is higher than most 'traditional' insulation materials such as rock, mineral or glass wool (Figure 3.2).

Clearly to achieve the PassivHaus U-value requirement of 0.15W/m²K using a fibre insulation, around 300mm of insulation would be required; whilst if a foam insulation is used thicknesses could be reduced to 150–200mm. For a conventional masonry wall with a 50mm air gap, this would require wall ties to bridge inner and outer elements 200–350mm apart. In the UK, a typical heavy duty wall tie is designed for a maximum gap of 150mm, although bespoke solutions have been achieved such as for the BedZed apartments in Beddington, England that incorporated 300mm of rockwool insulation into a 500mm masonry wall with no air gap (Figure 3.3).

However, for detached houses and for tall buildings (BedZED varies between one and three storeys) a 500mm thick wall starts to become very space

# Insulation Thicknesses for U values W/m2K

| Key to | Insulation Product Range | U value Range |
|---|---|---|
| | Cellulose Fibre | 0.2 W/m²K |
| | Glass Wool Fibre | 0.1 W/m²K |
| | Rock Wool Fibre | |
| | Expanded Polystyrene | |
| | Extruded Polystyrene Foam | |
| | Polyurethane Foam with $CO_2$ | |
| | Polyurethane Foam with Pentane | |
| | Phenolic Foam | |
| | Polyisocyanurate Foam | |
| | Phenolic Foam with Foil Face | |
| | Polyisocyanurate Foam with Foil Face | |
| | Aerogel Blanket | |
| | Vacuum Insulation | |

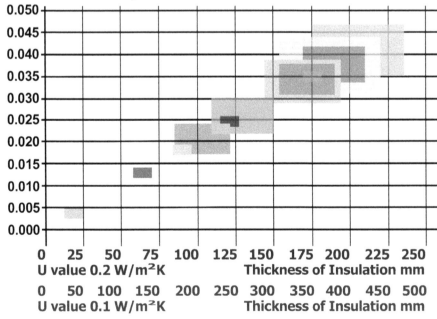

**Figure 3.2** *U-values and conductivity of different insulation types and thicknesses*

*Source*: www.bdonline. co.uk/story. asp?storycode=3140081; Ian Abley – www.audacity. com

greedy. For example, for a detached 2-storey house with a net internal floor area of 75m² the wall takes up a further 13.8m² (or 27%) of site area. Apartment buildings of 5 storeys or more may require a steel or concrete frame with infill panels or curtain walling that would create major challenges to anyone trying to incorporate 300mm of insulation.

Hence the race has been on to develop a 'super-insulation' that can achieve very low U-values in acceptable wall thicknesses. Currently the three major contenders are multi-foil, vacuum and aerogel insulants, discussed below.

## Multi-foil insulation

Multi-foil insulation is usually sold as a laminated panel comprising multiple layers of high emissivity foil and thin layers of foam. Since appearing on the market in the UK there has been a heated debate concerning its performance, since it does not perform so well under the Building Regulations approved test methods using the guarded hot box (BR443) as claimed by the suppliers. This has resulted in CLG commissioning a study that concluded that for a panel comprising 8 layers of reflective film with 4mm thick aerogel blankets between each and having a total thickness of 88mm including two 25mm air gaps 'the level of performance achieved even if the best possible materials were used in its construction, silver coatings with an emittance of 0.02 and foam core materials with a thermal conductivity equivalent to aerogel blankets (of) 0.012W/mK, they do not obtain a lower U value than 200mm of mineral wool insulation (as claimed by some suppliers)' (Eames, 2009).

**Figure 3.3** *BedZed super-insulated wall during construction*

*Source*: Lazarus, 2002; Bioregional

### Vacuum insulation

Vacuum insulation panels (VIP) use micro-porous fumed silica in powder form, vacuum packed into a metalized plastic membrane and heat-sealed to form a panel. As can be seen from Figure 3.2, the resultant conductivity is extremely low, typically less than 0.005W/mK and a U-value of 0.15W/m²K can be achieved with a panel thickness of 35mm or less. The resultant panel has the stability of vacuum packed coffee and if unprotected could easily be damaged. For obvious reasons on-site trimming to size is not possible. The technology has so far been applied mainly to fridges and freezers where it is protected by metal panels. A good example of its application to a building is a mixed office/residential building in Seitzstrasse, Munich by Martin Poole Architects completed in 2005 (Figure 3.4)

## Aerogel blanket

Aerogel consists of 2–5nm diameter particles which form a highly porous cluster having pores in the 1–100nm range. When based on silica, the aerogel produced is effectively transparent, although some light scattering occurs. The heat transfer occurs through gaseous conduction in the pores, conduction within the chains of particles that form the aerogel matrix and infra red radiation. Because the average pore size of 1–100nm is less than the mean free path of air molecules, (typically 70nm) and the silica matrix thermal conductivity is very low, silica based aerogels can achieve thermal conductivities

**Figure 3.4** *Office/ residential building in Munich with VIP insulation*

*Source*: Sascha Kletzsch, Munich

**Figure 3.5** *Example of VIP insulation by va-Q-tec*

*Source*: Ian Abley – www.audacity.com

of less than that of still air. To reduce long-wave radiation an opacifier may be introduced during manufacture, normally in the form of carbon particles. Thermal conductivities of silica-based aerogels as low as 0.005W/mK have been reported. However, the incorporation of reinforcing fibres into commercially available flexible insulation blankets to make them suitable for building applications increases the conductivity to around 0.012W/mK (Figure 3.2). Products have been developed that can be used as a substitute for glazing, with U-values as low as 0.6W/m²K and with good fire and impact resistance (EU Project HILIT, 2005). Although there have been some small installations in Sweden, the US and elsewhere, the technology has not entered into mass production at the time of writing.

## Dynamic insulation

'Dynamic' insulation is the term applied to air permeable insulation through which ventilation air for a building is driven by means of a pressure differential between the outdoors and the indoors. Outdoor air may be transferred into the building via wall and/or roof insulation such that the incoming air is heated during cold weather as it passes through the insulation so that the net heat loss through the fabric is reduced (see Figure 3.6). Air can either be drawn through the insulation by extract fans or driven by the stack effect, due to the difference in temperature between inside and out. Air is usually transferred through air inlets in the outer wall surface and at the eaves of a pitched roof and transferred from an outer air space through perforated board and fibrous insulation into an inner air space, from which it is distributed through the house, either through a negative pressure in the house or by a forced air supply that can be arranged to pick up further heat from the extract air via a simple recuperator.

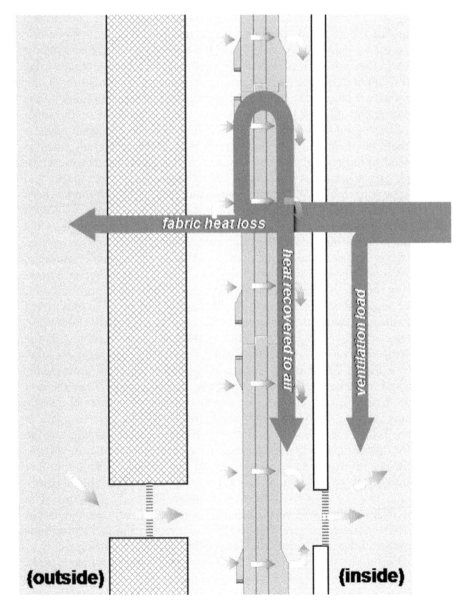

**Figure 3.6** *Principles of dynamic insulation*

Experience in Scandinavia indicates that measures to reduce air leakage are essential for extract only systems, since air can be drawn through gaps in the fabric rather than the dynamic insulation; hence a 'balanced flow' ventilation is more successful (Elmroth and Fredlund, 1996). Although in theory the U-values through the wall could approach zero, field studies have indicated that energy consumption is reduced by by 5–10 per cent compared to a house with insulation of the same thickness (Baker, 2003).

The same principle has been applied to the design of a number of commercial buildings including the McClaren Leisure Centre in Callander, Scotland and the Rykkinhalen basketball stadium in Norway.

## Thermal mass/phase change materials

All building materials allow thermal energy to flow along a temperature gradient. As the air temperatures either side of a building element fluctuate, the temperature of the material lags behind these changes and the resultant time lag means that changes in external temperature and incident solar radiation do not result in immediate changes at the internal surface, whilst the amplitude of the variations in energy flux at the outer surface are not so great at the inner surface – otherwise known as decrement. The time lag and decrement increase with the thermal capacity of the building element. Dense materials, such as concrete, tend to have a high thermal capacity, time lag and decrement, whilst lightweight materials, such as insulation, tend to have low thermal capacity and a rapid response to changes in incident energy flux. This phenomenon has been exploited for centuries by builders in North Africa, for example, who used thick high-mass flat roofs with such a long time lag that the high angle of the sun coincident with high day-time temperatures is not felt at the internal surface of the ceiling until during the night, when external air temperatures are much lower, whilst the radiation loss at night helps provide cooling of internal spaces during the day. A similar principle can be used for multi-storey buildings that are not occupied during the night by allowing cool night air to pass across high mass surfaces, such as the exposed soffit of a concrete slab, resulting in stored energy being available for the following day to provide cooling as the space air temperatures start to increase: usually referred to as 'night cooling'. This can be achieved with naturally ventilated buildings provided some means can be devised for opening windows at night without letting in either the rain or intruders. Normally this will require a degree of automation; for example by incorporating secure motor-actuated top-lights or vents into the facade immediately below the exposed soffits. Alternatively, night cooling can be achieved by operating an outdoor air supply mechanical ventilation system, however, the energy required to operate the fans may negate the benefit of free cooling. Similarly, if the building is likely to be in use for much of the night any night cooling may be dissipated by the resultant internal heat gains. The viability of night cooling will need to be assessed using computer simulation and WLC/NPV assessment (see Chapter 3.7 Computer Simulation of Built Environments and the case history for Great Glen House in that chapter).

Thermal mass can also be exploited for storing solar energy, for example in a conservatory or a Trombe wall. Both use solar energy to provide 'free' heating by using the 'greenhouse effect' that allows broad spectrum solar energy to pass through but absorbs much of the infra-red radiation from the internal heated

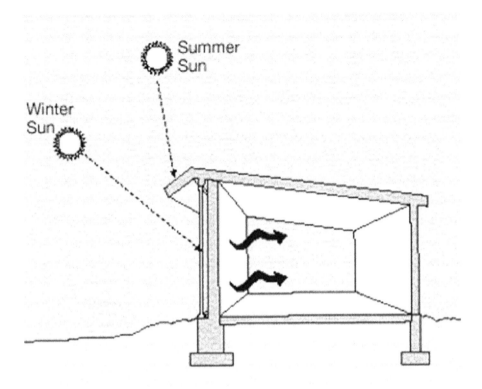

**Figure 3.7** *Principles of the Trombe wall*

surfaces. Although some of the trapped heat is conducted out through the glass, some can be allowed to enter the building via high level openings when required for heating.

Absorption of solar energy is increased by using surfaces with a very low emissivity.

One house in Australia, Birdwood House in Adelaide Hills, uses phase change material (PCM) instead of concrete. PCMs use the characteristics of certain substances, known as eutectics, to change from solid to liquid form at normal ambient temperatures. Products are available that use micro-encapsulated PCMs in board form that can be used in place of mass. The most commonly used materials are inorganic salt hydrates and organic materials such as paraffin wax. Salt hydrates generally have a high heat of fusion compared with organic materials, but can suffer from supercooling which can affect the heat transfer during the cooling phase and also tend to degrade with time. Organic materials are more stable but do not perform so well thermally. Although most organic PCMs are inherently flammable, this can be corrected using suitable additives. As the material cools it solidifies and when heated it liquefies and releases the energy absorbed as 'coolth'. The reverse effect occurs when used for solar heat absorption. For example, a typical PCM panel containing microencapsulated pure paraffin wax melts at 22°C and solidifies at 18°C. A 15mm thick PCM panel has a similar thermal storage performance to 90mm of exposed concrete. A number of commercially available products are available using paraffin wax with additives such as a copolymer for use in place of plasterboard, for example. Most salt hydrates melt at higher temperatures than are useful for night cooling, but have been applied to solar storage applications (Kelly, undated).

## Energy efficient building services

Building services that are associated with the emission of carbon dioxide and other greenhouse gases include heating, ventilation, air conditioning, refrigeration, domestic hot water services and lighting. Energy is also used in the treatment and supply of mains water. In addition, energy is consumed by a whole range of equipment and appliances used in human activities. For example, in homes gas or electricity is used in cooking, electricity is used to power washing machines, tumble driers, dishwashers, food preparation, televisions, computers and audio visual equipment. In offices electrical demand may arise from computer use and peripherals, photocopiers, telecommunications, etc., as well as a range of kitchen appliances. For some applications, such as certain factories and retail premises, process loads may dominate. This section will focus on the building services that fall within the remit of the design team. The design of energy efficient lighting design will be examined in Chapter 3.6 Light and Lighting.

### Environmental services

Detailed guidance for the energy efficient design of heating, ventilation and air-conditioning systems is provided in the Chartered Institution of Building Services Engineers Guide F: Energy efficiency in buildings (CIBSE, 2004). There is not sufficient space here to provide detailed guidance, however, some of the key principles can be set out as follows:

- Part-load performance can be more important than full-load and hence systems and equipment should not be over-sized, whilst control strategies should be designed for the full range of operating conditions likely to be experienced.
- A well-insulated building that has employed many of the passive design measures outlined above will operate for most of its life requiring very little space heating and hence the choice of heat source, heat emitter and control strategy should reflect the need to provide space heating for perhaps early morning pre-heat only.
- In the UK it has been a legal requirement to install condensing boilers since 2005 for gas-firing and since 2007 for oil-firing. Seasonal efficiencies must be more than 86 per cent, however, if a condensing boiler is designed to operate constantly with a flow temperature of 82°C and return of 71°C, which has been the norm for non-condensing boilers, efficiencies will be considerably lower. A condensing boiler achieves high efficiencies by incorporating a second heat exchanger in the flue gases in which it is intended that condensation should occur when exchanging heat with return water from the heating system. However, this water needs to be at 54°C or below for condensation to occur and maximum efficiencies occur if temperatures are below 45°C. This means that heat emitters must be designed to operate at lower than usual mean surface temperatures and

controls must allow flow temperatures to fall as outside temperatures increase (weather compensation). As can be seen below, this will result in efficiencies increasing with part-load operation, the opposite of conventional boilers. Condensing boilers are ideally suited to well-insulated buildings with low heating requirements and/or underfloor heating that requires low flow and return temperatures for comfortable floor surface temperatures. However, temperatures above 65°C will be required for domestic hot water, which for dwellings can be achieved through the use of a combination boiler having a separate heat exchanger for domestic hot water. Commercial applications will normally require a separate domestic hot water boiler.

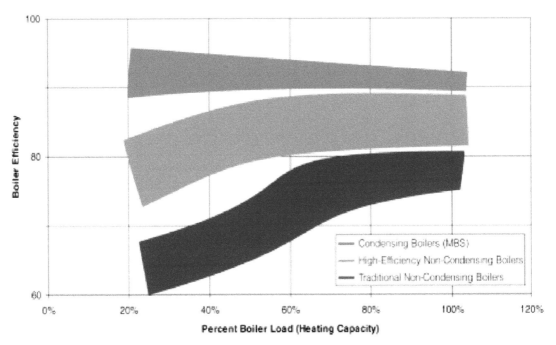

**Figure 3.8** *Typical seasonal efficiencies for different types of boiler at varying load percentages*

*Source*: www.bsdlive.co.uk/story_attachment.asp?storycode=3119440&seq=3&type=G&c=1.

- The design of environmental services must evolve to respond to the comfort requirements of occupants (Chapter 3.3 Thermal Comfort) to supplement the coarse climate control provided by the passive features of the building. Unless a process demands close climate control within very fine limits of temperature and humidity or the external climate is extreme, there should be no preconception that full air conditioning is mandatory. In a temperate climate such as that experienced in the UK, comfort conditions can be achieved for much of the year without recourse to artificial cooling or heating.
- If it is decided that a building has to be sealed to shut out noise and pollution or inclement external conditions, then a ducted air supply may be required utilizing electrically driven fans to circulate air through the building. This air may simply be used to provide outdoor air for occupants or replace air extracted from industrial processes, whilst temperature control may be from a separate system which could be a simple radiator-based heating system or

fan coil unit (FCU) based 'air-water' air conditioning; or the outdoor air may be fed into an 'all-air' air-conditioning system that provides summer cooling and possibly winter heating, such as variable air volume (VAV) or terminal reheat air conditioning (see CIBSE Guide B2 for detailed design guidance – CIBSE, 2001). A major advantage of an 'all-air' system is that when outdoor air conditions allow, the outdoor air can be used to maintain internal conditions with minimal heating or cooling. However, this may be offset by inherent inefficiencies in some systems such as concurrent cooling and heating, which is a particular problem for individual zonal control with terminal reheat systems and where tight humidity controls are required.

- Mixed mode systems allow a building to operate in natural ventilation mode whenever external conditions allow, whilst a simple air conditioning system can be used to provide 'load lopping' when a combination of heat gains and high external temperatures cause internal temperatures to exceed comfort levels. This is defined as a changeover complementary mixed-mode ventilation system in the categories devised by CIBSE (2000). This provides the maximum of flexibility and, when divided into zones, allows some tenants in a multi-tenanted office building, for example, to operate with full air conditioning, whilst others benefit from natural ventilation when feasible. The main disadvantage is one of cost, since to achieve a degree of automation motor-actuated windows will be required in addition to an air conditioning system sized to cater for peak demand.

- Low level supply and displacement ventilation: the method by which air is distributed through a space has a significant impact on energy consumption. For example, the distribution of air from a ceiling requires momentum to encourage heat exchange with the air below and the supply air temperature must be cooled to 10–14°C to deal with maximum heat gain. Delivering air with low momentum close to the floor, on the other hand, requires significantly lower fan energy and air must be supplied at 19–20°C to avoid draughts. This is known as displacement ventilation or buoyancy-driven air supply because the room air is displaced by the supply air, moves through the room by natural convection induced by the heat sources therein, and is extracted at high level. This encourages stratification in the room and a temperature gradient which is the limiting factor for design, since this must not exceed 3K between foot and head to avoid discomfort (CIBSE, 2001, Guide B2, para. 4.2.5). Displacement ventilation was originally developed in Scandinavia for industrial applications using large area perforated or fibre-faced outlets with large surface areas resulting in face velocities of 0.25m/s or lower. The systems have generally been designed to provide 100 per cent outdoor air, supplying significantly more air than is required to meet occupancy needs, but with efficient recuperators to recover heat from the exhaust air and exhaust temperatures of 30°C are not unusual. Floor to ceiling heights need to be slightly greater than normal, preferably no less than 3m, to allow for stratification above head level. A system using a similar principle was developed in Germany employing outlets recessed into a suspended floor. Air is supplied at a higher velocity with a slight twist, which encourages some mixing close to the floor and therefore allows slightly higher supply air temperatures. The advantage of this system is that it allows air distribution across a deep plan floor and with lower floor to ceiling heights. However, the latter may be offset by the need for a raised floor of sufficient depth to accommodate the air supply distribution (plenum or ducted).

- Variable air volume (VAV) air conditioning avoids the energy penalties of zonal reheat by using a damper system to vary the flow of air in response to a room thermostat. As the VAV dampers close, the supply and extract fans are automatically adjusted, typically by a pressure sensor in the ductwork or by a controller that receives information from the dampers calibrated to indicate the total flow required. Energy is saved provided the fans are controlled efficiently, with the best results being obtained using an inverter to vary the frequency of alternating current to a variable speed motor. Unfortunately some of the energy savings made from varying flow rates can be lost because VAV dampers require high inlet pressures to function properly, hence they tend to be used with high velocity ductwork and require high pressures to be developed at the supply fans. In order to avoid inadequate air movement, the control of VAV systems has to limit the turndown of the terminal dampers so that there is sufficient momentum in the supply air to penetrate the occupied zone. Similarly the amount of outdoor air in the supply air has to be kept at a level that satisfies the occupancy levels in all zones. This can be achieved through air quality sensors adjusting the amount of outdoor air brought into the building to satisfy the worst case.

- Heat recovery from lighting: recessed luminaires (light fittings) can be used as extract air points, commonly termed 'air handling luminaires', by creating a plenum above a false ceiling and channelling extract air along the full length of the lamps. A negative pressure is established in the plenum and heat is recovered at the air handling unit using a recuperator which transfers heat from the extract air into the intake air. The recuperator is bypassed when the heat is not required. A high quality grid is required for the ceiling tiles, installed with fine tolerances so that air leakage does not occur between them. The heat gain to the space below is reduced since approximately 50 per cent of the heat output from the lamps is in the convective heat that is carried away, whilst the extract air temperature may be as high as 28°C or higher if air is supplied at a low level via displacement ventilation. Air handling luminaires are more difficult to integrate into an exposed concrete ceiling and would have to be either connected into a surface-mounted ductwork system or through the concrete slab into a raised floor above, creating a fire spread risk that would necessitate fire dampers being fitted to each penetration of the slab.

- Cold water for space cooling can be obtained from aquifers or by circulating water through pipes drilled into the earth which is at a lower temperature than the air (see Chapter 2.6 Energy Strategy and Infrastructure). Alternatively, water can be cooled to a level close to the wet bulb temperature, which might be 22°C when the dry bulb is at 28°C for example, by circulating it through an evaporative cooling tower. Although the resultant water temperatures may be compatible with displacement low level supply systems (see above), they may not be low enough for 'full' air conditioning, including controlling space humidity in summer by dehumidification at a cooling coil. In which case refrigeration will be required, generating chilled water at perhaps 7–9 °C. This can be achieved through either vapour compression or absorption refrigeration. Vapour compression uses a refrigerant such as R134a which, although having a zero ozone depletion potential (ODP) has a global warming potential (GWP) 1300 times that of $CO_2$ (Table 3.1) and will be banned for certain types of new equipment from 2011.

**Table 3.1** *Global warming potential of common refrigerants*

| Refrigerant type | GWP | Refrigerant type | GWP |
|---|---|---|---|
| R11 (CFC-11) | 4000 | R32 (HCFC-32)* | 580 |
| R12 (CFC-12) | 8500 | R2407C (HFC-407) | 1600 |
| R113 (CFC-113) | 5000 | R152a (HFC-152a) | 140 |
| R114 (CFC-114) | 9300 | R404A (HFC blend) | 3800 |
| R115 (CFC-115) | 9300 | R410A (HFC blend) | 1900 |
| R125 (HFC-125) | 3200 | R413A (HFC blend) | 1770 |
| Halon-1211 | N/A | R417A (HFC blend) | 1950 |
| Halon-1301 | 5600 | R500 (CFC/HFC)* | 6300 |
| Halon-2402 | N/A | R502 (HCFC/CFC)* | 5600 |
| Ammonia | 0 | R507 (HFC azeotrope) | 3800 |
| R22 (CFC-22) | 1700 | R290 (HC290 propane) | 3 |
| R123 (CFC-123) | 93 | R600 (HC600 butane) | 3 |
| R134a (CFC-134a) | 1300 | R600a (HC600a isobutane) | 3 |
| R124 (CFC-124) | 480 | R290/R170 (HC290/HC170) | 3 |
| R141b (CFC-141b) | 630 | R1270 (HC1270 propene) | 3 |
| R142b (CFC-142b) | 2000 | R143a (HFC-143a) | 4400 |

*Indicates refrigerants which have a positive ODP and are banned for new equipment under the Montreal Protocol.

*Source*: BREEAM Offices Assessor Manual – BRE, 2008.

- Designers are being encouraged to explore alternatives to chillers that employ HFCs such as vapour compression machines which use hydrocarbons, including propane (R290) and butane (R600) (Table 3.1). With a Coefficient of Performance (CoP = condenser heat divided by energy input) of 3.5 and 4.5 an R290 based chiller is better than the equivalent R134a chiller[4] and, although the capital cost of plant is significantly higher at the time of writing, it is anticipated that as demand increases costs will fall to comparable levels.
- Vapour compression chillers require electricity to power the compressor whereas absorption chillers can use heat from any source to drive the absorption refrigeration cycle, which typically employs ammonia or lithium bromide and water. Lithium bromide (LiBr) is most commonly used for air

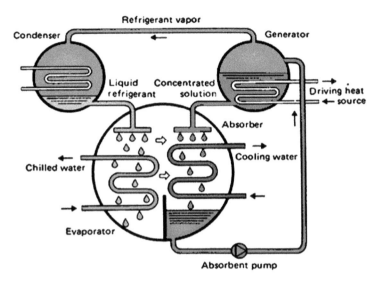

**Figure 3.9** *Absorption refrigeration cycle*

conditioning applications which use evaporation temperatures above 0°C LiBr is used as the absorbent and water as the refrigerant in a closed circuit that uses very low pressures in an evaporator so that chilled water returning from a cooling coil causes water to evaporate and be absorbed by the LiBr which produces heat that is rejected to a water circuit or directly to the air (see the simplified schematic in Figure 3.9). The solution is pumped at high pressure to the generator where heat is applied which separates the LiBr from the water and returns it to the absorber, whilst the water rejects its heat to the water circuit or air and is returned to the evaporator via an expansion valve which brings it back to the very low pressure required for evaporation.

• Although the CoP is typically less than 1.0, absorption chillers are very effective at utilizing waste heat, such as might be available from a combined cooling, heat and power plant (trigeneration) – see the section discussing On-site energy generation below – although the benefits may be marginal and a WLC and carbon saving calculation is required in each case before a decision can be made.

• One advantage of a refrigeration cycle is that the heat produced at the condenser can be used for heating purposes. For larger plant serving whole buildings, the chiller can be fitted with a double bundle condenser or valved circuitry to divert water to a heating circuit when required, so that cooling and heating can be provided simultaneously. Alternatively, and for smaller systems, the refrigeration circuit can be reversible so that the chiller can become a heat pump when demand requires. A variation on this has reverse-cycle water-source heat pumps distributed around a building and connected to a common water circuit, such that units that are experiencing a cooling demand reject heat to the water circuit and units that are operating as heat pumps pass this water through their evaporator coils ('Versatemp' system). Considerable energy savings are possible for applications where there are concurrent heating and cooling demands.

## Automatic control

The correct design and commissioning of automatic control systems have a major impact on the efficient operation of building services systems. In the introduction to their 2009 Guide H on Building control systems CIBSE suggests the following 'major contributions that the control system makes in reducing waste:

- the limitation of heating and cooling to the minimum period necessary; this usually includes the use of optimum start controllers and some form of occupancy detection to avoid excessive out of hours use;
- prevention of unnecessary plant operation and boiler idling;
- monitoring to give early warning of malfunction or inefficient operation.' (CIBSE, 2009).

A control strategy must be developed based on the complexity of the building and its services. At the heart of any automatic control system are time switches and thermostats. 'Optimum start' requires the start time to be adjusted automatically depending on the thermal response of the building and the amount of time required to bring the spaces therein up to comfort conditions. Room thermostats need to be located so as to accurately indicate the thermal environment being experienced by the occupants. Beyond this there are levels of sophistication that can be added to improve efficiency in more complex situations, such as air conditioning, taking into account the specific characteristics of the system. A Building Management System (BMS) or Building Energy Management System (BEMS) links all of these controls together with the monitoring of internal conditions and energy consumption giving a graphical output at a computer screen.[5]

## Electrical services

One of the most important reasons for energy wastage in electrical services is a low power factor (PF). The power factor is defined as the ratio of the useful power (kW) to the 'apparent' power in kVA. The 'apparent' power is that which is being drawn from the grid as measured at the incoming meter, whilst the useful power is that which is doing work. The reason why they are different is primarily due to a phase difference between the voltage and demand caused by an inductive load, such as occurs when there are induction motors and fluorescent lamp ballasts for example, and distorted waveforms, typically caused by inverters and variable speed drives. The PF can lie between zero and one, but should be as close to one as possible. PF correction is possible through a combination of suitably sized capacitors to compensate for lagging current and harmonic correction to correct for distorted waveforms.

## Lifts and escalators

The energy consumption of lifts can contribute to a significant percentage of the overall carbon emissions, particularly for tall buildings. A survey in Switzerland completed in 2005 reported that the standby energy consumption of lifts varied from 20 to 85 per cent depending primarily on utilization (Nipkow and Schalcher, undated). Standby consumption is due to lighting and ventilation fans. Some manufacturers incorporate controls that automatically switch off lighting and fans in standby mode.

Significant energy can be saved through the use of regenerative braking systems that use the forces generated during lift descent to generate electricity, resulting in potential savings in lift energy consumption of typically around 25 per cent.

## On-site energy generation

In Chapter 2.6 Energy Strategy and Infrastructure we examined alternative methods of generating energy on-site through CHP and fuel cell technologies. Plant that serves individual buildings tends to be called micro-CHP and many of the technologies explored can effectively replace boiler plant and supplement electrical supplies from the grid and renewable sources such as photovoltaics and wind turbines (see below). Some key issues are set out for consideration when integrating CHP or CCHP into a building's environmental services:

- Matching demand to maximize utilization is fundamental to successful design, particularly when attempting to combine CHP with renewable energy sources. For example, a CHP plant (i.e. no cooling) will generally rely on domestic hot water to provide a base-load during the summer months, hence CHP is not compatible with solar hot water which will supply at least 50 per cent of the domestic hot water demand in summer. This is less of an issue for a CCHP system generating electricity during summer coincident with the cooling demand in parallel with a PV installation or wind turbine, provided surplus electricity can be sold at a respectable feed-in tariff.
- A number of manufacturers have developed micro-CHP units suitable for use in dwellings.[6] In the UK the Government is encouraging their adoption by allowing 30,000 micro-CHP installations with electrical outputs of less than 2kWe to qualify for a feed-in tariff of 10p per kWh for 10 years, although these installations do not qualify for grants at the time of writing.
- The main types of micro-CHP available for residential use are based on either the external combustion engine (ECE), the organic Rankine cycle or the fuel cell.

Table 3.2 is an economic analysis for a commercially available micro-CHP with Sterling ECE based on UK prices at 1 November 2009 and including an allowance for feed-in tariff, but not for a £400 cash-back from the boiler scrappage scheme.

- The ECE Sterling units typically generate 1kWe electricity whilst producing 7kW of heat.
- The organic Rankine cycle is effectively a reversed refrigeration cycle not operating as a heat pump but as a steam turbine, using an organic fluid with a low boiling point, such as toluene, as a 'refrigerant'. The evaporation of the fluid drives a scroll expander or turbine. A 1kWe wall-mounted

**Table 3.2** *Economic analysis for WhisperGen micro-CHP*

| WhisperGen (1.0kWe/7kWt) micro CHP unit in an average sized family home with annual heat demand of 21,000kWh | | |
|---|---|---|
| Annual heat demand | 21,000 | kWh |
| Running hours | 3,000 | hours |
| Electricity generated | 3,000 | kWh |
| Own use of generation | 67 | % |
| Unit cost of avoided import | 10 | p/kWh |
| Value of avoided import | 200 | £ |
| Unit value of export | 4 | p/kWh |
| Value of export | 40 | £ |
| Total value of generation | 240 | £ |
| Additional gas consumed (compared with new condensing boiler) | 3,000 | kWh |
| Cost of additional gas consumed @ 2.5p/kWh | 75 | £ |
| Net saving = 240 − 75 | 165 | £ |
| Marginal cost of unit | 600 | £ |
| Simple payback | 4 | years |

*Source*: www.microchap.info.

version has been developed generating 10kW of heat and is expected to be highly competitive.

- The fuel cell technologies that have been found to be most suitable for residential use are proton exchange membrane (PEM) and solid oxide fuel cells (SOFCs). These technologies are being trialled globally, with the most experience being gained in Japan. Although still very costly, it is expected that commercial units will enter the residential marketplace globally in 2011.

## Building integrated renewable technologies

In Chapter 2.6 Energy Strategy and Infrastructure an introduction to renewable technologies was provided. Most of these technologies can be integrated into district or community energy systems or used for individual buildings. There follows a discussion of the relative merits of each technology for integration into individual commercial and residential buildings.

### Photovoltaics (PV)

Historically, and especially in the UK, the economic case for PV has been poor. However, a combination of rapidly reducing costs, grants/subsidies and beneficial feed-in tariffs (also a form of subsidy) means that it has become viable to install large areas of PV on suitably orientated surfaces. Crucially, with passive design and super-insulation the energy demand from both commercial buildings and dwellings is dominated by electricity, hence, since in most countries electricity has a higher carbon intensity than gas, it is logical to achieve carbon savings through the on-site generation of electricity.

## Solar hot water

Similarly, the dominant heat demand in well-insulated buildings is for domestic hot water, although this should be significantly reduced by the introduction of water saving measures (Chapter 3.9 Water Conservation). Solar panels have similar orientation requirements to PV panels and will compete for the same roof space. As stated above, they will also compete with CHP plant for base load in the summer. Domestic water requirements for office buildings are rarely sufficient for solar panels to be feasible unless there is a large restaurant and kitchen. Multi-occupancy residential use such as is found in apartment buildings, hotels and hospitals should provide sufficient demand, although close attention will have to be paid to the prevention of legionellosis due to the sometimes low temperatures leaving the panels. Vertical hot water cylinders will be required with heat exchangers installed close to the bottom that transfer the heat from the panels and with boiler-fed heat exchangers above that to ensure the flow temperature is at the required 65°C to minimize risk of the legionella bacterium multiplying (Chapter 3.5 Air Quality, Hygiene and Ventilation).

## Wind turbines

As we saw in Chapter 2.6 wind turbines are not suited to installation in sheltered locations and are difficult to integrate into building designs because of the noise and vibration generated. However, if these can be overcome it is possible to enhance the efficiency of wind turbines by installing them in exposed locations and using aerodynamically shaped elements to accelerate the approaching wind through the blades of the turbine (see Figure 2.17 Strata tower, London). However, the cost of achieving this may not be recoverable within the lifetime of the turbines.

## Aquifer Thermal Energy Storage (ATES)

Because of the cost of drilling boreholes into aquifers, ATES systems tend only to be viable for relatively large-scale projects. In order to ensure that the rock around the 'hot wells' stores enough of the heat rejected in the summer for use during the winter and similarly that the rock around the 'cold wells' cools to a temperature in winter that is useful for summer cooling, there should be a balance between the heating and cooling energy provided from the system.

It is very important that there is a minimal flow of water through the aquifer between the hot and cold wells; the horizontal velocity of flow should be no more than around 100m per year or else there will be thermal contamination of the downstream wells. The image in Figure 3.10 shows temperatures measured in an aquifer around two hot wells and one cold well showing a slight flow from top right to bottom left.

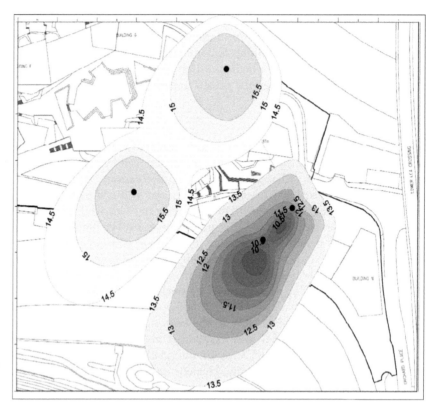

**Figure 3.10** *Typical isotherms measured in an aquifer around ATES hot and cold wells at end of summer*

*Source*: www.iftech.co.uk/index. php?option=com_content&task =view&id=19&Itemid=32, courtesy of IfTech Ltd

Flow rates through the aquifer and potential yield from boreholes can be determined from a site investigation requiring a number of test holes to be drilled into the aquifer.

Cooling water from the aquifer increases in temperature as the summer progresses, typically increasing from 7–8°C at the start of the cooling season to 12°C at the end, whilst the water from the hot well might fall from 22 to 18°C between autumn and spring. Ideally, it should be possible to use the cooling water without recourse to refrigeration, whilst the water from the hot well could be used in frost protection coils for applications requiring mechanical ventilation or be upgraded in a heat pump (see the section on Ground source heat pumps below).

## Closed loop geothermal and energy piles

Closed loop systems can use pipework buried in trenches, embedded in boreholes or integrated with piles. The amount of energy exchange achieved will depend on the surface area of pipework in contact with the earth, the conductivity of the surrounding earth and the ground temperature. Ground conductivities can vary considerably and hence a conductivity test is required before it is possible to size a system. Water temperatures available for cooling will be higher than for aquifer systems, so it will be usual to use the water in summer as condenser water in a reverse-cycle ground source heat pump (see below).

## Ground source heat pumps

For residential applications these can be used as an alternative to a boiler, with the option of chilled water being available for summer cooling. A typical operation is shown in Figure 3.11. If the heat pump is to be used for heating domestic hot water (DHW) as shown, then condenser heat must be available all year round. Normally condensing temperatures will not be high enough to heat water above that required to prevent legionella growth, so additional heating will be required from an immersion heater at the top of the cylinder (DHW tank).

## Micro-hydro

As mentioned in Chapter 2.6, historically small-scale hydroelectric installations have either been used for community electricity generation or DIY systems for home-owners with fast-flowing streams running through their land, which can be very cost effective. Guidance on these systems can be obtained from the British Hydropower Association.[7]

## Biomass and other biofuels

Biomass boilers have become quite sophisticated and the most recent developments are not much larger than an equivalent oil-fired boiler, apart from the space required for storage of course. One Austrian manufacturer offers a wood pellet fired product that is rated at 2–8kW and so could be used for low energy houses (see Figure 3.12). Wood chip is typically around half the price of the pelletized product, but the variation in size and moisture content of the material means that automation of boiler operation and feed is more complex. Typically, a larger helical screw feed is required to transport the wood

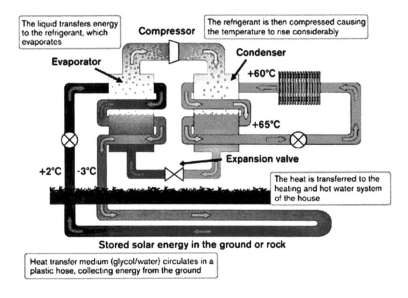

**Figure 3.11** *Ground source heat pump schematic showing summer operation*

**Figure 3.12** *Latest generation of wood pellet fired boilers for residential use*

*Source:* OkoFEN website at www.oekofen.co.uk/en/products.html.

chip from a hopper into the boiler to avoid blockage and a lamda probe is installed to measure the flue gas oxygen to adjust fuel feed and air intake volume depending on the moisture content and energy density of the fuel. A typical hopper-fed wood pellet boiler is shown in Figure 3.13.

A waste oil installation doesn't have the same storage and conveying challenges of biomass, but burners have to be suitable for the viscosity and constituents of the waste oil, which may require pre-heating and cleaning through sedimentation and filtration prior to combustion.

## Energy from waste

An energy from waste system that uses household waste as a fuel for incineration or gasification is an industrial process not suited to building integration and would be considered only for community energy schemes or major industrial sites. However, suitably processed pelletized waste material from household waste and sewage may be suitable for burning in conventional biomass boilers and CHP plant.

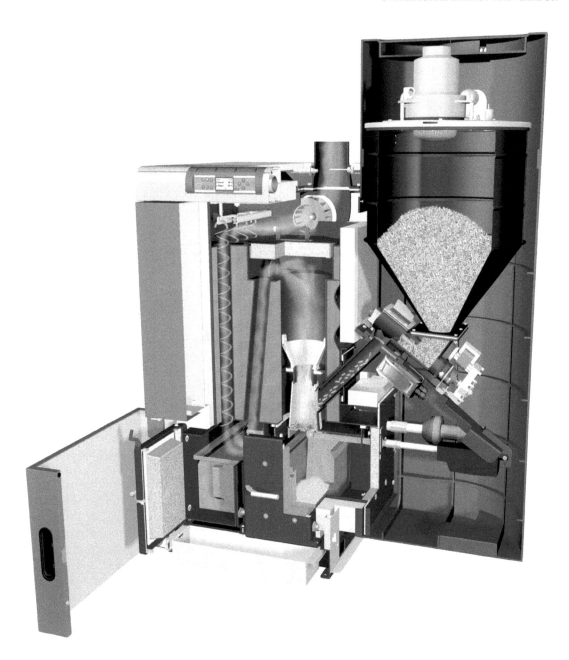

**Figure 3.13** *Internal view of wood pellet boiler with integral hopper*

*source:*Image supplied by Econergy (www.econergy.ltd.uk)

## Cost of carbon saving

It is very important to take into account the whole life economics of an energy strategy. When comparing different technologies a simple payback model can be used as shown for the Whispergen micro-CHP in Table 3.2. This should take into account capital and running costs, any subsidies or grants that can be guaranteed, including feed-in tariffs. Some authorities, such as the London Energy Partnership's Renewables Toolkit (LEP, 2004), have published capital cost per unit floor area and kg carbon saving per unit floor area for different combinations of renewable technology and building type which can be used to calculate the cost per kg of carbon saved. These should be used with great caution since they are based on the technologies and prices current at the time when the publication was prepared and they ignore the impact of subsidy and feed-in tariff on the economic case.

## Case histories

### Building integrated photovoltaics: OpTIC Centre, St Asaph, North Wales

The Welsh Development Agency (WDA) has provided a state-of-the-art research facility in St Asaph, North Wales. The OpTIC Centre is one of a series of facilities in the UK providing incubation for new businesses in key technological sectors. The Centre provides offices, laboratory accommodation and support facilities for new businesses and gives access to clean rooms, research and administrative facilities.

URS Corporation, in a consortium with architect Capita Percy Thomas and project manager/cost consultant Bucknall Austin, was awarded the commission for the design of the Centre during a competition among leading national and international teams. The winning solution was driven by a concept that combined striking architectural forms with innovative leading edge sustainable building design.

Working closely within the design team, URS was responsible for the development of sustainable features, including natural ventilation, a photovoltaic installation and rainwater harvesting, with a view to achieving an excellent rating under the BREEAM system. URS was also responsible for carrying out BREEAM assessments and design of the building services and structure.

The OpTIC Centre comprises 24 incubator units on two floors and a large laboratory facility containing clean rooms and workshops. Business support includes administrative offices, meeting rooms, lecture theatre and restaurant. Most rooms are comfort cooled or air-conditioned, whilst a double storey street between the incubator unit and laboratories is ventilated using unique solar siphons.

The most striking feature of the Centre is the 1200m² southern elevation that incorporates the largest copper indium diselenide thin film photovoltaic wall in Europe. The 1200m² wall comprises 2400 photovoltaic panels and forms an integral part of the design of the structure and building services, generating a peak output of 85kW. The photovoltaic installation generates approximately 70MWh of electricity each year, validated in a two-year trial that was funded by the Department of Trade and industry (DTI) and resulting in lower running costs and $CO_2$ emissions.

The installation was part-funded by a 70 per cent grant awarded under the DTI Building Integrated Photovoltaics (BIPV) Field Trial. The photovoltaic panels follow the line of the roof, forming a 70° inclined wall with a reflective

**Figure 3.14** *OpTIC Centre, St Asaph – note PV wall to right*

*Source*: Photograph by Paul Highnam

water feature at its base, which carries rainwater from the roof to an underground tank for recycling.

## Notes

1    SAP 2009 documentation can be viewed at: www.bre.co.uk/sap2009/page.jsp?id=1642).
2    When reading this section, refer also to the section entitled 'Energy efficient design at concept in Chapter 2.6 Energy Strategy and Infrastructure, which sets out a series of rules for low carbon design.
3    Details of the PassivHaus standards for the UK, the US and Switzerland can be found at: www.passivhaus.org.uk, www.passivehouse.us/passiveHouse/PHIUSHome.html and www.minergie.ch/tl_files/download_en/Faltblatt_Minergie_Standard_e.pdf.    The    AECB CarbonLite programme can be found at: www.carbonlite.org.uk/carbonlite/.
4    See www.hydrocarbons21.com/content/articles/2009-05-07-625kw-propane-chiller-keeps-westminster-cool.php.
5    For more information refer to CIBSE Guide H.
6    For example, www.microchap.info/ supplies information on the technology as well as having links to producers and suppliers.
7    www.british-hydro.co.uk/download.pdf.

## References

Baker, P., 2003. *Dynamic insulation for energy saving and comfort.* (BRE Information Paper 3/03 IP3/2003) Watford: BRE Press. Available at www.brebookshop.com/details.jsp?id=141647 (accessed 8 March 2010)

BRE, 2008. *BREEAM Offices Assessor Manual.* Watford: BRE. Available for licensed assessors from: www.breeam.org/page.jsp?id=17

BRE, 2009. *The Government's Standard Assessment Procedure for Energy Rating of Dwellings.* (Draft SAP 2009, version 9.90). Watford: BRE

CIBSE, 2000. *Mixed mode ventilation. Application manual AM13.* London: CIBSE

CIBSE, 2001 in CIBSE, 2004. *Guide B2 in Guide B. Ventilation and air conditioning and refrigeration.* London: CIBSE

CIBSE, 2004. *Guide F: Energy efficiency in buildings.* Available at http://pdfdatabase.com/cibse-guide-f-free-download.html (accessed 8 March 2010)

CIBSE, 2009. *Guide H: Building control systems.* London: CIBSE

CLG (Department of Communities and Local Government), 2009. *Consultation Simplified Building Energy Model* (cSBEM) v1.0.0 (18 June 2009). Available at www.2010ncm.bre.co.uk/page.jsp?id=2 (accessed 13/02/10)

Eames, P. C., 2009. *BD2768 : Multi-foil insulation.* London: CLG. Available at www.communities.gov.uk/publications/planningandbuilding/multifoil (accessed 8 March 2010)

Elmroth, A. and Fredlund, B., 1996. *The Optima-house: Air quality and energy use in a single family house with counterflow attic insulation and warm crawl space foundation.* Sweden: Building Science, Lund Institute of Technology

EU Project HILIT, 2005. *Super insulating silica aerogel glazing.* Eco Building Club. Available at www.ecobuilding-club.net/downloads/RTD/RTD_Aerogel_glazing.pdf (accessed 8 March 2010)

Kelly, R., undated. *Latent Heat Storage in Building Materials.* Available at www.cibse.org/pdfs/Latent%20heat%20storage.pdf (accessed 8 March 2010)

Lazarus, N., 2002. *BedZED: Toolkit Part 1: A Guide to construction materials for carbon neutral developments.* Bioregional Development Group. Available at www.bioregional.com/files/publications/bedZED_toolkit_part_1.pdf (accessed 8 March 2010)

LEP, 2004. *Integrating renewable energy into new developments: Toolkit for planners, developers and consultants.* London: London Energy Partnership. Available at www.lep.org.uk/uploads/renewables_toolkit.pdf (accessed 8 March 2010)

Nipkow, J. and Schalcher, M., undated. *Energy consumption and efficiency potential of lifts.* Available at: mail.mtprog.com/CD_Layout/Poster_Session/ID131_Nipkow_Lifts_final.pdf (accessed 5 August 2010)

# 3.3
# Thermal Comfort

## Introduction

It could be said that the demand for thermal comfort is a relatively recent invention that arose from a desire for improved living standards in the developed world. It blossomed following the invention of central heating and air conditioning and remains a concern primarily in industrialized countries. In the workplace it has become an expectation that comfortable conditions will be provided, supported in many countries by health and welfare legislation. There is a close correlation between maintenance of comfort conditions and productivity and hence it is a primary responsibility of the designer to create a building that is comfortable for occupants.

There is a significant link between thermal comfort and energy performance, since creating comfortable temperatures requires thermal energy. The tighter the tolerances of temperature demanded the more energy consumed, so it is important to understand the implications of relaxing those tolerances and what passive strategies are available as alternatives to mechanical temperature control.

UK guidance on thermal comfort can be found in Section 1 of the CIBSE Guide A: Environmental Design (CIBSE, 2006). US Guidance is provided in Chapter 9 of the American Society of Heating, Refrigeration and of the Airconditioning Engineers (ASHRAE) Handbook – Fundamentals (ASHRAE, 2009).

## Temperature and asymmetry

Perceptions of thermal comfort depend on a number of interdependent factors including air temperature, air speed, surrounding surface temperatures, humidity, clothing and the activity level of the person. The internationally recognized 'comfort' temperature is known as the operative temperature $(\theta_c)$. Because it is determined from air temperature and mean radiant temperature and takes into account convective and radiant surface heat transfer coefficients and air speed, it comes closer to representing the perceived temperature at the skin than the air temperature does. At air speeds below 0.1m/s it corresponds to the average of the air and mean radiant temperatures and is approximated to the temperature measured at the centre of a 25mm diameter matt black sphere. The temperature measured by a room thermostat with a dark finish gives a reasonable approximation of $\theta_c$.

Unless all of the surfaces in a room are at the same temperature and the air within is perfectly mixed, there will inevitably be temperature variations within the room. The radiant component of $_c$ is a function of the solid angles between the surfaces and the differences in surface temperatures, whereas variations in air temperature depend on a combination of the surface temperatures generating convection currents and air movement created by ventilation air or local fans.

## Air movement

The impact of air movement on the perception of thermal comfort will depend on the temperature of the air passing over exposed skin and the air velocity. The higher the temperature the greater the tolerance to higher velocities, whilst under typical comfort temperatures, as velocities increase beyond 0.1m/s discomfort from cooling of the skin increases. Figure 3.15 shows the elevation of $\theta_c$ required for increasing 'relative air speeds' (CIBSE, 2006).

Draught risk also increases with rapidly fluctuating or pulsing air movement. Studies reported in the CIBSE Guidance referred to above have indicated that if air speed fluctuates at a frequency between 0.3 and 0.5Hz then complaints of draught increase. It is possible to measure this variation and determine the standard deviation of speed changes which is used to calculate turbulence intensity ($T_u$) and hence a 'draught rating' (DR, %), which is an indication of the percentage of occupants likely to report discomfort from draught. CIBSE recommend a maximum value of DR of 15 per cent which, at the typical range of turbulence intensity found in most rooms of 30–50 per cent, corresponds with a maximum air speed of around 0.15m/s at an operative temperature of 22°C.

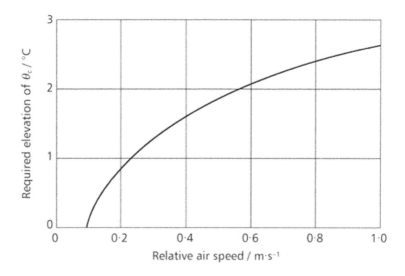

**Figure 3.15** *Correction to operative temperature to take account of air movement*

*Source:* Reproduced with permission of the Chartered Institution of Building Services Engineers

# Humidity

The humidity of the air is measured in absolute terms from the mass of water vapour contained in a kilogramme of dry air. This can be determined from the reading given by a wet bulb thermometer which is depressed below air temperature by the evaporation of water from a wick. It is useful to express this in relative terms by using the percentage saturation (%sat) or relative humidity (%RH), both of which indicate the amount of moisture that is associated with the dry air. One hundred per cent sat or RH means that the air can hold no more moisture or the 'dew point' has been reached, at which condensation or precipitation will occur.

Humidity only impacts on feelings of warmth or draught if the skin is wet – the lower the humidity the greater the rate of evaporation from the skin.

In moderate thermal environments humidity impacts on comfort in other ways; for example:

- Humidities below 30 per cent tend to be associated with a dryness of skin and mucosa and if there are high concentrations of airborne irritants an increased irritation of the nose and throat.
- Incidents of static electricity tend to increase as humidity levels fall below 40%sat.
- At high temperatures and humidities the body is more likely to overheat when sweating because the rate of evaporation decreases as %sat rises.
- High humidities tend to be associated with condensation onto surfaces at a lower temperature than the room air and there is a greater risk of mould and algal growth if this continues for long periods.
- Humidity in the range of 40–70%sat is generally considered acceptable for moderate thermal environments.

# Comfort conditions

There are a number of ways by which thermal comfort conditions can be defined and measured. For the purposes of designing and operating environmental services, it is important to define thermal comfort using parameters that can be measured by commissioning engineers and used to control the systems.

However, it is important to understand the relationship between controllable and measurable conditions such as $\theta_c$, parameters used to design services such as air speed, occupancy-dependent parameters such as clothing and activity-dependent parameters such as metabolic rate. BS EN ISO 7730 (2006), which has been adopted by both CIBSE and ASHRAE, uses statistical representations of the likely satisfaction of humans with their thermal environment in the form of the predicted mean vote (PMV) and predicted percentage dissatisfied (PPD). PMV is a parameter that was developed from a whole series of tests on human subjects exposed to varying temperatures wearing a wide range of clothing and at different levels of activity. A parameter was devised called the 'clo', the value of which is indicative of the insulating properties of the clothing. Similarly, activity levels were defined by the 'met' which is indicative of metabolic rate. Subjects were asked to report sensations of thermal comfort on a scale from

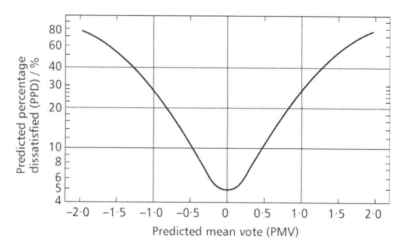

**Figure 3.16** *Predicted percentage dissatisfied as a function of predicted mean vote*

*Source*: Reproduced with permission of the Chartered Institution of Building Services Engineers

+3 (hot) to –3 (cold). The relationship between these reported sensations, expressed as a PMV, and PPD is shown in Figure 3.16.

As can be seen, a PMV ranging from +0.5 to –0.5 corresponds with a PPD of 10%. Since PMV and PPD cannot be used to control a heating or air conditioning system, thermal comfort conditions are normally defined in terms of operative temperature for specified activities and clothing and assume that room air speed will be below 0.15m/s and therefore not impact significantly on thermal comfort. Both CIBSE and ASHRAE give lists of recommended conditions for different applications. For example Table 1.5 of CIBSE (2006) recommends $\theta_c$ for offices in winter of 21–23°C, assuming an activity level of 1.2met (sedentary) and clothing of 0.85clo, whilst in summer the recommended temperatures are increased to 22–24°C because of the lower clothing insulation value at 0.7clo. For sedentary occupants the insulation properties of seating are taken into account when estimating clo. The summer value of clo is based on shirt and lightweight trousers for men and blouse, skirt and tights for women, both seated in office chairs. Winter clothing increases the thickness of trousers for men and adds a petticoat to the women's ensemble.

The ranges used by CIBSE are based on a PMV of +/–0.25, which corresponds to a PPD of 6%. Higher levels of predicted dissatisfaction may be acceptable, particularly for free-running buildings (see below).[1]

A maximum temperature gradient between head and feet of 3K is recommended, although this may need to be lower if an air speed greater than 0.15m/s is expected, for example with some displacement ventilation systems (see Chapter 3.2 Operational Energy and Carbon, the section on Environmental services).

In some circumstances the temperature of surfaces in proximity to occupants may be of importance. For example, floor temperatures should not exceed 28°C, which will have an impact on the design of underfloor heating systems. Similarly, where occupants are exposed to low surface temperatures from chilled ceilings or large windows, or high temperatures from radiant heating, an analysis of asymmetric thermal radiation may be required. This requires a calculation of the difference in radiant temperatures on either side of a reference point. CIBSE recommend that the dissatisfaction associated with this factor should be no more than 5 per cent, which corresponds to a 5K difference in radiant temperature for warm ceilings, 10K for cool wall/window, 14K for a cool ceiling and 23K for a warm wall. This calculation may be required for designing chilled ceilings or radiant heating and checking for likely discomfort from windows in winter.[2] It is interesting to note that CIBSE suggests that the 5K limit referred to above is unlikely to be economic for the design of ceiling radiant heating, and suggest a 10K limit corresponding to a 20 per cent dissatisfaction level, which some may consider to be too high. This will need to be factored into the decision on whether overhead radiant heating is an acceptable option.

Radiant temperature asymmetry does not take into account the impact of short-wave radiation from the sun passing through a window and directly heating someone located close to the window. This will not normally be an issue since in most cases people will have the option of closing a blind or moving away from the source of heat gain. For someone wearing typical summer clothing exposed to an incident solar radiation of 300W/m² at the internal surface of a window, it will feel around 4K warmer than the actual operative temperature.

## Adaptation, free-running and passive design

The summer time temperature range for $\theta_c$ of 22–24°C referred to above can only be achieved in most cases by the cooling of space temperatures in summer. If a strategy is to be devised that avoids mechanical cooling or air conditioning, then it must be accepted that temperatures above 24°C will occur for some of the time in the summer. When temperatures are uncontrolled in the summer, the building is said to be 'free-running'. Strategies for preventing an unacceptable rise in temperature during free-running are discussed in Chapter 3.2 Operational Energy and Carbon and the design of naturally ventilated free-running buildings is discussed in more detail in Chapter 3.4 Design for Natural Ventilation.

PMV and PPD are based on steady state conditions and fully adapted subjects and the operative temperatures derived assume the same level of clothing at all times. In real life occupants tend to have a number of adaptation strategies, which may include adjusting clothing (within the realms of decency) and, in naturally ventilated buildings, opening windows. Field studies have found that people tend to 'respond on the basis of their thermal experience, with more recent experience being more important' (CIBSE, 2006, para 1.6.4). Hence, as external temperatures increase, it is suggested that people will tolerate higher operative temperatures indoors as shown in Figure 3.17.

The differences between free-running and controlled environments is more to do with expectation than dissatisfaction, hence occupants of free-running buildings expect higher temperatures in summer. The running mean temperature is calculated by averaging the temperatures for a period in summer using weighting factors that decrease with each prior day. In the UK a summer rolling outdoor temperature would typically be around 20°C at the height of summer, indicating an upper limit for free running of 27.4°C.

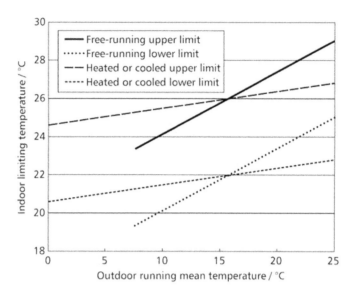

**Figure 3.17** *Operative temperatures against a running mean of outdoor temperature for free-running and controlled indoor environments*

*Source:* CIBSE, 2006, reproduced with permission of the Chartered Institution of Building Services Engineers

Note that these temperatures assume room speeds of 0.15m/s or below. An important adaptation strategy that this analysis does not assess is the potential for increasing air movement, by opening windows, operating desk or ceiling fans or personal air terminal devices. From Figure 3.15 it can be seen than an air speed of 1.0m/s, for example, is perceived as a reduction in temperature of around 2.6K. The limiting factor for elevated air speeds in offices is likely to be the inconvenience of loose papers being lifted from surfaces, which tends to happen with air speeds of 0.8m/s or more.

## Dwellings

Acceptable temperatures in dwellings are generally only quoted for winter operation, although CIBSE give summer operative temperatures for use in designing home air conditioning (CIBSE, 2006, Table 1.5). Generally it is accepted that people have a wider range of adaptation strategies for their own homes because there are a smaller number of people to be consulted. Surveys have also indicated that on average people keep their houses several degrees cooler in winter than the CIBSE recommended values, primarily because it is usually the householder who pays the energy bills.

Overheating can be a problem in summer, however, particularly for west-facing bedrooms exposed to the low summer sun at the end of the day. Field studies reported by CIBSE (CIBSE, 2006, para 1.6.4.3) have found that quality of sleep decreases as temperatures rise above 24°C, with an absolute maximum of 26°C, assuming that residents adapt by reducing bedclothes. Further adaptation by the operation of floor or ceiling fans can alleviate discomfort at higher temperatures.

# Notes

1   Values for met and clo are provided in CIBSE (2006, Tables 1.2 to 1.4).
2   See BS EN ISO 7730, 2006 for detailed methodology for calculating radiant temperature asymmetry.

# References

ASHRAE, 2009. *ASHRAE Handbook – Fundamentals*. Atlanta, GA: ASHRAE. Available at www.ashrae.org/publications/page/158 (accessed 8 March 2010)

BSI, 2006. BS EN ISO 7730: *Ergonomics of the thermal environment: Analytical determination and interpretation of thermal comfort using calculation of the PMV and PPD indices and local thermal comfort criteria*. London: British Standards Institute (BS1)

CIBSE, 2006. *Guide A – Environmental Design. Section 1: Environmental Criteria for Design*. London: CIBSE

# 3.4
# Design for Natural Ventilation

## Introduction

We have already seen that, along with fire, natural ventilation is the very oldest of environmental controls used by humans to make their shelters more habitable. Of course it was soon discovered that if you lit a fire indoors then smoke would travel upwards and it made sense to form an opening in the roof, whilst air would be drawn through openings at low level. This is what later became known as the 'stack effect' once humans learnt that combustion products could be carried away in stacks, both ejecting the pollutants clear of neighbouring buildings and increasing the amount of heat obtained from a combustion process.

The Romans exploited natural ventilation and thermal mass to create buildings that remained cool during the summer by using the combined force of stack effect and wind pressures, along with thick stone walls and narrow windows with deep reveals.

## Principles of natural ventilation

The forces associated with wind and natural convection form part of the microclimate in which every building sits. The designer can either chose to combat these forces by sealing the building and installing air conditioning, or capture and control them so as to provide 'free' cooling and ventilation for much of the year. However, if the designer chooses to give the building the response time of a cardboard box, then, when it is very hot outside, it will be even hotter inside. As we saw in the section dealing with 'Thermal mass/phase change materials' (PCMs) in Chapter 3.2 Operational Energy and Carbon, variations in external conditions can be attenuated by the incorporation of materials into the fabric that have a suitable time lag between absorbing and releasing energy.

Figure 3.18 shows an office building with three floors of open-plan offices opening onto a centrally located atrium with high level opening lights. The arrows indicate the likely passage of ventilation air driven by convection currents only (stack ventilation). Airflows are likely to be very different when a wind is blowing. Depending on the relative strengths of the wind and convective forces, as well as the direction of the wind, it is likely that there will be a cross flow of air between the upwind and downwind openings. Air entering the upwind side of the rooflight could either flow straight across to the downwind opening, inducing air from the atrium in the process, or be deflected

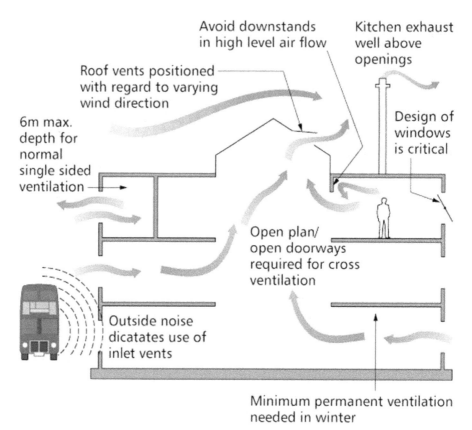

Avoid downstands
in high level air flow

Kitchen exhaust
well above
openings

Roof vents positioned
with regard to varying
wind direction

Design of
windows
is critical

6m max.
depth for
normal
single sided
ventilation

Open plan/
open doorways
required for cross
ventilation

Outside noise
dicatates use of
inlet vents

Minimum permanent ventilation
needed in winter

**Figure 3.18** *Simplified section through an office building showing principles of natural ventilation*

*Source*: CIBSE (2005), reproduced with permission of the Chartered Institution of Building Services Engineers

downwards against the convective flows, depending on the geometry of the openings and obstructions between them.

Natural ventilation should be designed so that wind direction and force do not reverse the flow of convection currents and prevent the removal of heat and pollutants from the occupied areas. The configuration in Figure 3.19 has been developed as a method for avoiding this.

The 'chimney' or 'solar siphon' has been designed to avoid the risk of downdraught, which can occur with wind blowing parallel with the long edge of a rectangular opening. An aspect ratio of 4:1 or less gives a good margin for error provided no upwind obstructions deflect wind downwards into the opening. A combination of convective and wind forces induce air from offices bordering the atrium. The glazing in the atrium faces south and the internal wall at a high level has a high thermal mass and dark finish; thus the heat from solar radiation continues to drive the stack effect for some period after the sun has disappeared. The chimney is designed so that rain entering the top is drained onto the roof to the north and a door is incorporated into the chimney that closes automatically when it is too cool to benefit from the high ventilation rates (or should there be a fire).

There are a number of products on the market that have been developed specifically to cap natural ventilation chimneys (Parker and Teekaram, 2005). Although these are sometimes called 'wind-driven natural ventilation terminals'

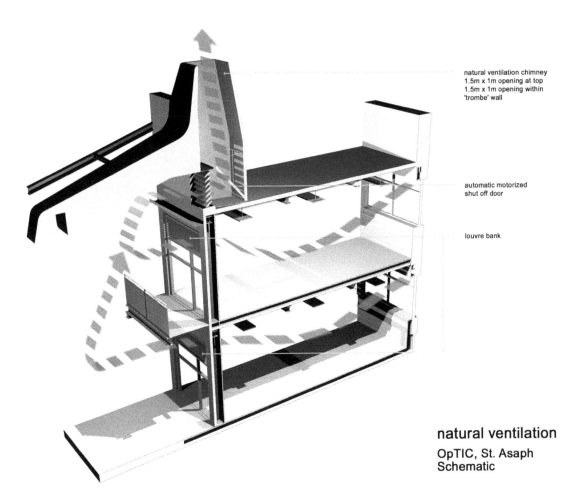

natural ventilation chimney
1.5m x 1m opening at top
1.5m x 1m opening within
'trombe' wall

automatic motorized
shut off door

louvre bank

**natural ventilation**
OpTIC, St. Asaph
Schematic

**Figure 3.19** *Natural ventilation strategy for OpTIC Centre, St Asaph*

*Source:* Matt Jones Capita Percy Thomas

in fact they use both the convective and wind forces described above. Some of these vents have been designed to act as inlets as well as outlets, using either separate passages exposed to positive upwind and negative downwind pressures or rotating cowls that incorporate sails that rotate the inlet into and outlet away from the wind (Figures 3.20 and 3.21).

Where no air passage can be provided across a building or into an atrium or chimney, single-sided ventilation prevails. The effectiveness of single-sided ventilation depends on a combination of floor to ceiling height and the location of openings in the external wall. Whereas for cross-ventilation a maximum wall to wall depth of 5 times floor to ceiling height ($h_R$) is recommended (CIBSE, 2005), for a single sided space with a vertical distance of at least 1.5m between the top of the upper opening and bottom of the lower opening the floor depth can be $2.5 \times h_R$. Hence cellular rooms require windows with top and bottom opening lights such as are provided by sash windows.

Natural ventilation in its simplest form can be controlled manually through the opening and closing of windows. However, where there are a number of

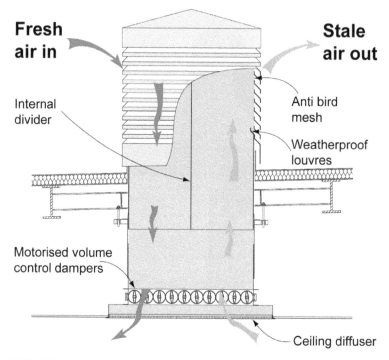

**Fresh air in**

**Stale air out**

Internal divider

Anti bird mesh

Weatherproof louvres

Motorised volume control dampers

Ceiling diffuser

**Figure 3.20** *Typical louvre-faced inlet/outlet roof vent*

*Source:* Monodraught Ltd

**Figure 3.21** *Rotating cowl inlet/outlet roof vents at BedZED*

*Source*: Bioregional

people affected, such as in an open-plan office, responsibility for these actions becomes more complicated and can lead to disputes and dissatisfaction. This is also true for temperature control and thermostat adjustment. Furthermore, the control of night cooling is difficult to achieve manually (see below). Hence it is important to develop a control strategy before embarking on the design process.

## Design process

As can be seen from Figure 3.18, it is not possible to design a building and then decide whether it is to be naturally ventilated or air conditioned. A natural ventilation strategy has to be integrated into the design approach from the start. The following seven sections represent the key stages in the design process and decisions that impact on the likely performance of the building.

### 1 Location

Proximity to a road or other source of noise and air pollution is very important since these can create unacceptable conditions for occupants, particularly for dwellings (Chapters 3.8 Noise and Vibration and 3.5 Air Quality, Hygiene and Ventilation). When analysing a site for its suitability to take a naturally ventilated building, a prediction of likely occupant exposure to noise and pollutants is essential. It may be necessary to develop a strategy that locates openable elements remote from potential sources of noise and pollution; for example from a courtyard at the centre of building or inlets located at roof level.

### 2 Heat gains and comfort

Even making use of night cooling there is a limit to the amount of heat that can be removed with natural ventilation whilst maintaining acceptable thermal comfort. Viability not only depends on heat gains but external summer time temperatures and the level of dissatisfaction with the thermal environment that can be tolerated. We saw from Chapter 3.3 Thermal Comfort that an upper limit for 'free-running' buildings of 27.4°C is acceptable, whilst if occupants can adapt further by increasing velocities across their bodies to perhaps 0.8m/s this could be increased to around 29°C. A useful approach to assessing the viability of natural ventilation employs the concept of temperature exceedances and has been used for assessment of summer overheating in the UK and elsewhere for some years. This uses the principle that an operative temperature of 28°C should not be exceeded for more than 1 per cent of the occupied hours in a year (CIBSE, 2006, Table 1.8). For a typical office occupancy this corresponds to around 25 hours. Predicting temperature exceedances will normally require computer simulation (Chapter 3.7 Computer Simulation of Building Environments).

### 3 Height

Because natural ventilation relies on the stack effect and stratification occurs above head level, room heights should be greater than in air conditioned spaces, preferably 3m or more, but with a minimum of 3m. Stratification can be a particular issue for the top storey of a multi-storey building opening onto an atrium, hence it is important that there is sufficient height within the upper zone of the atrium above the opening into the upper floor. Stratification can also be assessed through computer simulation, using a computer fluid dynamics tool to determine the likely temperature distribution under worst case conditions (Chapter 3.7 Computer Simulation of Building Environments).

The maximum number of storeys that can be naturally ventilated using an atrium or street is dependent upon a number of factors, including fire risk associated with an open atrium, noise from high wind velocities through openings in upper floors and the pressure and air speed created at low level openings by the stack height of the atrium, with a maximum of 10–12 storeys. Landmark tall buildings such as the Commerzbank in Frankfurt am Main and 30 St Mary Axe in the City of London (see the Case histories in Chapter 2.6 Energy Strategy and Infrastructure) have used a series of small atria or light wells. The Commerzbank has a triangular atrium running the full height of the 52 occupied floors with 4-storey 'winter gardens' spiralling every 120° for the full height and open to the atrium and external face (Figures 3.22 and 3.23). The atrium is divided into 12 storey-high compartments partitioned with glazed 'floors' to prevent excessive air speed at inlets due to the height of the stack. Office floors have opening windows via a double skin on the outer wall as well as opening into the atrium and winter gardens. Neither the Commerzbank nor St Mary Axe towers operate in natural ventilation mode at all times, with both offering load lopping through mixed-mode operation. Commerzbank report that their building operates for 9 months of the year in natural ventilation mode, the rest of the time air conditioning is required.[1] Foster & Partners, the architects for both Commerzbank and St Mary Axe are planning a 600m high naturally ventilated tower in Moscow having 118 occupied floors.

## 4 Shape, orientation and cellularization

The ideal configuration for natural ventilation is having a long axis running east to west, with 12–14m deep floors running either side of an atrium or street. A sloping atrium roof exposed to the south sun (in the Northern hemisphere) benefits from the greatest solar heat gain, heating surfaces at the top of the atrium and hence creating the greatest stack effect. Care needs to be taken to avoid overheating on the top floor and glare from sunlight for those occupants overlooking the atrium. Consideration should be given to installing vertical baffles/shades at high level to avoid this. If made from sound absorptive material, these can also be used to reduce some of the reverberation that can be experienced in an atrium (see the case history for Great Glen House in Chapter 3.7 Computer Simulation of Built Environment).

The ideal internal configuration for natural ventilation employs an internal atrium with open plan offices across the full floor depth. However, it may be necessary to cater for cellular offices, particularly for speculative development. Cross ventilation can only be achieved through cellular offices if openings are provided in partitions between offices and corridors that have pressure loss coefficients sufficiently low to allow the free passage of air with the pressures

**Figure 3.23** *Commerzbank winter garden*

*Source:* Gabriele Rohle

**Figure 3.22** *Commerzbank Frankfurt view*

*Source:* Mylius

available. Permanent openings may present privacy or nuisance problems due to cross-talk between offices and the corridor; attenuating louvres are likely to create excessively high pressure drops, although consideration might be given to white noise generation (see Chapter 3.8 Noise and Vibration).

## 5 Night cooling

As discussed in Chapter 3.2 Operational Energy and Carbon, exposed thermal mass or phase change materials can significantly reduce temperature exceedances and hence enhance the viability of natural ventilation. Peak day-time temperatures can be reduced by 2–4K compared with a building with conventional suspended ceiling tiles, delaying the time at which the peak occurs by up to 6 hours (Figure 3.24 and case history for BRE Environmental building).

Performance can be enhanced by locating openings for night-time ventilation at a high level in the external facade, whilst maximizing the surface area available for exposure to the night air. This can be achieved through the use of coffered or sinusoidal profile slabs, forming channels perpendicular to the outside wall that allow a smooth transfer of air into an atrium (Figure 3.25).

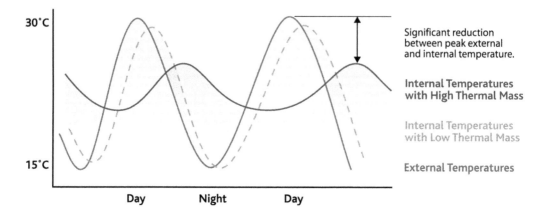

**Figure 3.24** *Stabilizing effect of thermal mass on internal temperature*

*Source:* From De Saulles (2006), Courtesy of MPA – The Concrete Centre

As can be seen from the image in Figure 3.25, these coffers can be designed to create reflectors for uplighters that can also be used to absorb sound, reduce reverberation and mitigate some of the risk of noise disturbance from neighbouring workstations (Chapters 3.6 Light and Lighting and 3.8 Noise and Vibration).

The facade design for night cooling is discussed below, whilst the control system for operation of the top lights at night needs to be designed so that they are held open for the optimum period during the night. This can be done using

**Figure 3.25** *Exposed concrete coffered slabs at Powergen headquarters*

*Source:* Bennetts Associates architects; photograph copyright Peter Cook/VIEW

a simple protocol that opens the top lights when the average temperature detected indoors ($T_S$) is 24°C, say, checking that the outdoor temperature has fallen below this value, and holds them open until the indoor temperature falls below 18°C, say ($T_L$). A degree of intelligence can be introduced to this system by incorporating an algorithm that each evening compares the maximum indoor temperature measured during the working day with $T_S$ and adjusting the temperature at which windows open depending on whether enough free cooling is being achieved. Software has been developed that uses data from weather forecasts to optimize opening and closing temperatures based on predicted weather conditions for the following day.[2]

The window and facade design should allow these top lights to remain open during rain storms, however, it may be necessary to incorporate rain and wind sensors to close the top lights during pre-defined extreme conditions.

## 6 Facade design

A facade for a naturally ventilated building is very different to an air-conditioned one. Normally the external windows will provide a route for air to enter and/or leave the building and there are certain key features that are needed for satisfactory performance:

- Windows should be at least 1.5m between the bottom of the lowest opening and the top of the upper opening in the external wall. This can be achieved with a single opening element but for security and control purposes it is usually necessary to have top and bottom lights that open separately, with the top light opened by an actuator linked to the night control system.
- Opening areas need to be confirmed using computer simulation but also must meet the requirement for purge ventilation set out in the 2010 version of Part F of the Building Regulations (CLG, 2009).
- Openable windows present potential routes for infiltration when closed, hence the quality of fit and seals must be excellent. This is particularly important for the upper storeys of tall buildings which are exposed to high wind pressures.
- Windows will need to incorporate trickle ventilation devices to meet winter ventilation requirements of Building Regulations Part F (CLG, 2009).
- Opening of windows located in the upper part of a tall building will need to be through mechanical means to avoid wind damage. Operating limits should be available from window manufacturers.
- Solar shading and daylighting strategies need to be adapted to cater for window opening (refer to the Passive design section of Chapter 3.2 Operational Energy and Carbon). Internal blinds are not compatible with openable windows, however, south facing windows can normally be shaded using external horizontal projections to cater for the high solar altitude angles in the summer. One option is to mount these below the top lights and extend them internally to form a 'light shelf' which reflects sunlight onto the ceiling and improves the daylighting when sunlight is available, bearing in mind the amount of sky 'seen' by occupants is reduced by the external shading (see Chapter 3.6 Light and Lighting and Figure 3.26).

**Figure 3.26** *Light shelf with external glazing*

*Source:* Kawneer Company Inc

East and west facing windows experience lower solar altitude angles than south facing windows at the beginning and end of the day, resulting in both potential glare problems and excessive heat gains. For this reason large areas of west-facing glazing should be avoided. Horizontal external shades are relatively ineffective, whilst external louvres covering the full window height can both interfere with the operation of windows and significantly impact on daylight, increasing the hours of operation of lighting and hence increasing both heat gain and energy consumption. One solution is to use opening windows with between-pane venetian blinds. These should be fully retractable, of light finish and be perforated, allowing some visibility of the outside world when drawn. For east facing windows, where low solar altitude is coincident with relatively low external temperatures, solar control glass with good light transmission properties may be the best solution, although between-pane blinds may be necessary for glare reduction.

## 7 Occupant satisfaction

The key to the success of a naturally ventilated building is to reduce the number of hours in the year when thermal conditions exceed levels at which dissatisfaction is unacceptable. It is recognized that occupant attitudes to natural ventilation are different to those towards air conditioning. Studies have found that many people appreciate being able to open windows and the resultant link with the outdoors. However, for open-plan offices, it has to be recognized this only applies to those whose workstations are located adjacent to the external wall. Indeed confrontations are possible over this, particularly if opening a window results in intrusive noise or draught, accepted by some but not others.

# Case history

## BRE Environmental Building, Watford, UK

The following case history has been adapted from the NatVent publication 'Natural ventilation for offices' (Perera, 1999).

The Building Research Establishment Environmental Building was designed by Feilden Clegg Design from a performance specification drawn up by the Energy Efficient Office of the Future (EOF) group and constructed on the BRE's Garston campus near Watford, England in 1997. The site is relatively quiet and pollution-free, with good exposure to wind and solar gain without being subject to extreme conditions. The 3-storey office wing has been built with a rectangular footprint having its long axis running east to west and a gross floor area of 2050m². There are 7.5m deep open plan offices with a south orientation and 4.5m deep cellular offices on the north elevation with a notional corridor between the two. There is also a single-storey wing housing a lecture theatre that creates an L shaped footprint. This area is mechanically ventilated from displacement ventilation outlets at low level.

The south and north elevations have approximately 50 per cent by area of double glazing with some manually operable elements. The first and second floor slabs are sinusoidal in cross-section providing exposed concrete channels in the ceiling below. Office heating is from pipework embedded in the floors.

The natural ventilation is driven from rectangular 'solar siphons' or stacks incorporated into the south elevation, faced with glass bricks and topped with circular ducts and cowls. Each of five vertical stacks is connected to the office space behind with windows having motorized elements for night cooling. Air is channelled through the troughs formed by the sinusoidal slab to maximize the exposure to thermal mass (Figure 3.28). The stacks incorporate simple axial

**Figure 3.27** *BRE Environmental building*

*Source:* Peter White, BRE

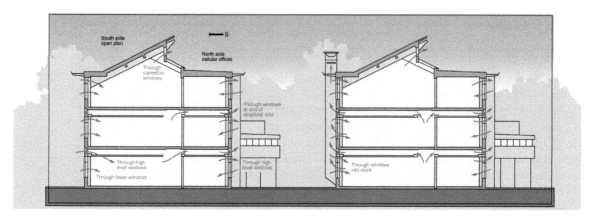

**Figure 3.28** *BRE Environmental building natural ventilation strategy*

*Source:* http://projects.bre.co.uk/envbuild/envirbui.pdf, image courtesy of BRE

fans that are intended to supplement the natural ventilation when there is inadequate pressure available from wind or stack effects.

The top storey is not connected to the solar siphons but instead uses clerestory rooflights to provide the draw for natural ventilation.

The graph in Figure 3.29 shows the results from monitoring carried out during the first summer of operation of the building. It can be seen that internal temperature, although rising daily over the six-day monitoring period, has a significantly lower amplitude than the external temperature plot and peaks at around 27°C despite external temperatures reaching 30°C for five consecutive days.

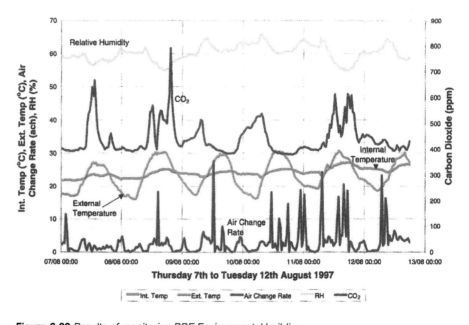

**Figure 3.29** *Results of monitoring BRE Environmental building*

*Source:* Perera (1999), image reproduced by permission of BRE

# Notes

1   www.commerzbank.com/en/hauptnavigation/konzern/engagement/oekologie/gebaeude
    management/gebaeudemanagement.html.
2   www.natvent.com/media/filebank/org/439-0905-UK.pdf and Wittchen et al, 2005.

# References

CIBSE, 2005. *Application Manual AM10: Natural ventilation in non-domestic buildings.*
    London: CIBSE
CIBSE, 2006. *Guide A1 Environmental Design, Section A1: Environmental criteria for
    design.* London: CIBSE
LG (Department for Communities and Local Government), 2009. *Proposals for
    amending Part L and Part F of the Building Regulations – Consultation.* London:
    TSO, Volume 3: Proposed technical guidance for Part F
De Saulles, T., 2006. *Utilisation of Thermal Mass in Non-Residential Buildings.*
    Camberley, London: The Concrete Society
Parker, J., and Teekaram, A., 2005. *Wind-Driven Natural Ventilation Systems (BSRIA
    Guide BG 2/2005)* Bracknell, UK: Building Services Research and Information
    Association (BSRIA)
Perera, E., 1999. *Natural Ventilation for Offices.* Watford, UK: BRE and NatVent
    Consortium
Wittchen, K. B., Løgberg, E., Pedersen, S., Djurtoft, R. and Thiesen, J., 2005. Use of
    weather forecasts to control night cooling. In: *Building Simulation 2005 – Proceedings
    of the 9th International IBPSA Conference.* Montreal, Canada

# 3.5

# Air Quality, Hygiene and Ventilation

## Introduction

Building occupants are exposed to a cocktail of airborne contaminants that includes chemicals and micro-organisms originating from sources both within and outside the building. Although in most non-industrial buildings the concentrations of these contaminants will be orders of magnitude below published occupational exposure limits, there may still be detectable odours present. In some instances, sensitized individuals may experience an allergic response to one or more airborne allergens, such as house dust mite excreta or fungal spores. Pollutants such as legionella bacteria, radon, lead or asbestos, all of which have no odour and are therefore undetectable by the human sensory system, may from time to time be present in sufficient concentrations to cause long-term health effects. In many countries smoking is no longer permitted in the workplace or places where the public congregate and so passive smoking is less of an issue for the designer or building manager.

Airborne substances enter the nasal cavity and are sensed by two largely separate detection systems: the olfactory sense, which is responsible for odour detection; and the general chemical sense, which is sensitive to irritants. The two senses do interact however; for example one sense can override or mask the other and some substances evoke sensations of both odour and irritation. Humans are known to adapt to odour with time, whilst irritation usually increases and potentially leads to sensitization.

## Design for good indoor air quality (IAQ)

The constituents of the air in a modern building will depend on a number of factors including:

- Whether the building is naturally or mechanically ventilated; i.e.:
  - If a building has openable windows, then IAQ will depend in part on the constituents of the outdoor air and the distance from window openings to sources such as traffic, chimneys, ventilation exhausts and other sources of pollution. Resultant indoor constituents and concentrations will vary, depending on emission rates, wind direction and weather events, such as inversions.

- If a building is mechanically ventilated, then the quality of the air supplied will depend on the location of intakes in relation to external pollution sources and the efficiency of the filtration and adsorption and desorption of pollutants within the air handling plant.
- Studies at the Technical University of Denmark in the late 1980s (Fanger et al, 1988) reported on the subjective assessment of volunteers of odour levels in a variety of naturally and mechanically ventilated spaces. Fanger discovered that the total strength of odour reported by the subjects was up to 6 times that which would be expected from the body odour of the occupants alone. He concluded that this was due to emissions from building materials, furnishings and air handling systems. Sources of odour within an air handling plant include fungicides used to treat filter material and thermal wheels, lubricants and fan belt materials. Further studies specifically on odour sources in poorly maintained air handling systems discovered that they emit odour and allergens from dirt and mould accumulated in filters, drip trays and other internal surfaces (Pejtersen et al, 1989).
- Malodorous internal sources depend on the usage, processes and equipment installed, but may include:
  - occupancy-dependent body odour;
  - organic materials used in the manufacture, decoration and cleaning of furniture, carpets, curtains and the building fabric;
  - ozone from photocopiers, laser printers and electric motors;
  - fumes from cooking;
  - volatile organic compounds (VOCs) emitted from micro-organisms such as fungi;
  - gases percolating through foundations from land-fill.

Good design practice is to adopt the following decision-making hierarchy to eliminate or reduce exposure to airborne contaminants (CIBSE, 2006 – para. 8.4 Air Quality and Ventilation):

1   eliminate contaminants at source;
2   substitute with substances that produce non-toxic or less malodorous emissions;
3   reduce emission rate of substances;
4   segregate occupants from potential sources;
5   improve ventilation, e.g. by local exhaust, displacement or dilution;
6   provide personal protection.

These are not mutually exclusive and some combination may be required – for example the specification of water-based rather than solvent-based paints – although good ventilation will be required in either case. The 2004 Volatile Organic Compounds (VOC) Paints Directive has been implemented across Europe, reducing the VOC content in paints and varnishes in two phases, the most recent of which came into force on 1 January 2010 (EC, 2004).

Substantial improvements in IAQ are possible if low emission materials are used in the building fabric, finishes, furnishings, ventilation plant and for cleaning. In the US, the Carpet and Rug Institute (CRI) have developed a green label for carpets and adhesives that involves testing products for 24-hour and

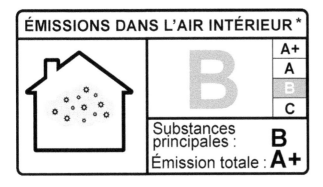

**Figure 3.30** *Proposed French label for construction and decorative products for use indoors*

*Source:* Image courtesy of Eurofins Product Testing A/S

14-day emission rates and test chamber concentrations for 'target contaminants'.[1] A voluntary American standard is also available for testing office furniture; this also uses a 14-day emissions test for a menu of VOCs (ANSI/BIFMA, 2007).

The use of VOCs and hazardous materials in the manufacture of wood-based furniture is addressed in a decision by the European Commission amending the 1980 Eco-label regulations (EC, 2009) and individual countries have already introduced proposals for their own mandatory labelling schemes, such as the French '*Emissions dans l'air interieur*' label (Figure 3.30).

Both BREEAM and LEED give credit for selecting 'low-emitting materials' (Chapter 1.3 Assessment Methodologies). BREEAM Credit Hea 9 is awarded on the basis that timber panels, structures and flooring, floor coverings and adhesives, ceiling tiles, wall coverings and adhesives, paints and adhesives have met relevant standards covering emissions of VOCs, including formaldehyde. LEED sets out a series of VOC content limits under EQ 4.1–4.4 covering adhesives and solvents, paints and coatings, carpet systems, and composite and agrifibre products.

## Health effects, hygiene and building-related illnesses

There is a fine line between discomfort caused by IAQ problems and building-related illnesses. The latter fall into two categories:

- those that have an identifiable cause, such as legionellosis, humidifier fever, and conditions resulting from exposure to known substances such as asbestos, lead and formaldehyde, for example; and:
- those that have no identifiable cause but can be described only by a group of symptoms – known as sick building syndrome (SBS).

As we saw in the section 'Strategy for avoiding sick buildings' in Chapter 3.1 Sustainability Strategy, poor IAQ can contribute to the symptoms that define SBS. Indeed most contaminants will have identifiable short- and/or long-term health effects if the concentrations are high enough. These effects are normally associated with occupational exposure to industrial pollutants and fall outside the scope of this book.

Beyond the industrial environment, health effects will normally be associated with exposure to allergens or, more rarely, pathogens, such as legionella bacteria. Most airborne allergens are organic particles and can be very small indeed, having a mean mass aerodynamic diameter of between 5 and 30 micron (µm), with some taking the form of aerosols or vapours. Allergens of this size tend to lodge in the bronchi causing bronchospasm and consequent breathing difficulties, whilst larger particles may collect in the nose and cause rhinitis in sensitized individuals and those that lodge in the eyes may cause conjunctivitis.

The concentration of allergens in most homes is higher than in a typical office, particularly in older homes with gardens and pets. The most common airborne allergens found in homes include house dust mite excreta, fungal spores, resins, insect detritus and animal proteins. House dust mites thrive in mattresses and other soft furnishings, while moulds thrive in damp conditions such as those found in bathrooms.

Legionellosis is a generic term that describes Legionnaires' disease and Pontiac fever. These are relatively uncommon infections, with 6280 cases of Legionnaires' disease reported in 35 European countries in 2006 (11.2 per 1 million of population),[2] although this represents a per capita doubling since 2000, with a fatality rate of 6.6 per cent. In the US the Occupational Health and Safety Administration (OHSA) estimates that there are more than 25,000 cases annually (between 8.3 and 10 per 1 million population), with around 4000 fatalities (12–16%).[3] However, globally the reporting rate is thought to be only 10 per cent because the symptoms are easily confused with other types of pneumonia.

Both Legionnaires' disease and Pontiac fever are caused by the inhalation of water droplets containing bacteria of the family Legionellaceae of which there are more than 40 species identified. The one most commonly associated with Legionnaires' disease is *Legionella pneumophila* serogroup 1. *L. pneumophila* has also been implicated in the 'flu-like illnesses Pontiac fever and Lochgoilhead fever.

### Box 3.4 Risk factors in outbreaks of legionellosis

- Water temperatures between 20 and 50°C.
- Nutrients available for growth, such as proteins and rust.
- Niches which protect legionella from the penetration of heat and biocides, such as limescale, sludge and algae.
- Generation of a fine, invisible aerosol such as that produced from taps, shower heads, cooling towers and spray humidifiers.
- Low water turnover which may create conditions during which temperatures drift into the risk zone, biocides decay and sediment precipitates to form a sludge.
- Water system open to ingress of animals, insects, dirt and sun: direct sunlight encourages algal growth which provides an ideal niche for legionella.
- Susceptible people exposed to aerosol, such as those with impaired lung capacity or immune system.

These bacteria are small and rod-shaped, penetrating deep into the alveoli where they cause infection. Symptoms tend to appear after an incubation period of between 2 and 10 days, with only 1 per cent of those who inhale the bacteria developing symptoms. It is thought probable that a significant proportion of the population inhale legionella bacterium at some time in their lives, especially those who work in high risk buildings such as hospitals.

Although legionella bacteria are found in nature they tend to only become a problem if they colonize building water systems that provide a suitable niche for the multiplication of the bacteria and create aerosols that can be inhaled by susceptible people (Box 3.4 summarizes the risk factors).

Legionnaires' disease is a pneumonia that, in its most severe form, can be fatal or leave the patient debilitated if it is not treated in time.

CIBSE and the Health and Safety Commission (since 2008 the Health and Safety Executive – HSE) in the UK and ASHRAE and Centers for Disease Control and Prevention (CDC) in the US provide advice on the design and operation of building services systems to minimize the risk of legionellosis (CIBSE, 2002; HSE, 2002; ASHRAE, 2000).[4]

Because the most serious outbreaks have been associated with contaminated cooling towers, there has been a tendency in recent years to use alternative plant, such as 'dry' cooling towers and air-cooled condensers. There are penalties to pay in terms of energy consumption and associated carbon emissions, however, as a 'wet' cooling tower uses evaporation to assist heat rejection, with significantly lower condensing temperatures and hence lower compressor power compared with either dry towers or evaporative condensers. Dry towers are also much larger than wet cooling towers, although the reduced water consumption is a significant advantage. Air cooled condensers are common for smaller applications and are frequently used in reversible heat pump configurations.

The key preventative measures for hot water systems are to ensure that water leaving the central storage cylinder is at 60°C or more and that this temperature is maintained in the distribution system through good insulation and either return circulation pipework taken to within a metre or less of the most remote outlet, or self-regulating tracer tape installed along the full length of the pipework.

The risk of legionella bacteria colonizing cold water storage tanks is frequently overlooked, but they have historically been oversized and located in tank rooms that overheat in the summer. They must be designed so that water flows through the full volume of the tank and there will be no significant temperature rise during periods of low use. Particular attention must be paid to connecting multiple tanks to ensure that the flows are balanced between them.

## External pollution and routes into a building

In an ideal world all buildings would be located in the middle of a field with no sources of pollution within several kilometers. However, sustainable brownfield sites tend to be located in close proximity to transport routes and industry. Furthermore the use of on-site biomass or waste to energy plant creates a further source of pollution within the site boundary. In addition, exhausts from kitchens and toilets create potential sources of odour close to the building.

BREEAM sets criteria under Credit Hea 8 that award points if in-takes for mechanical ventilation systems are more than 10m from exhausts and chimneys on the building under consideration or 20m from external sources of pollution, including roads that are in regular use, car parks and exhausts from neighbouring buildings. The rule is relaxed for naturally ventilated buildings with credit given when window openings are more than 10m from any pollution source. There are no equivalent criteria in the LEED scheme.

CIBSE provides useful guidance on minimizing the impact of external pollution indoors through an understanding of the behaviour of pollutants outdoors and estimating the effect of external pollutants on indoor air quality for both mechanically and naturally ventilated buildings (CIBSE, 1999).

A number of outbreaks of Legionnaires' disease have resulted from the dissemination of legionella bacteria in the aerosol produced by a cooling tower. A well-designed cooling tower will have high efficiency drift eliminators to remove these fine droplets. However, it is inevitable that some will escape and it is important that the plume from the cooling tower is well diffused into the atmosphere under all likely conditions of wind and weather so that a concentrated aerosol cannot enter open windows, ventilation inlets or external spaces where people congregate. If it is not possible to avoid cooling towers, for example on a tall building that has limited external plant space, then they should discharge at a high point well clear of openings into the building. Investigations into outbreaks have found that the legionella from a heavily contaminated cooling tower have spread through the streets surrounding a building, hence prevention at source is the priority (Brundrett, 1992).

## Ventilation criteria

We have seen above that, depending on location, outdoor air can be as much a source of pollution to a building as a means of diluting contaminants that are generated indoors. A naturally ventilated building should by definition be able to use the air that enters from outside as both a means of reducing summer time temperatures and ensuring good IAQ for all of the occupants. In the winter, however, the challenge is to allow the entry of sufficient outdoor air without a major energy penalty. In the UK, Part F of the Building Regulations (CLG, 2009) requires that naturally ventilated buildings have trickle ventilators in external facades, usually integrated into window frames. For 'buildings other than dwellings' Part F refers to the guidance in CIBSE AM10 which recommends 400mm$^2$ free area of trickle ventilator for every m$^2$ of (net internal) floor area, with at least 4000mm$^2$ in each room and located at least 1.75m above floor level.

The requirement in Part F for mechanically ventilated buildings with humans being the only source of pollution is 10 litres/s per person, which agrees with the most recent CIBSE Guidance (CIBSE, 2006, A1, Table 1.5). For both naturally and mechanically ventilated buildings local extract is required to deal with toilets, showers, baths, food/beverage preparation, and printers/photocopiers (with 50% utilization or more).

The US has similar standards, with the ASHRAE Standard 62.1 (ASHRAE, 2007) setting standards for outdoor air supply, however, it allows more flexibility than Part F in allowing adjustments to the outdoor air supply rate where high ventilation effectiveness is achieved through displacement ventilation or spaces are intermittently occupied, for example. The concept of ventilation effectiveness recognizes that not all of the outdoor air that enters at the air handling plant will reach every occupant at the design rate. This is because there is usually some short-circuiting of air due to the inefficiencies of air distribution. For example, some of the air supplied at a ceiling diffuser will stick to the ceiling (known as the ceiling or Coanda effect) and some may return into the extract air terminals before it has mixed fully with room air. This is a particular problem where there are supply air diffusers and air handling luminaires in close proximity.

It is common to adjust for this by supplying more than the requisite minimum amount of outdoor air and it has become common practice to use high efficiency recuperators to recover energy from the exhaust air and offset some of the resultant energy penalty. Where there are very low outdoor temperatures, however, the outdoor air will have a corresponding low moisture content, which can result in humidities in mid-winter below 30%sat (Chapter 3.3 Thermal Comfort). This can be overcome by incorporating steam humidification and/or recuperators such as thermal wheels that allow the recovery of latent as well as sensible heat. If possible steam humidification should be avoided since it would negate the energy benefit of recuperation and present a long-term maintenance burden.

## Notes

1    www.carpet-rug.org/commercial-customers/green-building-and-the-environment/green-label-plus/index.cfm. In the UK there is an independent advisory service at www.sustainable floors.co.uk/.

2    www.eurosurveillance.org/ViewArticle.aspx?ArticleId=753.

3    www.osha.gov/dts/osta/otm/legionnaires/disease_rec.html#risk.

4    www.cdc.gov/MMWR/PREVIEW/MMWRHTML/00045365.htm.

## References

J/BIFMA (American National Standards Institute / Business and Institutional Furniture Manufacturers Association), 2007. *M7.1-2007: Standard Test Method for Determining VOC Emissions from Office Furniture Systems, Components and Seating*. Grand Rapids, US. Available at www.bifma.org/standards/standards.html (accessed 14 February 2010)

ASHRAE, 2000. *Guideline 12-2000 ASHRAE STANDARD. Minimizing the Risk of Legionellosis Associated with Building Water Systems*. Atlanta,GA: ASHRAE. Available at www.lakoshvac.com/enewsimages/guide12.pdf (accessed 19 March 2010)

ASHRAE, 2007. *ASHRAE Standard 62.1-2007. Ventilation for Acceptable Indoor Air Quality*. Atlanta, GA: ASHRAE. Available at www.ashrae.org

Brundrett, G., 1992. *Legionella and Building Services*. Oxford: Butterworth-Heinemann

CIBSE, 1999. *Technical Memoranda TM 21: Minimising pollution at air intakes.* London: CIBSE

CIBSE, 2002. *Technical Guidance TM13. Minimising the Risk of Legionnaires' Disease.* London: CIBSE

CIBSE, 2006. *CIBSE Guide A – Environmental Design. Section 8: Health Issues.* London: CIBSE

CLG (Department for Communities and Local Government), 2009. *Proposed Changes to Part L and Part F of the Building Regulations: A Consultation Paper. Vol. 2: Proposed Technical Guidance for Part L.* London: CLG. Available at www.communities.gov.uk/documents/planningandbuilding/pdf/partlf2010consultation (accessed 21 February 2010)

EC, 2004. *2004/42/EC Directive on the limitation of emissions of volatile organic compounds due to the use of organic solvents in certain paints and varnishes and vehicle refinishing products and amending Directive 1999/13/EC.* Available at http://ec.europa.eu/environment/air/pollutants/pdf/eu_decopaint.pdf (accessed 19 March 2010)

EC, 2009. *EC 2009/894/EC. Commission Decision of 30 November 2009 on establishing the ecological criteria for the award of the Community eco-label for wooden furniture.* Available at http://ec.europa.eu/environment/ecolabel/about_ecolabel/pdf/ep_proposal.pdf (accessed 19 March 2010)

Fanger, P., Laurisden, J., Bluyssen, P., and Clausen, G., 1988. Air pollution sources in offices and assembly halls, quantified by the olf unit. *Energy and Buildings*, vol 12, pp7–19

HSE, 2002. *HSC-L8. Approved Code of Practice (ACOP) and Guidance – Legionella Bacteria in Water Systems.* Sudbury, Suffolk: HSE

Pejtersen, J., Bluyssen, P., Kondo, H., Clausen, G. and Fanger, P.O., 1989. Air pollution sources in ventilation systems. In: Kulic, E., et al (eds) *Proceedings of CLIMA 2000, Sarajevo. The Second World Congress on Heating, Ventilating, Refrigerating and Air Conditioning.* Sarajevo, vol 3, pp139–144

# 3.6
# Light and Lighting

## Introduction

Before the invention of effective artificial lighting the importance of daylight was recognized in the design of buildings with large windows and narrow floor plans. For most of our forebears work ceased after dark and productivity during the winter months was impacted accordingly.

The advent of electric lighting and cheap energy resulted in many deep plan buildings being constructed requiring artificial lighting throughout periods of occupation. However, following the energy crisis of the early 1970s and a number of surveys of occupant opinion in the 1980s, it was realized that not only does good daylighting save energy but that the majority of office workers like to be seated close to a window.

Artificial lighting typically represents around 18 per cent of carbon emissions for a typical air conditioned UK office building (see Figure 2.7), but with daylighting controls lighting energy can be reduced by 20–30 per cent for a narrow plan building which isn't overshadowed and is designed to best practice.

## Daylighting

It could be argued that providing good daylight is an art rather than a science and that this is furthest from hyperbole when applied to an art gallery or museum. The quality of daylight depends upon the orientation, location, shape and size of the windows in relation to the internal geometry of the space, the colours and textures of the surfaces therein and any internal or external obstructions to sunlight and a view of the sky.

It is interesting to note that the Workplace (Health, Safety and Welfare) Regulations (1992) in the UK require that, as well as 'every workplace (having) suitable and sufficient lighting', that lighting 'shall, as far as reasonably practicable, be by natural light'.

When considering the design of a workplace, it is the illuminance from daylight on the working plane that is the primary criterion, expressed as a percentage of the outdoor horizontal illuminance from a standard (CIE) overcast sky known as the Daylight Factor (DF) (see the section on Daylight, sunlight, overshadowing and glare in Chapter 2.3 Massing and Microclimate). The illuminance will vary across the working plane of a daylit room depending on the distance from the windows and obstructions to a view of the sky. For good quality daylighting, the variation in daylight should not be excessive.

BREEAM Offices awards a credit (Hea 1) based on an average daylight factor of 2 per cent or more (100lux in Britain) and with a uniformity of 0.4, which means the DF at any point in the room should not fall below 0.8 per cent, except where there is roof lighting, for which uniformity should not fall below 0.7. Additional credits can be awarded if an average DF of 3 per cent is guaranteed.

Higher standards than this, such as the mandatory German daylighting standard (Deutsches Institut für Normung, 1999), which requires an average DF of 4 per cent, have a significant impact on building design, resulting in narrower footprints and larger window areas. However, the actual operational DF will not be achieved if blinds have to be closed to eliminate excessive solar heat gain and discomfort glare. To mitigate this, solar shading strategies should be developed that maximize daylighting both when there is no direct sunlight and when solar incidence is at its greatest. In order to avoid having to close blinds to remove reflective glare in display screen equipment, the workspace should be designed to ensure that desks can be positioned with screens orientated away from windows. Surveys have found that for manually controlled blinds occupants tend to leave them closed, regardless of light conditions outside (Rea, 1984). Hence, ideally, solar protection should either be fixed or automated, although experience with external motorized blind systems indicates that they are vulnerable to damage and breakdown. CIBSE recommends the following techniques for the designer:

- give occupants a good view out as an incentive to raise the blinds;
- avoid overglazing, particularly where the occupants can see large areas of sky;
- use light shelves or overhangs to reduce the need for blinds;

...and for the building manager:

- maintain the blinds so that they are easy to operate, even when the windows open;
- ensure that occupants know how to operate the blinds;
- instruct security staff or cleaners to raise the blinds after or before each working day. (CIBSE, 1999)

As we saw in the section on facade design in Chapter 3.4 Design for Natural Ventilation, light shelves can be used to reflect sunlight onto a ceiling whilst the lower part of the window is shaded by external fins or retractable perforated between-pane venetian blinds. As discussed in that section, the optimum orientation for this configuration is southerly in the Northern hemisphere. However, studies indicate that 'Light shelves are found to be effective at increasing the light level at the back of a room, but only when the sun shines; otherwise they reduce the total amount of daylight in the room' (Littlefair, 1990). For east- and west-facing windows, the lower solar altitude angles can result in sunlight penetrating deep into the space above the light shelf and experience indicates that occupants will place articles on the light shelf to prevent glare. The solar geometry of the window, shading and shelf design should be assessed to reduce the potential for abuse. Ideally light shelves should be located out of reach thus avoiding the temptation to use them for storage.[1]

For good daylighting, buildings should not be located too close to large obstructions. Many city centre buildings suffer from poor daylighting on the lower floors because of this. The amount of daylight that falls onto the working plane in an office depends on having an unobstructed 'view' of the sky, reflection from external and internal surfaces and the light transmittance of the glass. Daylight factors can be calculated manually using the techniques provided in the CIBSE Lighting Guide LG10, the IESNA Lighting Handbook or the relevant National standard (CIBSE, 1999; IESNA, 2000). However, for a full assessment of daylighting in a workplace, including an analysis of the energy/carbon saving from daylight controls for artificial lighting, a computer simulation is required (see Chapter 3.7 Computer Simulation of Building Environments).

Daylighting is important for dwellings primarily for its quality rather than quantity. In the UK the Code for Sustainable Homes (CSH) awards credits for designs that achieve an average DF of 2 per cent in kitchens and 1.5 per cent in living rooms and studies. These are adapted from BS 8206, Part 2 (BSI, 1992) that also specifies a minimum standard for bedrooms of 1 per cent. The Homes and Communities Agency may require compliance with the CSH daylight criteria under their 'enhanced standards' for specific programmes under the National Affordable Homes Programme (NAHP).

Although the CSH/BS 8206 criteria should be achievable for most housing developments, they will have a significant impact on the design of apartment buildings. This is because the desire to achieve an efficient use of floor space has led to architects designing individual apartments with internal kitchens or living rooms with inboard kitchen areas. The rationale behind the 2 per cent DF criterion for a kitchen comes from it being considered the focal point of activity for a family home, and the room most likely to be occupied during daylight hours. In reality, most one and many two bedroom apartments will be occupied by individuals or couples who are not at home during hours of daylight, apart from the weekends, and will probably spend little time in the kitchen during the day.

## Energy efficient lighting

Lighting levels in offices have been steadily falling since peaking in the late 1960s, when there was a trend to installing lighting to achieve general levels of 1000 lux or more at the working plane. Not only did this consume excessive electrical energy but, as the use of display screens increased, there was a commensurate problem with glare. CIBSE and the Society of Light and Lighting (SLL) now recommend a maintained illuminance of 300–500 lux for the workplace, with task lighting if necessary to suit the closeness of the task and visual acuity required (CIBSE, 2005; SLL, 2005). In order to avoid a gloomy and oppressive visual environment, the guidance includes a requirement to provide an illuminance on the ceiling of 30 per cent and on the walls of 50 per cent of that on the working plane. This demands luminaires that are designed to illuminate the ceiling and walls and in most instances are suspended below the ceiling as shown in Figure 3.31.

**Figure 3.31** *Office lighting with upward component at Powergen headquarters, Nottingham*

*Source:* Bennetts Associates architects; photo copyright Peter Cook

Lamp technology has also evolved over the last 40 years, although fluorescent lamps are still used for most general lighting in offices, luminous efficacies (light output in lumens over electrical power consumed in Watts – lm/W) have steadily improved, with the latest T5 lamps achieving a luminous efficacy of around 104lm/W compared to 75lm/W for tube fluorescent lamps in the 1970s (see Figure 3.32).

As can be seen, the luminous efficacy (lamp efficiency) for most technologies has improved, but incandescent lamps, including tungsten halogen, remain at less than half that of fluorescent lamps. Metal halide lamps have slightly better luminous efficacy than fluorescent, but are used primarily for floodlighting and downlighting applications. Apart from White-SON, the sodium based lamps

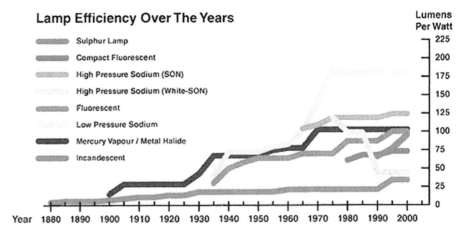

**Figure 3.32** *Lamp efficiency trends during 20th century*

*Source:* http://americanhistory.si.edu/lighting/tech/chart.htm, image by Hal Wallace, National Museum of American History

**Figure 3.33** *LED lighting at Federal Reserve, Washington US*

*Source:* John Harrington, Cree

give a yellow light which, although offering a high luminous efficacy, has found application only to street lighting.

Light emitting diode (LED) lighting offers the opportunity for luminous efficacies of between 120 and 150lm/W. In fact one American manufacturer has recently reported laboratory results for a 'white power LED' lamp of 208lm/W.[2]

Recently the Pentagon has had fluorescent lamps in 4200 of their luminaires replaced with LED lamps resulting in a 22 per cent reduction in electricity consumption and 140 tonnes per annum saving in $CO_2$ emissions. See Figure 3.33 for a typical LED installation at the Federal Reserve in the US. At the time of writing, however, many of the products available for replacing fluorescent tubes have luminous efficacies no better than T5 lamps, but at much greater cost.

LED lamps run cooler than fluorescent lamps and hence the heat gain and associated impact on summer time temperatures or air conditioning loads is lower. On the other hand they are more sensitive to high temperatures, which have a drastic impact on light output and life. Life expectancy is expected to be greater than 50,000 hours, compared to between 15,000 and 30,000 hours for fluorescent lamps, depending on wattage. LED lamps are also better suited to dimming than fluorescent lamps and switch immediately to full light output from cold. The current generation of LED tubes comprises a cluster of individual LED lamps within the tube which emit light directionally and hence rely on reflectors and diffusers to provide the combination of upward and downward illumination required by the current UK guidance for offices (SLL, 2005).

## Intelligent lighting controls

The three main methods of controlling lighting in the workplace are photoelectric switching and dimming; absence detection and occupancy sensing; and local dimming. (CIBSE, 1999)

1   Photoelectric switching uses light sensitive cells (photocells) that are calibrated to switch lamps according to a pre-determined illuminance level which corresponds to the maintained illuminance that the electric lighting has been designed to achieve. Generally speaking photoelectric switching is not appropriate for areas having a daylight factor below 2 per cent. There has to be a 'dead zone' between the set point for switching lamps on and off to avoid 'hunting' of the lamps, which will be annoying and shorten lamp life. More subtle control of illuminance levels and integration between artificial and natural lighting can be achieved with photoelectric dimming, whereby daylight is supplemented with artificial lighting when external levels drop below that required to maintain illuminance at the necessary level. Fluorescent lamps can be dimmed through electronic high frequency ballasts which also provide a 'soft start', increasing lamp life.

    The photocells can either be located externally, usually in a shaded spot on the roof, separately programmed for each elevation; attached to the underside of ceilings at key locations in each perimeter zone of the building, but not in direct sunlight; or attached to or integrated into luminaires. The first option is only suitable for applications where shading strategies are identical for each elevation, whilst the others respond to internal daylight conditions and can be arranged to control a cluster of luminaires together or provide separate control for each luminaire.

2   Absence detection and occupancy sensing are two sides of the same coin. Both use infra-red sensors to detect the presence of a human body within a preset distance of the sensor. Absence detection relies on manual or timed switching of the lighting and switches lamps off when no human is detected, and is the preferred option for office lighting. Occupancy sensing switches lighting on only when someone is detected, switching off again when they are out of range, and can be a good option for intermittently used areas such as corridors and warehouses.

3   Local dimming is very useful in auditoria and living rooms, but can also be applied in offices through local controls at individual workstations, either through personal computers or telephones. This can be combined with other local controls such as temperature (Chapter 3.2 Operational Energy and Carbon).

A Canadian study into the energy savings associated with the three measures described above found that the electrical energy required for lighting fell by around 69 per cent with the introduction of all three measures (Galasui et al, 2007). Each individual measure was assessed separately, indicating a saving of 35 per cent for the absence detectors, 20 per cent for the photoelectric switching and 11 per cent for individual dimming. These results were influenced by the building design, which has facades that are not overshadowed, containing large windows which provided good daylighting when the venetian blinds were open. The researchers found that blind occlusion was 55 per cent during the survey, presumably to reduce glare and overheating from direct sunlight.

# Health and lighting

## Display screen equipment

The importance of light in relation to the health of workers is recognized in the Health and Safety (Display Screen Equipment (DSE), Regulations (1992) and Workplace (Health, Safety and Welfare) Regulations (1992). These regulations require lighting to be suitable and sufficient for the task in hand and the DSE Regulations require 'an appropriate contrast between the screen and the background environment' and that 'possible disturbing glare and reflections on the screen … shall be prevented by coordinating workplace and workstation layout with the positioning and technical characteristics of the artificial light sources.' It also requires that 'sources of natural light … cause no direct glare and no distracting reflections on the screen' and consequently that 'windows shall be fitted with a suitable system of adjustable covering to attenuate the daylight that falls on the workstation.'

The use of combination luminaires that provide say 300 lux at the workstation with 90 lux at the ceiling and 150 lux on the walls, along with suitable diffusers to shield lamps from view, should significantly reduce the risk of reflective glare in display screens from artificial lighting. Integrating the requirements of the DSE Regulations with a desire to maximize daylighting is more problematic and is best dealt with by ensuring that display screens can be orientated away from windows, since even a reflected view of a light-coloured venetian blind can be distracting. Reflection is far less of a problem if DSEs have low reflectance screens, although this is outside the influence of the designer.

## Seasonal Affective Disorder (SAD)

Seasonal Affective Disorder (SAD) is a depressive illness caused by a biochemical imbalance in the hypothalamus associated with a lack of daylight and sunlight impacting on the human circadian system. It is generally associated with winter months in the high latitudes when days shorten to such an extent that there is very little exposure to daylight. Although with some individuals the symptoms of depression can be extremely debilitating, it is thought that approximately 17 per cent of the UK population suffer from milder symptoms, sometimes referred to as 'winter blues'.[3] It has been found that exposure to broad spectrum lighting that simulates daylight, but without the ultraviolet component, is effective in alleviating symptoms.

Interference with human circadian rhythms are also implicated in sleep disorders, broadly termed 'circadian rhythm sleep disorders' (CRSD) including delayed sleep phase syndrome, which can also be alleviated with exposure to 'artificial daylight'.

### Flicker

The light from a fluorescent lamp pulses at twice the frequency of the mains alternating current, or 100Hz in the UK and 120Hz in the US. Humans can only perceive flickering light at up to 50Hz, however, studies have indicated that many people are sensitive to the higher frequencies, reporting that there was a 50 per cent reduction in complaints of headache and eye strain when people were exposed to lighting with high frequency ballasts rather than those supplied at mains frequency (Wilkins et al, 1989). Modern electronic high frequency ballasts operate at a frequency of 30kHz.

## Light pollution

The images in Figures 3.34 and 3.35 do more than a thousand words to illustrate the phenomenon of night-time light pollution in the developed world. The issue is not restricted to external lighting but to luminaires left operating

**Figure 3.34** *Image of the Earth from space at night*

**Figure 3.35** *City of London at night*

*Source:* Viosan

in buildings at night, which also has a considerable implication for carbon emissions. A recent study by the Carbon Reduction in Buildings (CaRB) Programme has predicted that some one million tonnes of $CO_2$ could be saved annually if offices in the UK switched off their lights at night.[4]

The BREEAM Schemes include credits that reward good design practice for external lighting based on the Institution of Lighting Engineers 'Guidance notes for the reduction of obtrusive light' (ILE, 2005) as well as their guidance on illuminated advertisements (ILE, 2001). These limit the average upward light ratio, the illuminance from external lighting at neighbouring buildings, the intensity of each external light source in potentially obtrusive directions and the luminance of floodlit buildings. In order to qualify for this credit, BREEAM also requires that non-essential external lighting be switched off between 2300 and 0700 hours and that security lighting be kept to a minimum.

The LEED Schemes have similar requirements but also require designers to limit the amount of light that can escape through windows, or incorporate controls that switch off internal lighting during non-business hours. LEED has different requirements depending on location, divided into four categories: rural, residential, commercial/industrial or dense residential and major city centres. For each case LEED limits the amount of light crossing site boundaries and the lumens emitted above the horizontal plane.

## Notes

1    Refer also to www.lrc.rpi.edu/programs/daylighting/dr_windows.asp for a summary of relevant research papers on daylighting controls and shading.
2    www.cree.com/press/press_detail.asp?i=1265232091259.
3    www.sada.org.uk/what-is-SAD.html.
4    www.carb.org.uk/.

## References

BSI (British Standards Institute), 1992. *BS 8206, Part 2,1992. Code of practice for daylighting*. London: British Standards Institute

CIBSE, 1999. *CIBSE Lighting Guide LG10,1999. Daylighting and window design*. London: CIBSE

CIBSE, 2005. *CIBSE Guide A. Environmental Design. Section A1 Environmental criteria for Design, 1.8 – Visual environment*. London: CIBSE

Deutsches Institut für Normung. 1999. *Daylight in Interiors – General Requirements*. DIN 5034-1:1999-10 (E). Germany: Deutsches Institut Fur Normung E.V.

Galasui, A., Newsham, G., Suvugau, C. and Sander, D., 2007. *NRCC – 49498: Energy saving lighting control systems for open-plan offices: A field study*. Canada: Institute for Research in Construction. Available at http://irc.nrc-cnrc.gc.ca (accessed 20 March 2010)

*Health and Safety (Display Screen Equipment (DSE), Regulations, 1992*. Norwich: HMSO

IESNA (Illuminating Engineering Society of North America), 2000. *Lighting Handbook*. 9th ed. New York: IESNA

ILE (Institution of Lighting Engineers), 2001. *Technical Report TR05: The brightness of illuminated advertisements*. London: ILE

ILE, 2005. *Guidance GN01: Guidance notes for the reduction of obtrusive light.* London: ILE

Littlefair, P., 1990. Innovative daylighting: review of systems and evaluation methods. *Lighting Research & Technology*, vol 22, no 1, pp1–17

Rea, M., 1984. Window blind occlusion: A pilot study. *Building & Environment*, vol 19, no 2, pp133–137. Available at www.nrc-cnrc.gc.ca/obj/irc/doc/pubs/nrcc23592.pdf (accessed 20 March 2010)

SLL (Society of Light and Lighting), 2005. *Lighting Guide LG7 – Office Lighting.* London: SLL. Available at www.lg7.info/index.html (accessed 20 March 2010)

Wilkins, A., Nimmo-Smith, I., Slater, A. and Bedocs, L., 1989. Fluorescent lighting, headaches and eye-strain. *Lighting Research & Technology*, vol 21, pp11–18

*Workplace (Health, Safety and Welfare) Regulations, 1992.* SI 1992/3004. Norwich: HMSO

# 3.7

# Computer Simulation of Building Environments

## Introduction

Each new building is a prototype and as such its performance has to be predicted from standard assessment methodologies and compared against appropriate benchmarks. The integrated design process requires optimization through a series of option studies. It can be seen from the other chapters in this book that the factors to be considered and the options available are numerous. When it comes to the environmental design of a building, the viability of natural ventilation, penetration of natural light and consequent impact on human comfort and carbon emissions cannot readily be tested using manual techniques. Since these depend on climate-dependent variables, modelling has to be based on dynamic simulation of the building, its services and environment.

Dynamic simulation models are not the same as the steady state models such as the UK's SAP and SBEM (Chapter 3.2 Operational Energy and Carbon), which are used for predicting building carbon dioxide emissions and assessing compliance with Part L of the Building Regulations, although outputs from most dynamic simulation models can be used to assess compliance with the Part L targets.

We are going to examine three types of computer model for this chapter:

1  Dynamic simulation modelling (DSM) of building thermal performance to predict temperature exceedances to assess the viability of natural ventilation.
2  Computational fluid dynamics (CFD) modelling of airflows and temperatures to assess the impact of stratification in an office space and atrium on thermal comfort of occupants.
3  Prediction of daylight factors and interaction with photoelectric lighting controls and annual carbon dioxide emission modelling.

At the end of the chapter there is a case history that demonstrates the processes involved in the dynamic simulation of a naturally ventilated office building with atrium.

## Dynamic simulation modelling (DSM)

This procedure determines the likely operative temperatures due to the airflows, heat fluxes and thermal mass in a free-running building over the whole of the year, enabling a prediction of the number of hours per year that specified indoor operative temperatures will be exceeded. For example, to test against the CIBSE criterion that stipulates that an operative temperature of 28°C should not be exceeded for more than 1 per cent of occupied hours per year (see the section on Heat gains and comfort in Chapter 3.4 Design for Natural Ventilation).

There are a number of validated DSMs which can be purchased for use by trained personnel, or the modelling can be sub-contracted out to specialist consultancies. These models use the equations of thermodynamic heat transfer to simulate the thermal performance of buildings with input from a database of the thermodynamic properties of building materials, heat transfer and air flows associated with building services equipment, dimensions of building elements, meteorological data and solar geometry.

The materials database includes U-values, admittance coefficients (Y – thermal storage capacity), solar gain factors (S) and solar energy transmittance for windows (G-values), shading coefficients, etc. Meteorological data are available in a form suitable for input to DSM from a number of suppliers, but care needs to be taken that these account for likely increases in external temperatures due to global warming. For example, CIBSE publish Future Test Reference Year and Design Summer Year (TRY/DSY) data for 14 sites across the UK for the years 2011–2040, 2041–2070 and 2071–2100 that take into account the four UKCIP02 climate change scenarios between Low to High $CO_2$ emission rates.[1]

## Computational fluid dynamics (CFD)

CFD can be used for modelling air flow distribution and velocities, temperature profiles, contaminant distribution and the movement of fire and smoke indoors. It can also be used for predicting the wind environment and contaminant distribution outdoors (Chapter 2.3 Massing and Microclimate).

CFD is based on two- or three-dimensional modelling for a snapshot in time by dividing the cross-section or space to be analysed into control areas or volumes and solving the equations governing air flow, thermal transport and contaminant transport, as required. The input data required is the same as for DSM described above.

The resultant software solves the necessary governing equations which are used to model forced and free convection, including buoyancy-driven flows, differentiating between low velocity laminar flow and turbulent flow conditions and predicting the velocity profiles that occur at solid boundaries (boundary conditions). If contaminant transport is to be modelled, then the location of sources and sinks needs to be established along with the rate of emission, dilution and absorption of the contaminants.

## Daylight simulation models

The ideal daylight simulation model uses the building and solar three dimensional geometry data from DSM to produce daylight illuminance levels and daylight factors, along with photo-realistic visualization internally, complete with sun shadow casting. The output from the illuminance models, based on the likely availability of sunlight for internal illumination, should feed back to the DSM for predicting the operation of photoelectric lighting controls and hence the energy consumption of lighting and the likely impact on temperature exceedances or air-conditioning cooling energy consumption and associated carbon dioxide emissions. The models can also use input data on luminaires in order to model the lux distribution and visualize the effect of alternative lighting strategies.

# Case history

## Scottish Natural Heritage offices, Great Glen House, Inverness, Scotland

When Scottish Natural Heritage decided that they needed a new headquarters building that rationalized their office space in Scotland, they decided to aim for an exemplary environmental performance. With this in mind they employed the Building Research Establishment to develop performance requirements and employed them to monitor their implementation. They also decided to use an integrated design and build contract with a highly competitive tender process that attracted bids from Scotland's leading contractors and designers.

**Figure 3.36** *Great Glen House south end of atrium*

*Source:* Photograph by Michael Wolchover

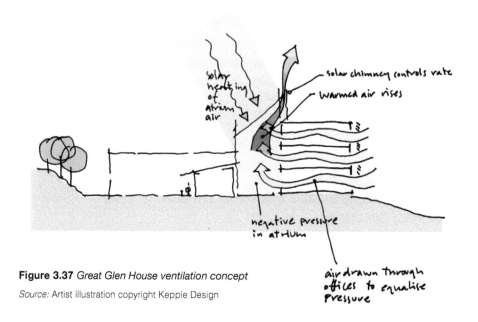

**Figure 3.37** *Great Glen House ventilation concept*

*Source:* Artist illustration copyright Keppie Design

The winning team, led by the developer arm of the Robertson Group and architect Keppie Design, included a Sustainability Advisor (URS Corporation Ltd) as well as building services and structural engineers from the outset.

It was decided from the first project team meeting to aim for a fully naturally ventilated building and an Excellent BREEAM rating. The concept of using an atrium with solar siphons to drive ventilation across open plan offices was developed by the Sustainability Advisor and architect during the early brainstorming resulting in a sketch similar to that in Figure 3.37 (Stark, 2006).

From this sketch and initial layouts, a preliminary dynamic simulation model was built using the proprietary Tas software developed by Environmental Design Solutions Limited (EDSL). This allowed a number of different options to be developed over the following months, including:

- the use of exposed concrete soffits compared with suspended ceilings;
- the impact of using a flat soffit compared with profiled coffer soffits;
- different solar siphon (chimney) configurations, with and without additional openings in the wall of the atrium;
- external shading louvres compared with tinted glass, which impacted on daylighting, glare and operation of the lighting controls;
- impact of internal shading 'banners' at high level in the atrium;
- single sided ventilation using 'herringbone' wall configuration for meeting rooms.

In the absence, at the time, of a climate model that predicted the impact of global warming on external temperatures in Inverness, data for Edinburgh were used, adding 2 or 3 degrees Celsius to temperatures. Criteria were adopted

**Figure 3.38** *Great Glen House East elevation*

*Source:* Photograph by Michael Wolchover

from the 1999 CIBSE Guidance (now superseded) that operative temperatures should not exceed 25°C for more than 125 and 28°C for more than 25 of the occupied hours during the design year. The model indicated that exposed soffits with windows at high level in the office walls opened automatically at night reduced the period that the top floor was likely to exceed 25°C by up to 40 hours. Although a coffer slab improved these results slightly, it was ruled out because of the cost implications. Openings at high level in the atrium were required to improve the performance of the solar siphons. The model also demonstrated that although external shading on the dominant east wall was effective at removing solar gain it also removed daylight, thus requiring the artificial lighting to be operating for most of the day, hence it was decided to use tinted glass.

DSM also showed that external shading on the west wall of the atrium was essential in reducing summer time temperatures and glare for the office spaces overlooking the atrium.

The modelling indicated that the top floor would experience the highest summer time temperatures, but that, even taking into account global warming, the operative temperature is unlikely to exceed 25°C for more than 40 hours of the year.

It has since been reported that 'the systems responded well to the warmer than normal weather during July 2006 (the first summer of operation). With no air conditioning, and an external temperature of 29°C, internal daytime temperature in open plan offices was 22°C.' (Stark, 2006).

Not only was the BREEAM Excellent rating achieved but, at 84 per cent, in 2006 the design achieved the highest score awarded to an office building from its inception in 1991 and was presented with the inaugural BREEAM Offices Award.

## Note

1    www.ukcip.org.uk/index.php?id=161&option=com_content&task=view. Refer also to CIBSE TM48 on the use of climate change data for building simulation (CIBSE, 2009).

## References

CIBSE, 2009. TM48: *The use of climate change data for building simulation*. London: CIBSE
Stark, D., 2006. *Great Glen House*. Edinburgh, UK: Keppie Design

# 3.8
# Noise and Vibration

## Introduction

Noise and vibration are an omnipresent part of human existence and, although individual sensitivity to noise varies considerably, there are statistically derived thresholds for discomfort, annoyance, interference to speech intelligibility and hearing damage that are widely accepted. Noise is generated across a wide range of frequencies, ranging from very low frequency bass notes through to high frequency sound which is not audible to humans but can be heard by other animals such as dogs and bats. In fact the human auditory system typically responds to frequencies in the range of 20–20,000Hz, although the response is not linear and tends to deteriorate at the top end, particularly with age.

The ear responds to pressure waves in the air, expressed in the logarithmic decibel scale (dB), the strength of which depends on the power of the noise source, the acoustic properties of the space in which the source is located and the distances between the source, ear and nearby surfaces. The same source located indoors and outdoors will sound very different, depending on the reverberation indoors. The amount of reverberation that occurs depends on the absorptive properties of the surfaces that are impacted by sound waves. Sound in a reverberant room, that is a room with many hard surfaces, will follow a number of different length paths to the ear and hence individual sounds will be followed by a series of decaying echoes. In highly reverberant 'live' spaces this can make speech unintelligible. At the other end of the spectrum an anechoic chamber has surfaces that are entirely sound absorbent or 'dead', so the sound follows only one route to the ear, which is close to the conditions in an exposed location outdoors, apart from reflection from the ground.

## Acoustic design of office buildings

An office is a complex acoustic environment in which it should be possible to concentrate on the task at hand without too many distractions. The key to a successful acoustic design is to identify all potential sources and develop a strategy for either separating them from noise sensitive activities, attenuating them at source or providing a barrier between source and receptor.

### Noise and vibration criteria

The first step is to identify noise sensitive tasks and define the acoustic environment required to reduce likely dissatisfaction to an acceptable level.

A number of noise criteria have been developed that are intended to express the variation in human sensitivity to different frequencies in a single index. The oldest of these is the dBA measure, still in wide use today, that corresponds approximately to the response of the ear. For example, because the human ear is more sensitive to noise at 1000Hz than 31.5Hz it assumes that a sound at the lower frequency would have the same loudness as one 40dB greater at 1000Hz.

When it comes to assessing the noise resulting from building services, a number of competing criteria have been developed, the most common of which are Noise Rating (NR), Noise Criterion (NC) and Room Criterion (RC). These all use a series of curves (or straight lines in the case of RC) to represent the response of the human ear against a range of sound frequencies at octave band centre frequencies ranging from 31.5Hz to 8000Hz (CIBSE, 2006, A1, Section 1.9). Historically, comfort criteria and performance of room equipment have been expressed in any one of the above criteria, although they were all originally developed for use with air-conditioning systems. The favoured criterion for situations where there are different types of source is the statistically derived criterion $L_{Aeq,T}$, which is the A-weighted sound pressure level (dBA) of a continuous steady sound having the same energy as a variable noise over the same time period (T). The designer has to take into account the noise from all sources in order to assess whether design criteria are being met. This will normally include a baseline noise survey and/or a prediction of likely noise levels from traffic and other external sources once the building is completed. This will be particularly important for a building having openable windows, but may also be required to determine noise transmission through a sealed facade.

BREEAM Offices Hea 13 awards a credit if noise levels in office spaces achieve the criteria shown in Table 3.3.

Note that in the absence of a significant contribution from low frequency noise dBA is normally between 4 and 8dB greater than NR, hence an $L_{Aeq,T}$ of 40dB would be approximately equivalent to one between NR30 and NR35.

Vibration may be one of the mechanisms by which noise is transmitted through a structure from plant, construction machinery or road traffic. However, it may also impact on the human body separately and people may be sensitive to vibration at levels that are barely above levels of perception. Guidance is provided by CIBSE and in more detail in BS6472 on levels at which

Table 3.3 BREEAM criteria for acceptable noise levels in office spaces

| Room type | $L_{Aeq,T}$ |
| --- | --- |
| Single-occupancy offices | 40dB or less |
| Multi-occupancy offices | 40–50dB |
| Rest rooms | 40dB or less |
| Seminar/lecture rooms | 35dB or less |
| Canteen | 50dB or less |

there is 'a probability of adverse comments' for both continuous and intermittent exposure for different applications (BSI, 1992; CIBSE, 2006, Section 1.10). Humans are more sensitive to vibration that is transmitted from one side of the body to the other or back to front compared with from foot to head. These criteria indicate that humans are approximately three times more sensitive to long-term exposure at night in their homes than in offices, and six times more sensitive to intermittent exposure with up to three occurrences.

## Reverberation time

As we saw above, the reverberation time is dependent upon the volume of the space and the sound absorption characteristics of room surfaces. It is defined as the time taken for the sound level to decay by 60dB at specified frequencies, normally 500Hz and 1000Hz, and can be approximated from the Sabine formula: reverberation time (RT) = (0.16 × room volume m³) (V) divided by (the total room surface area m²) × (the area weighted average absorption coefficient) ($\alpha_{av}$).

In the UK there are statutory requirements for reverberation only for the 'common internal parts of buildings containing flats or rooms for residential purposes' and for schools as set out in Sections 7 and 8 respectively of the Approved Document E of the Building Regulations (ODPM, 2003). Section 7 does not specify RT as such but provides a technique for establishing the sound absorption coefficient ($\alpha$) of suitable material for an exposed surface in a common area (usually a ceiling) based on the properties of the other surfaces at octave band mid-frequencies of 250, 500, 1000, 2000 and 4000Hz. Section 8 refers to the reverberation times specified for various uses in school buildings in Section 1a of Building Bulletin 93 (DfES, 2003). For example, an RT of 0.8 second is required for secondary school classrooms, 1.0 second for offices and 1.5 seconds for atria and circulation spaces.

There are no mandatory requirements for other types of office space, although good practice RTs of between 0.5 and 0.9s have been recommended for offices up to 1000m², with figures between 0.3 and 0.55s for large open plan office spaces (Woods, 1972). RTs this low may be very difficult to achieve in offices that have exposed soffits, which are likely to require sound absorptive material integrated into the suspended luminaires. Reverberation in atria may also be an issue, particularly where office space opens into the atrium. Incorporation of absorptive material into the walls and suspended from the ceilings should be investigated. Absorption coefficients for some building materials are provided in Approved Document E (ODPM, 2003), and for specific products should be available from most manufacturers.

## Noise and vibration from building services

The noise and vibration from building services can be divided into the following categories:

- Noise and vibration transmission to occupied spaces through the building structure from plant and equipment within a plant room or mounted externally.
- Noise transmitted from plant to occupied spaces via ductwork.
- Noise transmitted from plant rooms or externally mounted plant to the external environment.
- Noise transmitted from equipment located within the space.

Noise sources in plant rooms or externally mounted include fans, refrigeration plant, air cooled condensers, heat pumps, pumps, boilers, CHP plant, cooling towers, stand-by generators, cooling towers, lifts and escalators. All of these incorporate rotating or reciprocating machinery that generates both airborne noise and vibration. Figure 3.39 from CIBSE (2002) summarizes the transmission routes from building services plant.

Noise can find its way into the occupied space either by transmission from a plant room or by transmission into a duct or pipe that is used to carry fluid into the occupied space. A plant room may contain a combination of the items listed above, each of which will contribute to the overall noise level within the plant room. This may vary as items are switched on and off or have their speed varied. If there is one particular item which has a much higher sound power level than the others then this may dominate the sound field. In general all rotating plant will need to be mounted on suitable anti-vibration mountings (AVMs) and may require suitable inertia bases to reduce vibration and noise transmission directly to the structure. Where ductwork passes through walls into the building, sound attenuators (silencers) should be installed so that plant noise cannot be transmitted back into the ductwork between the silencer and the penetration through the plant room wall. Similar precautions may be required for penetrations to outside. Large pumps will also have to be mounted on AVMs and connected to pipework via flexible connections.

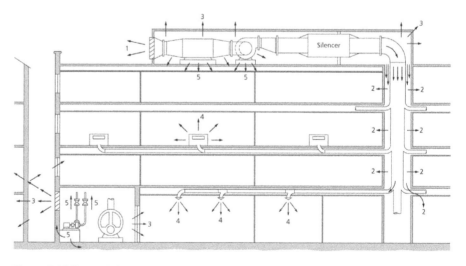

**Figure 3.39** *Transmission paths for roof-top and ground level plant rooms*

*Source:* CIBSE (2002). Reproduced by permission of the Chartered Institution of Building Services Engineers

The building services engineer should select plant, equipment, silencers and AVMs based on achieving the specified noise criteria for the spaces served. This will need to take into account the noise transmitted from the plant through ductwork and the noise generated by the air passing through the ductwork and at terminal devices. If the terminal devices incorporate fans or variable air volume (VAV) dampers, these may dominate the noise generated through this route.

When deciding on locations for plant and plant rooms, the potential for transmission to neighbouring spaces should be taken into account and noise sensitive spaces not positioned adjacent to significant sources of noise. There tend to be correlations between noise generation and size and noise reduction and cost. Hence it is generally more cost efficient to house large and potentially noisy plant either in separate buildings or clustered together with other service areas of a building such as car parking, rubbish storage, etc. Due to the substantial inertia provided by the foundations and earth, the vibration from large plant may also be easier to deal with in plant rooms that sit on the ground rather than on upper floors.[1]

## Cross talk and white noise

Generation and transmission of noise within an office building is not restricted to building services. The occupants talking on phones or in meetings can generate significant sound power levels which have the potential to be extremely distracting for their neighbours. This can be an issue both for people working in open-plan offices and for neighbouring cellular offices and meeting rooms. To avoid interference the sound level from the intruding speech should be 5–10dB below the sound level in the receptor space.

The amount of cross talk between neighbours in an open-plan office will depend on the height and acoustic properties of the partitions between workstations, the reverberation time of the space and particularly the sound absorption of the ceiling between neighbouring workstations, and the overall sound level in the space due to other sources (masking). Spaces which are subject to broad spectrum noise generated by the air conditioning and that benefit from sound absorption from textured ceiling tiles with associated low reverberation times will have far less intrusion from cross talk than a space with no noise from air conditioning or mechanical ventilation and exposed concrete ceilings and consequent higher RT.

A similar thermal performance could be achieved with a suspended ceiling manufactured from a PCM that will provide greater sound absorption than exposed concrete (Chapter 3.2 Operational Energy and Carbon). Alternatively, sound absorptive panels can be integrated with the suspended lighting.

With cellular offices there are a number of potential routes for sound transmission between neighbours:

- via gaps between partitions and the external wall or ceiling;
- via the common ceiling void above the neighbouring rooms, if partitions are not carried upwards to the underside of the slab, which can be exacerbated if there are openings in the suspended ceiling through light fittings or extract grilles;
- via openings into common ductwork serving neighbouring rooms.

Ideally partitions should be carried upwards and sealed at the underside of the floor slab as well as at external walls. However, this may not be compatible with the flexibility required to cater for future changes in office use. Cross talk through the ceiling void can be reduced if the underside of the slab is lined with sound absorbent material, whilst cross talk through ductwork can be reduced by a combination of laying out ductwork so that sound has to travel through a convoluted route between neighbouring rooms and by lining ducts with sound absorbent material.

Unfortunately most sound absorbent material is textured and can form a hidden trap for dust, making the internal cleaning of ductwork very difficult.

The noise generated by an air-conditioning system into the occupied space will have a sound characteristic that depends on the velocity of air at the diffusers, the noise regenerated in the ducts and dampers, the noise generated from air moving through the ductwork, the noise generated by terminal equipment such as fan coil units and the amount of noise that is transmitted from central plant, not attenuated through silencers or energy loss in the ductwork. This is likely to have a fairly broad spectrum and provide a degree of masking of conversations between neighbouring workstations. In the absence of such masking, it is possible to generate masking noise through speakers distributed throughout the occupied space.

Sound masking originated in the US in the 1960s following the widespread introduction of open-plan offices. The original systems incorporated speakers installed above suspended ceilings generating broad spectrum 'white noise', but the volumes required to mask speech were so high that discomfort and complaints resulted. In the 1970s systems were developed that simulated the spectrum of human speech and an annoyance threshold of 48dBA evolved. The problem with plenum mounted speakers was that due to the amount of reflection off ductwork and structural elements in the ceiling void, the noise generated into the space was not uniform. Hence direct systems were developed, using an array of speakers within the space, which are ideally suited to spaces with exposed soffits. These systems typically generate a noise level of 45dBA.[2] The speakers can also be employed for emergency announcement and tannoy use if required.

## External noise and window opening

By definition a naturally ventilated office building will allow noise from outside to enter. It is one of the links with the outside world that many office workers find appealing about natural ventilation. However, if the noise generated externally disturbs occupants they will close windows and the natural ventilation strategy has failed (Chapter 3.4 Design for Natural Ventilation). According to CIBSE (1999), any type of window when open will reduce the sound level entering the building by between 10 and 15dB compared with the level at the external surface of the window. If that building were located 20m from the edge of a motorway, the 18h average of noise levels at the building facade ($L_{Aeq,18h}$) is typically around 77dB, which corresponds to 67dB for someone sitting close to an open window. This compares with the BREEAM requirement for cellular offices of 40dB or less. Put simply, the $L_{AeqT}$ at the external openings of a naturally ventilated building should be less than 50dB from all sources and close proximity to all but the quietest of roads will mitigate against the viability of a natural ventilation strategy.

There are a number of computer models designed to assess the noise generated by predicted traffic flows when combined with noise survey data (see the section on Prediction of transported-related emissions in Chapter 2.7 Integrated Sustainable Transportation Planning). This assessment should be carried out for the noise impact assessment as part of the EIA (Chapter 1.4) and should be integrated into the design process for the building.

## Noise and dwellings

### External noise

In the UK local planning authorities make their decisions on applications for new homes based on the 1994 Planning Policy Guidance 24 (CLG, 1994), which requires a noise impact assessment to determine the likely noise levels as $LA_{eq.T}$ values at 1.2m above the ground in an open site at the location where the dwellings will be constructed. The criteria vary slightly depending on whether the dominant source is from road traffic, rail traffic, aircraft, or a mix of sources with no dominant component. Decisions are based on whether the predicted levels fall within specified Noise Exposure Categories (NEC) on a scale from A to D, where for NEC A noise will not be a determining factor, for NEC B permission is possible but conditional, for NEC C permission is unlikely and for for NEC D permission will be refused. Separate criteria are provided for day-time (0700–2300) and night-time (2300–0700) exposure. Day-time levels are based on 1980 WHO guidelines for an external $L_{Aeq}$ of 55dB 'required to prevent any significant community annoyance', which corresponds with the NEC A criterion. Night-time levels are based on the same WHO guidelines which stated that 'a level of less than 35dBA is recommended to preserve the restorative process of sleep'. Taking into account a 13dB attenuation from an open window and 3dB increase due to reflection of noise from the ground, a NEC A of 45dB is given in PPG24 (CLG, 1994). Predicted noise levels above these will require mitigation. To mitigate against day-time noise, bearing in mind the level has to be achieved at the external elevation of the dwelling, mitigation will have to be from barriers or increasing the distance between source and receptor. For night-time noise, mitigation can be from a combination of barriers, relocation, attenuation and ventilation. Part F of the Building Regulations (CLG, 2009) requires that dwellings be provided with continuous ventilation through either trickle ventilators or balanced mechanical ventilation. Trickle ventilators provide an opening through which noise can pass, although passive attenuating trickle ventilators are available that can provide sound attenuation of typically between 40 and 42dB (Figure 3.40).[3]

It is interesting to note that the World Health Organization revised their recommendations for acceptable noise levels to avoid sleep disturbance in their 1999 'Guidelines for community noise' which recommends an $L_{Aeq.8h}$ of 30dB, rather than 35dB with a 'small number of noise events' having an $L_{Amax}$ of 45dB, preferably during the latter part of sleep, since disturbance is most likely during the first part of the night (WHO, 1999).

**Figure 3.40** *Sound attenuating trickle vent*

*Source:* R W Simon

One external noise source that has become an issue of particular concern since the publication of PPG24 is the wind turbine. A report prepared by an 'expert panel' for the American and Canadian Wind Energy Associations expressed the view that noise from wind turbines is not unique, with no specific adverse physiological or health effects, and therefore should be considered alongside other noise sources (Colby et al, 2009). In the UK the criteria that would be applied to a planning application for a wind farm referred to in PPS 22 Renewable Energy[4] (ODPM, 2004) are published in an Environmental Technology Support Unit report prepared for the Department for Trade and Industry in 1996 which refers to a sound level exceeded for 90 per cent of a 10min period ($L_{90,10min}$) of 35dBA when the wind speed is 10m/s at 10m from the ground (ETSU, 1996).

## Transmission between rooms

In the UK the Building Regulations Part E sets minimum standards for the sound insulation between neighbouring flats and houses via walls and floors and for the transmission of impact noise between flats via floors. Sound insulation that can reduce incident sound levels by 45dB is required, whilst impact noise measured in the space below should be no more than 62dB. Internal floors and walls within new houses and flats should achieve a minimum sound insulation value of 40dB. These values have to be validated prior to handover using test procedures set out in the relevant parts of BS EN ISO 140 (BSI, 1998).

Credit Hea 2 of the CSH has four points available for improving on the standards for party walls and floors described above. A maximum score is awarded for an improvement of 8dB on both insulation and transmission properties, that is party walls and floors providing 53dB sound reduction and party floors that reduce impact noise to 54dB to the receptor.

The Regulations and CSH allow for methods of construction to be used that follow Robust Details.[5] Significant enhancements are required to achieve the maximum score under CSH, including resilient floors and suitable air gaps. At the time of writing there are no robust details for floors that achieve four points (8dB enhancement) and only one wall construction, whilst there are three floor types and 9 wall types that achieve a 5dB enhancement and therefore three points under Hea 2.

## Notes

1   For more detailed guidance see CIBSE Guide B5 (CIBSE, 2002).
2   www.speechprivacysystems.com/voicearrest-sound-masking-systems/the-evolution-of-sound-masking-systems/.
3   More information is available on the attenuation provided by windows and ventilation products in BRE Digest 338 and BRE Information Paper 4/99 (BRE, 1998; White et al, 1999).
4   www.greenspec.co.uk/documents/drivers/PlanningState22.pdf (Accessed 21 March 2010).
5   www.robustdetails.com/. The Robust Details Handbook, including a subscription service with regular updates is available for purchase from this website.

## References

BRE, 1998. *Digest 338. Insulation against external noise.* Watford: Building Research Establishment

BSI (British Standards Institute), 1992. *BS 6472:1992. Guide to evaluation of human exposure to vibration in buildings (1 Hz to 80 Hz).* London: British Standards Institute

BSI, 1998. *BS EN ISO 140:1995. Acoustics. Measurement of sound insulation in buildings and of building elements.* London: British Standards Institute

CIBSE, 1999. *Lighting Guide LG10. Daylighting and window design.* London: CIBSE

CIBSE, 2002. *Guide B5. Noise and Vibration Control for HVAC.* London: CIBSE

CIBSE, 2006. *Guide A – Environmental Design. Section A1. Environmental criteria for design.* London: CIBSE

CLG (Department for Communities and Local Government), 1994. Planning Policy and Guidance 24. Planning and Noise. Available at www.communities.gov.uk/documents/planningandbuildings/pdf/156558.pdf (accessed 21 March 2010)

CLG, 2009. *Proposed Changes to Part L and Part F of the Building Regulations: A Consultation Paper. Vol. 2: Proposed Technical Guidance for Part L.* London: CLG. Available at www.communities.gov.uk/documents/planningandbuilding/pdf/partlf2010consultation (accessed 21 February 2010)

Colby, W., Dobie, R., Leverthall, G., Lipscomb, D.M., McCunney, R.J., Seilo, M.T. and Søndergaard, B. 2009. *Wind Turbine Sound and Health Effects: An Expert Panel Review.* American Wind Energy Association/Canadian Wind Energy Association. Available at www.awea.org/newsroom/releases/AWEA_CanWEA.SoundWhitePaper_12-11-09.pdf (accessed 21 March 2010)

DfES (Department for Education and Science), 2003. *Building Bulletin 93: Acoustic design of schools – A design guide.* London: Department for Education and Science

ETSU, September 1996. *The Working Group on Wind Turbine Noise, The Assessment and Rating of Noise from Wind Farms.* Harwell UK: New and Renewable Energy Enquiries Bureau, ETSU. ETSU-R-97.

ODPM (Office of the Deputy Prime Minister), 2003. *Approved Document E of the Building Regulations 2000: Resistance to the passage of sound.* London: Office of the Deputy Prime Minister

ODPM, 2004. *Planning Policy Statement PPG22: Renewable Energy.* London: Office of the Deputy Prime Minister

White, M., McCann, G., Stephens, R. and Chandler, M. 1999. *BRE Information Paper 4/99: Ventilators: ventilation and acoustic effectiveness*. Watford: Building Research Establishment

Woods, R., 1972. *Noise Control in Mechanical Services*. Colchester: Sound Research Laboratories Ltd

World Health Organisation (WHO), 1999. *Guidelines for Community Noise*. Geneva: WHO

# 3.9
## Water Conservation

## Introduction

Globally there is a huge variation in the availability of water for consumption and the per capita demand. As with energy, the greatest demand is in developed countries. For example, in 2006 the water consumption per head in the US was 575 litres/day and in Australia 493 litres/day, whilst in the UK it was 150, China 86 and Rwanda, Ethiopia and Uganda 20 litres/day (UNDP *Human Development Report*, 2006; see Figure 3.41). These figures are just the tip of the iceberg, however, since they indicate domestic household use only, which when considered globally only represents some 8 per cent of total water consumption, whereas irrigation makes up 70 per cent and industry 22 per cent.[1]

The massive consumption figures for the US and Australia indicate endemic waste behaviours, which is of considerable concern when it is realized that some of the driest regions on the planet exist in these countries.

Although the UK may appear to do well in comparison, there remains an enormous amount of waste, particularly in large cities such as London, which historically has been losing some 30 per cent of its water supply due to leakage from underground pipes. Thames Water has, however, replaced some 1600km of mostly Victorian cast iron pipework over the past few years to bring this figure down.

The production of potable water is resource-hungry, due to the energy used for treating and conveying water and sewage and the chemicals used for treatment, all of which use natural resources and carbon associated with their manufacture. Where there is a heavy reliance on desalination, using such techniques as high pressure reverse osmosis to remove salt, the carbon footprint is even greater.

## Water consumption

Regulation 17K of the Building Regulations (HMSO, 2010) now limits water consumption for new dwellings in England and Wales to 125 litres/person per day based on a demand prediction carried out using a standard Water Efficiency Calculator (CLG, 2009) which is also used for Credit Wat 1 of the CSH. This credit awards a maximum of five points if it can be demonstrated that water consumption can be brought down to 80 litres/person per day, which is also a mandatory requirement to qualify for a CSH level 6 score. CSH levels 1 and 2 require a maximum water consumption of 120 litres/person per day, which is

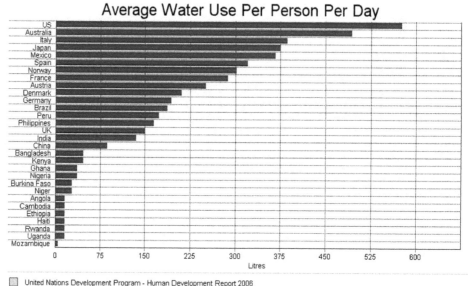

Figure 3.41 *Global water use statistics*

*Source:* Graph reproduced by permission of UNDP, Human Development Report 2006 (http://hdr.undp.org/en/media/HDR06-complete.pdf) and www.data360.org (www.data360.org/dsg.aspx?Data_Set_Group_Id=757)

in line with Regulation 17K that includes a fixed allowance of 5 litres/person per day for external water use, not included in the CSH benchmarks.

The Water Efficiency Calculator takes into account the water used in flushing water closet (WC) cisterns, bidets, wash basins, sinks, baths, showers, dishwashers, washing machines, water softener backwash and waste disposal units. It excludes water consumption from jacuzzis that are not used as baths and swimming pool top-up, maintenance and operation. It takes into account savings in demand from the mains that are achieved through grey water recycling and rainwater harvesting, but not from using private boreholes. This is because the latter does not reduce the burden on the municipal drainage and treatment systems.

CSH deals with external water consumption separately through Wat 2 which rewards rainwater harvesting for irrigation use.

There is currently no legal requirement for water metering in the UK, although it has become standard practice for new development.

Regulation 17K only applies to dwellings and there are currently no targets for water use in non-residential development. The various versions of BREEAM reward water saving measures and BREEAM Offices includes an escalating reward scale for predicted water consumption based on a calculator that is built into the protocol. It awards from a maximum of 3 points for internal water use of 1.5m³/person per year to a minimum 1 point for between 4.5 and 5.5m³/person per year. Separate credits are available for the provision of a water meter with a pulsed output suitable for remote analysis and billing, major leak detection and sanitary supply shut-off (see below).

In the US the Energy Policy Act of 1992 introduced maximum water use standards for sanitary appliances, such as the 9.5 litre/min standard for showers referred to below. It also introduced a requirement for labelling and provided funding for state and local incentives. Recently this was augmented by the American Recovery and Reinvestment Act which was adopted on 17 February 2009, providing $4 billion for the Clean Water State Revolving Fund (SRF), $2 billion for the Drinking Water SRF and, among others, $126 million for water recycling projects through the United States Bureau of Reclamation.

LEED rewards the use of appliances that improve on the Energy Policy Act requirements by 20 and 30 per cent, as well as the use of design features that reduce water demand such as zero irrigation landscaping and waterless urinals. LEED also gives credit for the use of rainwater harvesting and grey water recycling.

## Water saving appliances

As can be seen from Figure 3.42 for the typical British household, personal washing and toilet flushing dominate water consumption. Hence the attention on water saving has logically been on reducing water use from WCs and baths or showers.

The challenge with the development and installation of water saving appliances is to achieve adequate performance at lower flow rates. For example, there is little point in installing a low flush WC which has to be operated two or three times to clean the bowl. Similarly a low flow shower that provides only a dribble will both prolong the length of time required for an effective wash and create dissatisfaction and frustration amongst users.[2]

There follows a discussion on some of the technologies available for reducing water use. In most cases the technologies are the same for residential and non-residential applications.

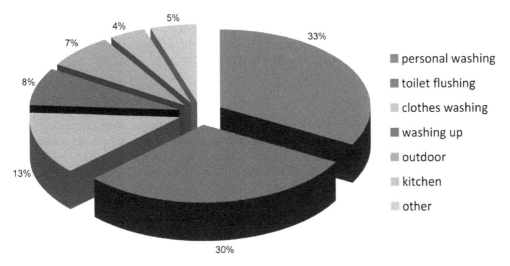

**Figure 3.42** *Proportions of water use for a typical British home*

*Source:* www.waterwise.org.uk/images/site/products/toilet%20flushing%20product%20catalogue.pdf.

## WCs

The main methods for reducing the water consumption of a WC are to either reduce the volume of water used for flushing or use the solid waste for compost. Dual-flush WCs have become the norm across Europe, including the UK, and with different combinations of flushing volumes for solid and liquid waste. However, where there are urinals present, there is very little point in installing dual flush toilets because there should be no need to use them for flushing liquids alone.

In the UK the performance requirements of WCs are governed by the Water Supply (Water Fittings) Regulations 1999 WC Suite Performance Specification (DEFRA, 2000) which sets out 'the Regulators' Specification for WC Suites delivering a single flush of 6 litres maximum or a dual-flush of 6 litres maximum and reduced flush of no greater than 2/3 of the maximum flush'. The minimum allowable flush rate for the main flush is 1.85 litres and 1.6 for the reduced flush.

The most common dual flush toilets available use either 6 litre and 3 litre flush (6/3), 4 litre and 2.5 litre flush (4/2.5) or even 4/2 litre flushes. All products on the market in the UK must pass the flush tests for solid and liquid waste set out in the above Specification. Research into residential use of a 6/3 dual flush WC found that water consumption fell by 18 per cent compared with a 6 litre flush. The smaller flush volumes give greater water savings, but not proportionately so, presumably because there are more incidences of double flushing.

Delayed action inlet valves prevent the cistern starting to fill before it is empty which research has shown can increase water consumption by 17 per cent (Environment Agency, 2007). The same report points out the drawback of drop valve cisterns, usually associated with push button flushes, which have been allowed in the UK since an amendment to the Water Supply Regulations in 2001. These have been shown to leak into the cistern after a while, something which can easily go undetected. Siphon-operated cisterns do not suffer from this problem but require lever operation rather than a button and hence are less frequently used for dual flush operation.

Composting toilets are unlikely to find application other than for remote dwellings located beyond the reach of municipal sewers. They have been successfully used in remote locations for many years.[3]

## Urinals

Conventional urinals also require periodic flushing and this can either be carried out continuously, automatically or manually. Water Supply Regulations stipulate that no more than 7.5 litres/h should be used for each stall, however continuous flushing systems set to this rate frequently drift over time. Automatic systems use a variety of methods including operating central cisterns from a timer or a door switch, from water pressure fluctuation when a hot water tap is operated and from an infra-red presence detector. The water consumption from flushing conventional urinals can be reduced by ensuring that flushing matches demand for individual stalls. The most effective methods of doing this are for either each stall to be fitted with an infra-red presence detector or, less hygienically, a lever operated valve. Continuous and timed systems have no feedback and are likely to consume more water accordingly.

Alternatively, waterless urinals are available that use a liquid which is lighter than water to provide both a seal that caps the outlet once the urine has drained away and masks the odour from the drain. These are obviously dependent upon good and regular housekeeping to ensure odour free operation, although this is also true of water-flushed urinals (Figure 3.43). There are also obvious advantages in reducing plumbing and the potential for leaks and scale accumulation.

## Taps

The type of tap that offers the greatest water saving will depend on how it is to be used. A tap over a kitchen sink or bath will usually be used for filling things and hence the flow rate is probably not very important. Taps used primarily for washing hands however can perform well with low flow rates, provided their wetting capability is maximized. Hence aerated or spray taps are ideally suited to hand washing and can save up to 80 per cent of water consumption compared with a conventional bib tap. However, in a residential application and with a flow rate of typically around 1.7 to 2 litres/min, a single spray tap will not generate sufficient flow through a combination boiler for it to operate.

Products are available that operate as a spray tap when the handle is partially turned then open to full bore for basin filling when operation is continued. Where there is a risk of taps being left running, which applies to most taps used for hand washing outside the home, then a degree of automation is worth considering. This can be achieved with infra-red operation or self-closing push button taps, although the reliability of the latter is varied, with large variations in delay before closing possible. Infra-red has the advantage of hygienic operation, although batteries or electricity supply will be required.

**Figure 3.43** *Waterless urinals*

*Source:* Kohler Steward L waterless urinal (see www.tuvie.com/waterless-urinal-from-kohler-avoid-splashing/).

## Baths and showers

In the UK it is generally considered a minimum requirement to provide both a bath and a shower, either with the shower over the bath or in a separate cubicle or room. Hence the designer is not usually in a position to dictate which is to be used for bathing. Although there are various bath sizes available, the standard size in the UK is 1700mm × 700mm with a capacity of 230 litres, which is the maximum allowed by Water Supply Regulations, and corresponds to approximately 160 litres of water when occupied. An efficiently designed bath as shown in Figure 3.44 can reduce these volumes to 140 litres and 70 litres respectively.

Showers should offer a significant saving on the water used in taking a bath. However, a lot depends on the shower type and the average length of shower taken. For example, a 'power shower' might use 16 litres/min which corresponds to 160 litres for a 10 minute shower. However, there are low flow shower heads which operate at between 4 and 8 litres/min and with aerating of the spray can provide excellent wetting and a high level of satisfaction amongst users. Table 3.4 compares a number of different combinations with a low water volume 73 litre bath (Critchley and Phipps, 2007).

The potential for combination boilers not to operate at low flow conditions has been examined in Grant (2007) and found only to be an issue with 35kW boilers that only modulate down to 30 per cent of maximum duty and boilers of 48kW or larger.

**Figure 3.44** *Low water volume bath*

*Source:* Ideal Standard

**Table 3.4** *Comparison of water use between different shower types and a 73l bath*

| Appliance | Flow rate (l/min) | Duration (min) | Volume/ event (litres) | Energy/ event (kWh) | Cost/ event (p) | Water use/yr (litres) | Energy use/yr (kWh) | kgC/yr |
|---|---|---|---|---|---|---|---|---|
| Electric shower | 3.9 | 5.8 | 22.6 | 0.96 | 20 | 14,000 | 580 | 249 |
| Mixer shower | 8.0 | 5.8 | 46.4 | 2.80 | 26 | 28,000 | 1720 | 327 |
| Mixer shower | 8.0 | 9.0 | 72.0 | 4.30 | 40 | 44,000 | 2650 | 503 |
| Power shower | 12.0 | 9.0 | 108.0 | 6.50 | 60 | 66,000 | 3980 | 756 |
| Bath | N/A | N/A | 73.0 | 4.90 | 43 | 35,000 | 2330 | 443 |

Mixing valves serving showers should achieve a comfortable temperature as quickly as possible with minimal water wastage. This can best be achieved with mixers that have a separate thermostatic temperature adjustment that can be left at the desired setting whilst the water valve is opened or closed.

## Piped distribution

Hot water 'deadlegs' should be kept as short as possible in order to reduce the amount of water that has to be run-off in order to attain the desired temperatures. This is also a requirement to reduce the risk of legionella bacteria multiplication (Chapter 3.5 Air Quality, Hygiene and Ventilation).

For non-residential or multi-residential buildings meters should be provided with a pulsed output for linking with a Building Management System (BMS). For individual dwellings, meters should either be located so as to be easily read by the occupier or form part of a 'smart metering' system that also includes output from electricity and heat or gas meters.

Major leaks can be sensed through the pulsed output from a meter linked to the BMS that includes a trend analyser that is able to tell when flows exceed the norm and hence raise the alarm. For small buildings with no BMS, bespoke leak detectors are available that can be installed in the main water supply.

Small leaks in toilet areas or blocks that might go undetected overnight or at the weekend can have their impact minimized through the installation of proximity detectors positioned to sense occupation that operate solenoid valves in the water supply to the area/block.

Pulsed metering, leak detection and sanitary supply shut-off all gain credits under the BREEAM scheme.

## White goods

A new washing machine uses approximately 50 per cent of the water and energy that is used by a typical 10-year old machine. Most new washing machines use between 40 and 50 litres of water per 6kg wash. However, most washer-dryers use mains water as part of the condensing cycle which increases water consumption to between 90 and 170 litres per wash and dry, but with a maximum of 5kg of clothes for drying.

Dishwashers are also becoming more water and energy efficient, with the most efficient machines using as little as 12 litres to wash 12 place settings.

It should be noted that for the European market manufacturers have to quote water consumption on the Ecolabel for each product.

## Water recycling and rainwater harvesting

Water that has been used for washing purposes can be collected, cleaned to some extent and re-used for flushing toilets and to supply washing machines, if the quality is good enough. Generally speaking, water is collected from showers, baths and wash hand basins, and possibly from washing machines and dishwashers, although the cleaning system will have to be sufficiently robust to handle the additional detergent and grease loads from these appliances. The simplest arrangement is shown in Figure 3.45, which uses wastewater from a bath and shower for toilet flushing. Stored water is held for a time to allow

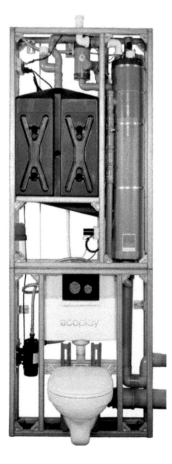

**Figure 3.45** *Packaged grey water system for single dwelling*

*Source:* Copyright CME Sanitary Systems Ltd (see www.ecoplay-systems.com/content.aspx@id=2.htm).

sludge to collect and surface scum to be removed automatically and both flushed away. The whole system is drained and flushed through if there is no demand for a pre-determined period.

For larger installations a water treatment plant can be installed on similar lines to a municipal plant, but the additional carbon emissions associated with water treatment will have to be justified by the water savings achieved.

Rainwater harvesting installations range from a simple water butt storing rainwater from the roof of a house for irrigation purposes to large underground storage tanks which use rainwater for toilet flushing and irrigation.[4]

An example of a large-scale installation is shown in Figure 3.46. The storage tank can comprise simple sectional water tanks buried underground or located in a basement, oversized drainage pipes or the flexible cellular system shown in the figure.

The storage tank is normally sized from the amount of rainwater likely to be available, which depends on the surface area of the roof and the slope and type of roof surface. For example typically only 50 per cent of rain that falls on a flat roof is available for harvesting (i.e. a drainage factor of 0.5). This proportion falls to even lower levels for green or landscaped roofs, which do not make an ideal catchment area for rainwater harvesting. Smooth sloping roofs can have a drainage factor as high as 0.9. The amount of water stored is usually calculated from 5 per cent of the total annual rainfall adjusted for the

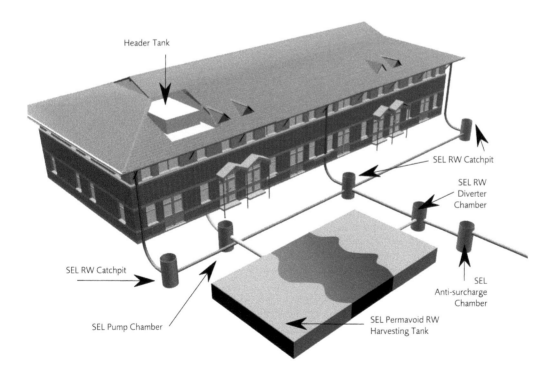

**Figure 3.46** *Typical arrangement for large-scale rainwater harvesting*

*Source:* Althon, Norwich (see www.althon.co.uk/products/althon-suds/sel-rainwater-harvesting-systems/)

drainage factor of the roof and the filter efficiency of the installation (available from the manufacturer, but typically 0.9). Coarse filters are required to remove leaves and other large solids. Beyond this the level of filtration and water treatment provided depends on the appliances being supplied. It may be considered that coarse filtration is sufficient for irrigation and even supplying WC cisterns, although there have been some problems with staining of WC basins from residues in the harvested water. Some installations have incorporated additional filtration and ultra violet biocidal control, although the cost, energy consumption and additional maintenance from these must be factored into the feasibility study for the installation.

Rainwater harvesting products are listed in the UK enhanced capital allowance directory[5] whilst grey water systems have to go through a certification process. For further guidance refer to CIBSE KS01 Reclaimed Water (2005)

## Notes

1   see UN-Water Statistics at www.unwater.org/statistics_use.html.
2   For a list of water conserving appliances that benefit from enhanced capital allowances, refer to www.eca-water.gov.uk/.
3   For more information on composting toilets, see www.compostingtoilet.org/.
4   www.ukrha.org/home.
5   www.eca-water.gov.uk/. For further guidance on rainwater harvesting and grey water systems refer to CIBSE KS01 Reclaimed Water (2005).

## References

CIBSE, 2005. *KS01: Reclaimed Water*. London: CIBSE

Critchley, R. and Phipps, D., 2007. *Water and Energy Efficient Showers: Project Report.* United Utilities. Available at http://stakeholder.unitedutilities.com/downloadfile. aspx?ID=15 (accessed 12 August 2010)

CLG (Department Communities and Local Government), 2009 *The Water Efficiency Calculator for New Dwellings*. London: Communities & Local Government

DEFRA (Department for Environment, Food and Rural Affairs), 2000. *Water Supply (Water Fittings) Regulations 1999: WC Suite Performance Specification*. London: DEFRA. Available at www.defra.gov.uk/environment/quality/water/industry/wsregs99/ documents/waterregs99-wcspec.pdf (accessed 21 March 2010)

Environment Agency, 2007. Conserving Water in Buildings. Bristol: EA. Available at www.environment-agency.gov.uk/static/documents/Leisure/geho1107bnjree_ 1934318.pdf (accessed 21 March 2010)

Grant, 2007. *Combination boilers and low flow fittings. Elemental Solutions*. [Available at www.elementalsolutions.co.uk/downloads/combis%20and%20low%20flows.pdf (accessed 21 March 2010)

HMSO, 2010. Approved Document G. *Sanitation, Hot Water Safety and Water Efficiency*. London: HMSO. Available at www.planningportal.gov.uk/uploads/br/ 100312_app_doc_G_2010.pdf (accessed 29 October 2010)

UNDP, 2006. *Human Development Report*. New York; United Nations Development Programme

# 3.10

# Design to Reduce Vehicle Impacts

## Introduction

This chapter focuses on what measures can be incorporated into the design of individual buildings to encourage occupants and visitors to use alternatives to the motor vehicle. The strategic issues for planning and masterplanning are covered in Chapter 2.7 Integrated Sustainable Transportation Planning.

## Travel plans

Planning guidance mandates local authorities in the UK to require developers to submit a green travel plan alongside the Transport Impact Assessment (TIA) with a planning application for most types of development (PPG13 – CLG, 2001). PPG13 suggests that these plans include the following sustainable transport objectives:

1   reductions in car usage (particularly single occupancy journeys) and increased use of public transport, walking and cycling;
2   reduced traffic speeds and improved road safety and personal security particularly for pedestrians and cyclists; and
3   more environmentally friendly delivery and freight movements, including home delivery services.

BREEAM Offices provides a useful checklist as part of the requirement for Credit Tra 4, which requires that a transport plan includes:

(a) where relevant, (analysis of) existing travel patterns and opinions of existing building or site users towards cycling and walking so that constraints and opportunities can be identified;
(b) travel patterns and transport impact of future building users;
(c) current local environment for walkers and cyclists – accounting for visitors who may be accompanied by young children;
(d) disabled access – accounting for varying levels of disability and visual impairment;
(e) public transport links serving the site;
(f) current facilities for cyclists.

Examples of measures that might be included in a travel plan are given, including:

(a) providing parking priority spaces for car sharers;
(b) providing dedicated and convenient cycle storage and changing facilities;
(c) lighting, landscaping and shelter to make pedestrian and public transport waiting areas pleasant;
(d) negotiating improved bus services, e.g. altering bus routes or offering discounts;
(e) restricting and/or charging for car parking;
(f) criteria for lobby areas where information about public transport or car sharing can be made available;
(g) pedestrian and cycle friendly (design) – for all types of user regardless of the level of mobility or visual impairment – via the provision of cycle lanes, safe crossing points, direct routes, appropriate tactile surfaces, well-lit and signposted to other amenities, public transport nodes and adjoining off-site pedestrian and cycle routes.

The TIA should demonstrate how the measures in the green travel plan will impact on the number of vehicle movements and the reduction in carbon emissions and pollution associated with the smaller number of vehicles on the road. Every development should demonstrate that it is catering for those who want to cycle or walk and that consideration has been given to accommodating car clubs, with facilities for low emission vehicles, such as charging points for electric vehicles. Residential development should cater for those who wish to work from home and ideally include a proportion of units that allow for establishing a studio or workshop as part of a live/work unit.

**Figure 3.47** *Example of 'compliant' communal bicycle shelter in Belfast*

*Source:* Ardfern

# Facilities for cyclists and pedestrians

We have discussed the importance of access to local amenities and transport hubs in Chapter 2.7 Integrated Sustainable Transportation Planning, including the provision of secure routes for cycles and pedestrians through such measures as the Home Zone. In order to encourage bicycle ownership and use, BREEAM, CSH and LEED all include credits for providing storage for cycles.

BREEAM Offices Tra 3 is a fairly complex requirement which rewards the provision of secure and sheltered cycle storage, and some combination of showers, changing facilities, lockers and clothes drying. The minimum requirement for one credit is for 10 per cent of staff (not including visitors) to be provided with 'compliant storage' up to a maximum of 500 staff. Beyond this and up to 1000 staff, 7 per cent are to be provided with storage and from 1001 upwards the percentage drops to 5 per cent. 'Compliant storage' should be no more than 100m from the main entrance, be overlooked, well lit, secure and sheltered, with no more than one open side (Figure 3.47). It should be possible to secure both a wheel and the frame to a fixed stand. The 'Sheffield' design is both compliant, robust and simple, comprising a sturdy tubular frame as shown in Figure 3.48.

Stands need to be fixed at 1 metre centres and 500mm at either side to ensure easy access.

BREEAM Tra 3 has a further credit for either provision of showers for 10 per cent of the cycle storage provision, space for cyclists to change, including suitable lockers, or drying space. In order to gain this credit two of these must be provided and:

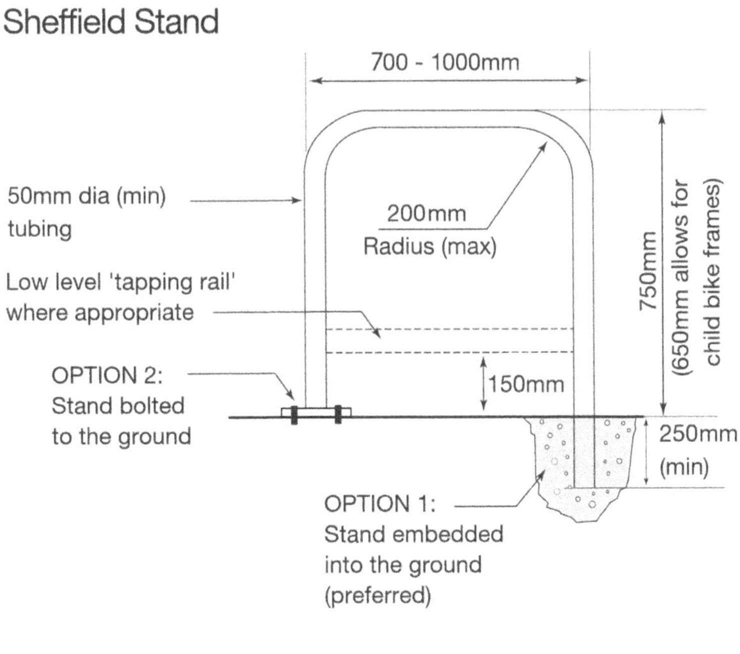

## Sheffield Stand

700 - 1000mm

50mm dia (min) tubing

200mm Radius (max)

Low level 'tapping rail' where appropriate

OPTION 2: Stand bolted to the ground

750mm (650mm allows for child bike frames)

150mm

250mm (min)

OPTION 1: Stand embedded into the ground (preferred)

**Figure 3.48** *Sheffield Stand*

*Source:* Image courtesy of Sustrans (www.sustrans.org.uk/assets/files/guidelines/Signing%20parking%20etc.pdf)

- the showers must be accessible for both men and women;
- at least 1m$^2$ of changing space must be provided for each shower, up to 4m$^2$ per cluster;
- lockers must measure at least 900mm high by 300mm wide by 450mm deep; and:
- drying space must have provision for heating and ventilation and be designated for drying clothes only, allowing cyclists to hang clothes in a boiler room would not be acceptable, for example.

BREEAM allows half the above provision if the proposed offices either have very good access to public transport or are located where the shortest commute for staff would be ten miles or more, making commuting by cycle more difficult.

The requirements for LEED NC SS 4.2 are simpler and include provision of secure racks/storage located within 200 yards of the building entrance for 5 per cent of building occupants and shower provision for 0.5 per cent of full-time equivalent (FTE) staff.

The CSH also rewards provision of cycle storage, dependent upon the number of cycles that can be accommodated for one-bed homes, two/three-bed homes, or homes with four beds or more. A maximum score is awarded for the provision of suitable accessible, sheltered and secure storage for one, two and four cycles respectively. These can be provided in communal stores, garages or garden sheds provided there is space for the primary use and an area of 2m × 0.75m for one bike, 2m × 1.5m for two and 2m × 2.5m for four.

Some local authorities include specific requirements for cycle parking, the most detailed of which have been developed by Transport for London (TfL, 2004), reproduced as Table 3.5. Policy 3C.22 of the 2008 London Plan (Mayor of London, 2008) requires London Boroughs to include these in their Local Development Frameworks and 'the Mayor will use/apply these standards in considering applications for strategic developments'.

We have already discussed the design of safe cycle and pedestrian routes as part of the masterplanning process (Chapter 2.7 Integrated Sustainable Transportation Planning). BREEAM and CSH require that cyclists should have direct safe access from the place where they park or store their cycle to the nearest cycle route. Cycle lanes should be designed to meet Sustrans National Cycle Network (NCN) Guidelines and Practical Details and their Design and Construction Checklist.[1]

Shared cycle and pedestrian routes should be at least 3m wide, whilst paths used for cycles only should be 2m wide or more and for pedestrians only 1.5m. If the cycle lane forms part of a highway then it can be reduced to 1.5m wide. Lighting for these routes needs to be designed in accordance with good practice as set out in CIBSE/SLL Lighting Guide 6 (SLL, 1992).

## Car clubs and sharing

Designs should incorporate dedicated spaces for the use of cars that can have multiple users, either as part of a community or commercial car club or a car sharing scheme for the commuters. Incentives can be provided in the form of allocating reserved spaces close to entrances and/or offering discounted or free parking.

**Table 3.5** *Cycle parking requirements for new developments in London*

| Land use category | Location | | Cycle parking standard |
| --- | --- | --- | --- |
| A1 | Shops | Food retail | Out of town 1/350m²* |
| | | | Town centre/local shopping centre 1/125m²* |
| | | Non-food retail | Out of town 1/500m²* |
| | | | Town centre/local shopping centre 1/300m²* |
| | | Garden centre | 1/300m²* |
| A2 | Financial and professional services | Offices, business and professional | 1/125m²* |
| A3 | Food and drink | Pubs, wine bars | 1/100m²* |
| | | Fast food takeaway | 1/50m²* |
| | | Restaurants, cafes | 1/20 staff for staff + 1/20 seats for visitors |
| B1a | Business | Business offices | 1/250m²* |
| B1b | | Light industry | 1/250m²* |
| B1c | | R&D | 1/250m²* |
| B2–B7 | General industrial | | 1/500m²* |
| B8 | Storage and distribution | Warehouses | 1/500m²* |
| C1 | Hotels | Hotels | 1/10 staff |
| | | Sui generis hostels | 1/4 beds |
| C2 | Residential institutions | Hospitals | 1/5 staff + 1/10 staff for visitors |
| | | Student accommodation | 1/2 students |
| | | Children's homes, nursing homes, elderly people's homes | 1/3 staff |
| C3 | Dwelling house | Flats | 1/unit |
| | | Dwelling houses | 1/1- or 2-bed dwelling, 2/3+-bed dwelling |
| | | Sheltered accommodation | 1/450m² |
| D1 | Non-residential institutions | Primary schools | 1/10 staff or students |
| | | Secondary schools | 1/10 staff or students |
| | | Universities/colleges | 1/8 staff or students |
| | | Libraries | 1/10 staff + 1/10 staff for visitors |
| | | Doctor, dentist, health centre, clinics | 1/50 staff + 1/5 staff for visitors |

**Table 3.5** *Cycle parking requirements for new developments in London* (Cont'd)

| Land use category | Location | | Cycle parking standard |
|---|---|---|---|
| D2 | Assembly and leisure | Theatres, cinema | 1/2 staff for staff + 1/50 seats for visitors |
| | | Leisure, sports centres, swimming pools | 1/10 staff + 1/20 peak period visitors |
| Transport | Train station | A Central London termini | 1/600 entrants |
| | | B Zone 1 interchanges | 1/1000 entrants |
| | | C Strategic interchanges | 1/600 entrants |
| | | D District interchanges | 1/200 entrants |
| | | E Local interchanges | Upon own merit |
| | | F Zone 1 non-interchanges | 1/200 entrants |
| | | G Tube termini/last 3 stations | 1/150 entrants |
| | | H Other | Upon own merit |
| Transport | Bus stations | | 1/50 peak period passengers |

*Note:* * minimum 2 spaces

*Source:* TfL (2004).

## Catering for low emission vehicles

Similar incentives can be provided for those who use electric or hybrid cars for commuting, with spaces set aside for electric vehicles having suitable charging points. One option is to establish a car club fleet which uses primarily electric vehicles, with hybrid cars available for longer journeys if necessary.

BREEAM does not include any credits that cover low emission vehicles, whilst LEED NC SS 4.3 includes three options for encouraging the use of low emitting and fuel efficient vehicles. Option 1 involves providing these vehicles to 3 per cent of the FTE workforce, along with preferred parking. Option 2 requires preferred parking for occupants using low emission vehicles for at least 5 per cent of the car parking spaces. Option 3 rewards the provision of refuelling facilities to cater for at least 3 per cent of the car parking capacity. The last option is referring to alternative fuels such as hydrogen for fuel cells.

## Home working and live/work units

As we saw from Chapter 2.7 Integrated Sustainable Transportation Planning, designs should encourage working from home and this is recognized in the CSH through Ene 8 which rewards the provision of a 'home office'. It requires sufficient space for a desk, filing cabinet and chair whilst allowing the primary use of the space to be retained. The credit specifies that at least 1.8m of wall space containing a minimum of two double electrical sockets and phone socket be provided for connection to broadband or a double socket for separate phone and modem connections. The space must achieve a minimum average daylight factor of 1.5 per cent and have suitable ventilation. In dwellings with three or more bedrooms the space provided should not be in the main living room, master bedroom or kitchen, whereas in smaller dwellings the living room or either bedroom can be used.

Live/work units take the concept of home working one step forward by providing a purpose-designed facility that incorporates an apartment and workspace into the same envelope. It is important to understand both the likely planning and funding requirements whilst catering for the possible marketplace, which may be very location specific. For example, the 2008 London Plan suggests that clusters of live/work units could provide flexible and affordable space for creative industries and help with local regeneration (see the section on Local facilities and jobs in Chapter 2.7 Integrated Sustainable Transportation Planning).[2]

## Notes

1    www.sustrans.org.uk/assets/files/guidelines/appendix.pdf.
2    Some further guidance on the design of live/work units is provided at www.liveworkhomes. co.uk/content/view/332/295/designing-live-work%3A-have-you-ticked-right-boxes.htm.

## References

CLG (Department for Communities and Local Government), 2001. *Planning Policy Guidance 13: Transport (PPG13)*. London: TSO. Available at www.communities. gov.uk/documents/planningandbuilding/pdf/155634.pdf (accessed 21 March 2010)

Mayor of London, 2008. *The London Plan. Spatial Development Strategy for Greater London. Consolidated with Alterations since 2004*. Available at www.london.gov. uk/thelondonplan/docs/londonplan08.pdf (accessed 21 March 2010)

SLL (Society of Light and Lighting), 1992. *Lighting Guide LG6 – The Outdoor Environment*. London: SLL

TfL (Transport for London), 2004. *Cycle Parking Standards: Proposed Guidelines*. London: Transport for London. Available at www.tfl.gov.uk/assets/downloads/ corporate/Proposed-TfL-Guidelines.pdf (accessed 21 March 2010)

# 3.11
# Waste Management and Recycling

## Introduction

In 2007 the UK buried approximately 73 million tonnes of waste from all sources in landfill sites, which includes household waste of around 320kg per person. This compares with 100 million tonnes in 2000, when only about 12 per cent of household (municipal) waste was recycled. The proportion recycled in 2007–2008 was 34 per cent (DEFRA, 2009). The US has been recycling similar percentages of municipal waste since the mid 1990s but the per capita waste that was sent to landfill in 2005 was around 780kg.[1]

The emphasis from the World Health Organization (WHO, 2007) and the European Commission is on 'the protection of human health and the environment against the harmful effects caused by the collection, transport, treatment, storage and tipping of waste' (EC, 2008). This section sets out the issues that need to be addressed by designers to enable future building occupants to minimize the amount of operational waste that is taken to landfill whilst storing and conveying waste in an efficient and sanitary manner.

Strategies for minimizing and recycling site waste during demolition and construction are dealt with in Chapter 4.3 Construction Waste Management.

## Waste strategy

It is possible to transpose the strategy set out in the EU Waste Directive (EC, 2008) to the design process for individual developments. In summary it requires Member States to encourage:

- prevention or reduction of waste and its harmfulness;
- waste recovery through recycling, re-use or reclamation;
- recovering energy from waste.

The designer's responsibility for reducing waste and minimizing its harmfulness is shared with that of the contractors who procure and assemble the construction materials. The choices available to the designer in the specification of materials and building products are covered in Chapter 3.12 Materials Specification, whilst reducing waste and its harm during construction is discussed in Chapter 4.3 Construction Waste Management.

The specification and procurement of materials and products should pay attention to the opportunities for the re-use of elements, components and materials from existing or demolished buildings, the recycled content of materials and of the products and to the wastefulness of the extraction of raw materials and manufacturing processes.

The end user of the proposed development will need to create and implement their own waste management strategy and it is the responsibility of the designer to ensure that they have the opportunity to separate different types of waste for recycling and disposal. Where compostable materials are produced on site then opportunities for on-site composting or collection should be catered for. Some projects may lend themselves to storing this waste for anaerobic digestion (see below and Chapter 2.6 Energy Strategy and Infrastructure).

The UK Waste and Resource Action Programme (WRAP) provides tools for both the concept and detailed design stages of a construction project that can be downloaded from their website.[2]

WRAP's Design out Waste Tool for Buildings helps the designer:

- identify opportunities to design out waste in buildings projects;
- record design solutions pursued to reduce material consumption or wastage;
- calculate the impact of these solutions, including savings in project costs, waste to landfill and embodied carbon;
- compare the performance of different projects/alternative designs; and
- provide an indicative waste forecast for the Site Waste Management Plan (SWMP).

## Recycling and composting

It is important that sufficient space is provided in a suitable location for bins dedicated to storing recyclable waste alongside the bins, compactors and balers that are required for general waste. The various versions of BREEAM have specific space requirements, for example BREEAM Offices Credit Wst 3 requires $2m^2$ per $1000m^2$ of floor area up to a maximum of $10m^2$, with double this where catering is provided on-site. These must be readily accessible for building users and no more than 20m from the service entrance to the building. These bins must be accessible for waste collection vehicles and located so as not to create a hazard or nuisance for building occupants during collection.

The number of separate receptacles for recyclable waste will depend on the way in which the local authority (LA) or waste contractor separates recyclable materials. For example, some separate glass from other recyclables, others offer separate collections for paper/cardboard and compostable food waste.

The CSH has complex requirements for waste management which cover arrangements for the storage and collection of both recyclable and non-recyclable household waste. This includes a mandatory requirement for storage volumes to comply with BS 5906 (BSI, 2005) based on 100 litres for a one-bed dwelling and an additional 70 litres per bed for larger homes. It is also mandatory that waste storage be accessible to disabled occupants in accordance with Building Regulations Approved Document M (ODPM, 2004), which is also covered by BS 5906 (BSI, 2005).

Access for wheelchair users should be direct, free from obstructions and raised thresholds, have a suitable smooth surface (as defined in Approved Document M, section 6.9), and allow easy manoeuvring of a wheelchair.

Where changes in level are unavoidable, suitable ramps should be specified having a minimum width of 900mm, a slope no greater than 1 in 15 and with landings top and bottom at least 1.2m long plus the length required to accommodate the door swing (Approved Document M, section 6.15). Where appropriate a turning circle of at least 1500mm diameter should be provided to allow a wheelchair user to turn and return in the opposite direction. For communal bin stores, signs, lighting and information should be provided for visually impaired people. Bins, hoppers, light switches, etc. must all be at a height accessible to wheelchair users.

Detailed design guidance is provided in BS 8300: 2009 (BSI, 2009) which covers all aspects of designing buildings for a variety of disabilities.

Credits are also available under CSH for facilities for the storage of recyclable waste internally and externally for various scenarios. If there is no LA collection of recyclable waste and it is decided that no external storage will be catered for, then 2 points can be won by providing each dwelling with storage for at least three recyclable materials (paper, card, glass, plastics, metals or textiles), providing a total capacity of at least 60 litres with the smallest being 15 litres or more (Figure 3.49).

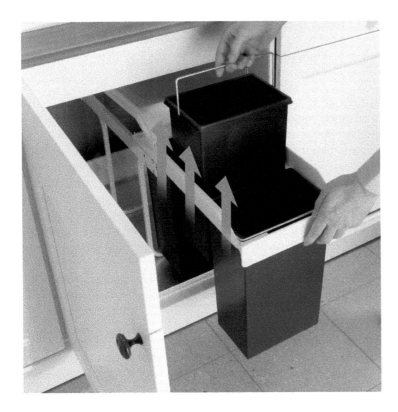

**Figure 3.49** *Kitchen waste integrated recycling bins*

*Source*: Photograph courtesy of Laundry Company Ltd and Hailo

This storage has to be provided as part of the fit out – standalone bins do not comply.

Alternatively, four points are available if there is either a LA collection scheme for recyclable waste at least fortnightly or adequate storage is provided externally as well as internally. If the LA scheme provides for sorting of waste after collection, then a single 30 litre bin for recycled waste can be provided indoors. If the LA requires the householder to sort waste before collection, then three bins should be provided as above, but with a total capacity of at least 30 litres and a minimum bin size of 7 litres. Alternatively credits will be given if there is an automated waste collection system collecting separated waste from each dwelling (see below).

If there is no LA collection of recyclable waste then the credits will be awarded if there is adequate storage externally comprising a total of 180 litres for a house, with a minimum bin size of 40 litres, located no more than 30m from the door. For apartment buildings the amount of space provided externally will have to be agreed with the recycling scheme operator. Bins will have to be easily identifiable for the type of waste for which they are intended and located no more than 30m from the common entrance. Bin stores should be covered, have well lit access and be located on a level hard-standing.

Composting is a natural process which involves the conversion of organic waste into a soil-like material through the action of bacteria and micro-organisms. The composting process is also supported by larvae, wood lice, beetles, worms and other such creatures. Composting avoids the production of methane gas and toxic leachate that occurs if organic materials are sent to landfill. However, this process may be reproduced in a controllable environment for the production of methane from anaerobic digestion (bio-digestion).

Separate facilities need to be provided for compostable waste. This is also covered by a credit in CSH (Was 3) which is awarded for compliance with two different scenarios:

- Home composting: using a bespoke composting bin to process kitchen and/or garden waste. Such containers should not be sited in close proximity to openable windows, doors, or ventilation intakes for habitable areas within the dwelling or neighbouring buildings.
- A local authority or community run composting scheme. Communal or community composting is where a group of people share a composting system. The raw materials are provided by all who take part in the scheme, and the compost is then used in the community, either by individuals in their own gardens, or for use on local community projects. As with other waste facilities, the storage must be accessible by disabled persons (see above). The distance between the site entrance and the communal/community containers should not exceed 30m. The waste conveying systems described below can be used for conveying separated organic waste if dedicated inlets are provided.

LEED includes a prerequisite that requires measures to reduce the amount of operational waste going to landfill including the provision of an easily accessible area that serves the entire building and is dedicated to the collection and storage of non-hazardous materials for recycling, including (as a minimum) paper, corrugated cardboard, glass, plastics and metals.

# Evacuated waste conveying

The concept of piped waste conveying was developed for a hospital in Sweden in the late 1950s as a logical extension of the central vacuum cleaning systems already in use at that time for large buildings. The principle is simple, but developing a system that was reliable, hygienic and did not suffer from periodic blockage was quite a challenge.

There are now hundreds of underground vacuum waste conveying systems installed across Europe, many of them serving entire communities or large mixed use developments such as the first installation in the UK serving the Wembley City complex in north London (see case history below). Figure 3.50 shows a three-dimensional representation of a typical installation and Figure 3.51 a typical bin cluster for general and recyclable waste.

Users throw their waste into inlets which may be located indoors or out, and the bags are stored temporarily above a closed storage valve. Once the inlet is full, a valve opens under the dictates of a central control system that only allows waste of the same type into the underground pipes at any one time.

When the control system senses it is time to empty the inlets, the central fans are started creating a vacuum in the pipework and a supply air valve is opened in order to allow transport air to enter the system. The storage valves beneath the inlets are opened one at a time allowing the waste to fall into the underground pipework and be evacuated away to the collection station where it is drawn through a cyclone and separated from the transport air. The waste then falls into a compactor where it is compressed and transferred into a sealed container designated for the specific waste type. The transport air is discharged to atmosphere after having passed through filters and silencers.

Although expensive to install, these systems offer significant advantages over the normal approach to waste collection. For example, refuse vehicles are kept away from most of the site, reducing nuisance from noise, odour and spilt

**Figure 3.50** *Underground vacuum waste conveying system*

*Source*: Envac Group (www.envacgroup.com/web/Stationary_vacuum_systems.aspx)

**Figure 3.51** *Typical waste collection points for underground conveying system*

*Source*: Envac Group (www. envacgroup.com/web/ Envac_UK.aspx)

waste as well as reducing the fuel consumption, carbon emissions, pollution and risk of accidents associated with the operation of these vehicles. The carbon emissions associated with the fans operating four or five times a day for an average of 30 minutes is usually significantly lower than from the fuel consumption of refuse vehicles handling the same volume of waste. There is also far less manual lifting of waste bags, especially if inlet points are provided on every floor of apartment buildings.

Not everything can be handled by these systems. The size of waste handled is limited by the size of the inlet openings and underground pipework, which are typically 300–400mm diameter. Clearly it is possible that there could be some abuse of the system, particularly those designed for access by the public. Blockages could be caused if large, highly absorbent articles are disposed of and exposed to moisture, whilst a separate system would be required to handle large volumes of slurry from kitchens. Problems could also arise from depositing heavy items such as builder's rubble or scrap metal into the system. The system is not intended for the disposal of dead animals, faeces, highly acidic or alkaline solutions, paints and adhesives. Appropriate signage and community information is therefore very important.

There is also a small, but real, risk that someone could deposit an explosive device into one of the inlets. However, this is true of any waste bin and the underground system is more likely to contain the explosion and limit the damage than a street-level bin would. When installed as part of a private community, it is possible to introduce a measure of security by installing lockable inlets accessed with a swipe card.

## Energy from waste

The UK Waste Strategy of 2007 includes a number of initiatives to promote the adoption of energy from waste by energy companies and private developers through:

Using the Private Finance Initiative, Enhanced Capital Allowances and, where appropriate, the proposed banding system for Renewable Obligation Certificates to encourage a variety of technologies of energy recovery (including anaerobic digestion) so that unavoidable residual waste is treated in the way which provides the greatest benefits to energy policy. (DEFRA, 2007)

The strategy goes on to suggest that 'Energy from waste is expected to account for 25% of municipal waste by 2020 compared to 10% today' and sets out the following initiatives for achieving this:

- Putting in place an operational protocol for anaerobic digestate ...' which has been developed by the Environment Agency and was published in 2009 (WRAP/EA, 2009).
- Developing collection arrangements and the energy market for wood waste which cannot be re-used or recycled'.

We saw in Chapter 2.6 Energy Strategy and Infrastructure a number of the technologies that are available for large-scale energy from waste plants. For individual buildings it is unlikely that there will be sufficient waste generated on-site for use as a fuel, however, the market already exists for pelletized products manufactured from waste biomass, used cooking oil and biogas from anaerobic digestion or landfill.

## Case history

### Wembley City Underground Waste Conveying System

The new Wembley City is a development by Quintain Estates of 85 acres of land surrounding the new national stadium. The development comprises 4200

**Figure 3.52** *Computer image of proposed Wembley City development*

*Source:* Copyright Wembley City 2010 (www.wembleycity.co.uk/news/31.html)

new apartments, a designer outlet shopping facility, a new Hilton hotel plus retail, leisure and entertainment venues.

Early in the design process it was decided that waste management at this landmark development would form an important part of its infrastructure. The Envac underground vacuum waste conveying system was chosen because of the predicted reduction in vehicle movements and associated carbon savings, along with the potential for a cleaner and safer environment where people can live and work. It has been predicted that rubbish collection will generate around 400 fewer tonnes of $CO_2$ a year than conventional refuse collection at the Wembley City site.

The Envac system, comprising 252 inlets and 2500m of underground pipework, became operational in December 2008. Four waste fractions are collected which are non-recyclables, dry recyclables, organic recyclables and cardboard. During the first year of operation an average recycling rate of 41 per cent was achieved, which is five times higher than the average recycling rate for high density development in London. To save space, a bridge crane has been installed to lift the full containers straight on to the collection vehicle. The loading operation is in the collection station behind closed doors, no sound or waste odour will be detectable outside the station. The roof of the collection station will feature a green roof garden when the development is fully built.

## Notes

1    www.zerowasteamerica.org/Statistics.htm.
2    http://nwtool.wrap.org.uk/.

## References

BSI, 2005. *BS 5906:2005 Waste Management in Buildings – Code of Practice.* London: BSI

BSI, 2009. *BS 8300: 2009. Design of buildings and their approaches to meet the needs of disabled people. Code of Practice.* London: BSI

DEFRA, 2007. *Waste Strategy for England.* London: Defra

DEFRA (Department for Environment, Food and Rural Affairs), 2009. *Sustainable Development Indicators.* London: Defra. Available at www.defra.gov.uk/sustainable/government/progress/defra-resources/sdiyp.htm (accessed 22 March 2010)

EC, 2008. *Revised Waste Directive.* Available at http://eur.lex.europa.eu/LexUriserv/lexUriServ.do/uri=OJ:L:2008:312:0003:0003:EN:PDF (accessed 22 March 2010)

ODPM (Office of the Deputy Prime Minister), 2004. *Approved Document M: Access to and Use of Buildings.* London: ODPM. Available at www.planningportal.gov.uk/uploads/br/BR_PDF_ADM_2004.pdf(accessed 22 March 2010)

WHO (World Health Organisation), 2007. *Population, Health and Waste Management: Scientific data and policy options.* Rome: World Health Organization

WRAP/EA, 2009. *Anaerobic digestate: End of waste criteria for the production and use of quality outputs from anaerobic digestion of source-segregated biodegradable waste.* London: Environment Agency

# 3.12
## Materials Specification

## Introduction

'Materials in construction make up over half of our resource use by weight. They account for 30% of all road freight in the UK. The construction and demolition industries produce over 4 times more waste than the domestic sector, over a tonne per person living in the UK. The environmental impacts of extracting, processing and transporting these materials and then dealing with their waste are major contributors to greenhouse gas emissions, toxic emissions, habitat destruction and resource depletion.' (Lazarus, 2002a)

Each of the components of a building has to go through a process prior to construction, which may vary from simple adaptation of a re-used component through to a complex industrial process involving extraction of numerous raw materials, processing, assembly and delivery. Once installed, some components will need periodic maintenance or replacement using additional materials and processes. Eventually the whole assembly may need to be demolished and disposed of, or adapted for another use.

## Life cycle environmental impact

In order to make raw materials suitable for use in a building or its landscape, there is the potential for a whole array of environmental impacts which occur from 'cradle to grave', that is through the life cycle of each material from extraction of the raw materials through to demolition and disposal. The BRE has incorporated these impacts into a life cycle rating system which they refer to as the Green Guide.[1] The environmental issues that are covered by this rating system are summarized in Box 3.5.

---

**Box 3.5 Environmental issues covered by the BRE Green Guide and Ecopoints systems**

- Climate change – embodied carbon expressed as $kg\ CO_2$ equivalent greenhouse gas emissions giving global warming potential (GWP) over 100 years.
- Water extraction – cubic metres of mains, surface and ground water extraction.

- Mineral resource extraction – tonnes of virgin irreplaceable materials such as ores and aggregates.
- Stratospheric ozone depletion – chlorinated and brominated gases that destroy the ozone layer as kg of CFC-11 equivalent.
- Human toxicity potential* (HTP) – based on the full life cycle of the material.
- Ecotoxicity* to freshwater – maximum tolerable concentrations in water for ecosystems.
- Ecotoxicity* to land – maximum tolerable emissions to land.
- High level nuclear waste – volume of high level waste requiring a minimum of 10,000 years storage before it may be safe, expressed in $mm^3$.
- Waste disposal – tonnes of solid waste going to landfill or incineration in terms of loss of resource.
- Fossil fuel depletion – energy content in tonnes of oil equivalent ($toe_{eq}$).
- Eutrophication – over-enrichment of watercourses causing algal growth and oxygen depletion, in kg of phosphate equivalent.
- Photochemical ozone creation – emissions of $NO_x$ and VOCs that convert to ozone in presence of sunlight expressed as kg ethene equivalent.
- Acidification – emissions of $SO_2$ and $NO_x$ that lead to acid deposition (acid rain) expressed as kg sulphur dioxide equivalent.

Note: * Expressed as kg of 1,4 dichlorobenzene (1,4-DB) equivalent.

The BRE has looked at most of the main elements of a building and, using a formula that they have devised, calculated 'Ecopoints' and Green Guide ratings for the commonly used composite elements for internal and external walls, windows, roofs, ground and intermediate floors, insulation, floor finishes, landscaping and boundary protection.

The Green Guide takes the overall impact of the best and the worst results for a particular element and building type, and divides the gap between them into six equal bands, E to A+. The other results are then fitted into these bands. The Green Guide ratings are used in the BREEAM Offices and CSH Materials calculators, for example, to determine the rating for the main building elements, weighted according to the area of each.

The Ecopoint is a single score that measures the total environmental impact of a product or process over a 60-year period as a proportion of overall impact occurring in Europe – 100 Ecopoints is equivalent to the impact of a 'European Citizen'. Green Guide ratings are derived by sub-dividing the range of Ecopoints/$m^2$ achieved by all components of a building element according to a formula devised by BRE.

For some elements, such as separating walls, windows or commercial floor construction, the difference between the best (A+) and the worst (E) is relatively small in absolute terms with an E rating having approximately two to three times the impact of an A+ rating and the range being about 0.5 Ecopoints/m². In other elements, such as roofing or surfacing for heavily trafficked areas, the range is more than 1.5 Ecopoints/m² and an E can be over four times worse than an A+.

BRE offer to determine Green Guide ratings for specific products as a consultancy service to manufacturers. They provide 'Environmental profiling certification' for specific products using a Type III environmental declaration model in accordance with ISO 14025 and labelling in compliance with ISO 14024.[2]

Green Guide ratings provide an indication of the environmental performance of the average composite element available in the UK marketplace installed anywhere in the UK. It does not reward the designer for specifying materials that are entirely native to the development location, for example, unless an environmental profile is available from a company who is responsible for the manufacture of the whole building element. However, BRE is introducing a calculator for use by licensed BREEAM/CSH assessors that will enable the calculation of Green Guide ratings for elements not covered by the online Green Guide library.[3]

In parallel with BRE, the National Institution of Standards and Technology (NIST) in the US has developed the Building for Environmental and Economic Sustainability (BEES) LCA model which uses a very similar protocol to the Green Guide (Figure 3.53).

BEES provides a rating system for the environmental performance of building products by using the LCA approach specified in the ISO 14040 series

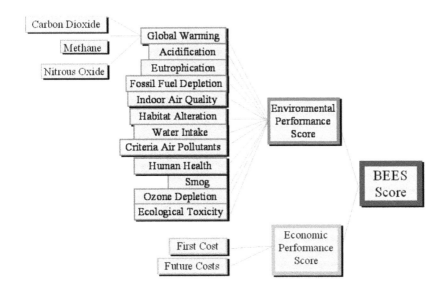

**Figure 3.53** *BEES structure*

*Source*: NIST (www.bfrl.nist.gov/oae/software/bees/bees.html)

of standards. As with the Green Guide rating, all stages in the life of a product are analysed from raw material acquisition, manufacture, transportation, installation, use, potential for recycling and waste management. Economic performance is measured using the American Society for Testing and Materials (ASTM) standard life cycle cost method, which covers the costs of initial investment, replacement, operation, maintenance, repair and disposal. Environmental and economic performance is combined into an overall performance measure using the ASTM standard for Multiattribute Decision Analysis.[4] For the entire BEES analysis, building products are defined and classified according to the ASTM standard classification for building elements known as UNIFORMAT II (ASTM E1557-09 Standard Classification for Building Elements and Related Sitework).

An example of a BEES rating comparing various wall constructions is shown in Figure 3.54.

BEES also develops a lifetime cost score for the building element which provides a total environmental and economic score for comparing alternative constructions (Figure 3.55).

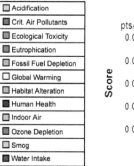

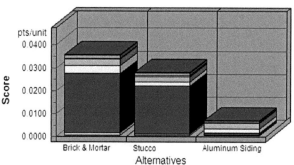

**Figure 3.54** *BEES rating for alternative wall constructions*

*Source*: NIST

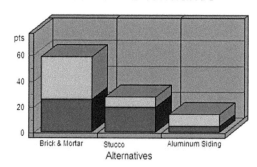

**Figure 3.55** *BEES total performance of alternative wall constructions*

*Source*: NIST

Despite a number of feasibility studies and lobbying by NIST (Scheuer and Keoleian, 2002), neither BEES nor LCA techniques in general had been incorporated into LEED at the time of writing.

Many of the chemical components that are used in the manufacture of construction materials are covered by the EC Registration, Evaluation and Authorisation of Chemicals (REACH) Regulations which require the registration of substances providing a dossier setting out the potential health and environmental impacts of all chemicals that are manufactured or supplied into the EU at a rate greater than one tonne per year. This is likely to result in the phasing out of a number of substances and their substitution in a large number of construction products.[5]

# Embodied carbon, global warming and ozone depletion

Embodied carbon dioxide emissions are of course incorporated into both the BRE Green Guide rating and BEES under the heading of Climate Change or Global Warming. These are determined as kg $CO_2$ equivalent, which includes all emissions that impact on global warming, including methane and a variety of refrigerants and blowing agents for insulation. See Table 3.1 in Chapter 3.3 Operational Energy and Carbon. Methane has a GWP over 100 years that is 25 times more powerful than $CO_2$, whereas some common refrigerants have a very high GWP. R407C, for example, has an equivalent GWP of 1600. At one time CFCs, HCFCs and HFCs were all used as blowing agents for foam insulation until the ozone-depleting chlorinated products were phased out under the Montreal Protocol. Globally HFC blown insulation is still widely used, particularly for non-residential applications. However, in Europe this insulation material is steadily being replaced with hydrocarbon blown foams, typically using pentane, isobutene, air or $CO_2$. While $CO_2$ has a GWP of unity, the value for pentane is quoted between 3 and 11.[6] It should be noted that the authorities differ on this and many do not quote the period over which the GWP is estimated, which should be 100 years. For example, BRE quote pentane (iso-pentane, cyclo pentane or n-pentane) blown insulation as an example of compliant material having a GWP of less than 5 as required by Credit Pol 1 in their guidance to CSH 2009.

Currently the above LCA rating systems are primarily used to evaluate and compare construction materials and not building services systems, plant or equipment. We saw in the renewable section of Chapter 2.6 Energy Strategy and Infrastructure that the embodied impact of carbon saving measures can be greater than the life cycle savings. This can be just as important when comparing insulation strategies as for renewable technologies. One Canadian study from 2007 has determined the impact of the leakage of high GWP gas blown products on the carbon payback, which can be over 100 years for high GWP halocarbon blown products compared to 10–50 years for low GWP non-halocarbon blown insulation (Harvey, 2007).

Globally no country demands that total life cycle embodied $CO_2$ emissions must be calculated in order to meet building regulations or planning requirements. However, in the UK because of the focus on reducing operational carbon to zero by 2016 for dwellings, some pioneering projects, such as the Beddington Zero Energy Development (BedZED) have been the subject of greater attention than others. There has been a concern that the quest for zero operational energy and its association with high thermal mass will lead to increased embodied carbon.

One assessment demonstrated that embodied carbon in the BedZED scheme was some $675kgCO_2/m^2$, compared with an average of $550kgCO_2/m^2$ for 'normal' dwellings (Lazarus, 2002b). However, this latter figure is based on work from BRE and others on dwellings designed prior to the 2002 Building Regs Part L revisions – hence with significantly smaller quantities of insulation and associated embodied carbon. Elsewhere values that vary from 500 to $1200kgCO_2/m^2$ are given in a publication dating from 1999, hence it appears that there are massive differences in estimated values for dwellings (Crane Environmental for Sustainable Homes, 1999).

The differences lie not just in the amount of insulation used but also in the method of construction. It is no accident that apartment blocks lie at the top end of the embodied carbon range. This is partly due to the amount of concrete and other dense materials employed in larger buildings – to hold them up, to prevent the spread of fire and to reduce transmission of noise through party walls and floors.

For a 'normal' dwelling during a 60-year lifespan and built to 2006 Building Regulations targets – embodied carbon represents between 13 and 18 per cent of total carbon emitted. For a low energy dwelling such as those found at BedZED or built to the German Passivhaus specification, total annual carbon emissions associated with heating may be 10–15 per cent of Part L 2006 targets. Hence embodied carbon can represent 50 per cent or more of lifetime carbon emissions. Whilst for any dwelling that ultimately reaches the zero carbon 'Mecca', embodied carbon will represent 100 per cent of lifetime carbon emissions.

For anyone wishing to estimate the total carbon footprint of a project there are numerous web-based tools available that use simplified approaches to estimating embodied energy.[7] Many of these have been developed for carbon offsetting or for personal use. However, the designer requires more detail than these methods allow and there are now a number of databases of embodied carbon for different composite constructions. The University of Bath has produced an 'Inventory of Carbon & Energy' that covers most of the common building materials and provides estimated 'cradle to grave' energy and $CO_2$ emission rates per kg for each material, rather than composites (Hammond and Jones, 2008).[8]

Ozone depletion potential is also integrated into the Green Guide rating, although for the entire developed world there should no longer be any products available that contain ozone depleting substances (see the discussion of the Montreal Protocol in Chapter 1.2 Policy, Legislation and Planning).

## Re-use and recycling

BREEAM (but not CSH) rewards elements that are re-used in-situ in five ways. First they will be awarded an A+ Green Guide rating by default and second a separate credit rewards projects that make use of existing facades that constitute more than 50 per cent of the external envelope of the building by area and 80 per cent by mass. Similarly if more than 80 per cent of the volume of primary structure is re-used without significant reinforcement or alteration an additional credit can be won. Fourth re-used elements achieve a Tier 1 rating under the Responsible Sourcing credit (see below). Lastly the credit dealing with construction waste management benefits from the re-use of existing building materials (Chapter 4.3 Construction Waste Management).

LEED does not have the equivalent to a Green Guide rating and has different approaches in the New Construction (LEED NC) and Homes protocols. LEED NC has a series of credits that reward various levels of fabric re-use and recycled content, including 'rapidly renewable' materials such as bamboo, wool insulation and cork. LEED Homes rewards material efficient framing and environmentally preferable products, which includes a number of the issues covered under the 'responsible sourcing' credits in BREEAM and CSH (see below).

BREEAM rewards the use of recycled materials either from the demolition of existing buildings on the site or within 30km of the site, or suitable non-construction waste, to supply at least 25 per cent (by weight or volume) of 'high grade' aggregate uses in the construction of the development. It is common practice to use crushed demolition materials as fill for landscaping. This reduces the cost of transport, landfill taxes and bought-in aggregates. However, for a compact site it may be difficult to accommodate a large crushing machine within the boundary (see Figure 4.5) and it may be necessary to transport material off-site for processing.

BREEAM defines high grade aggregate uses as concrete for the structural frame and floor slabs, paved areas and roads, granular fill and capping, sub-bases and foundations, pipe bedding and gravel landscaping. Suitable non-construction waste covers a wide range of post-consumer and post-industrial wastes that cannot be re-used locally. These might include anything from colliery spoil to plastics and tyres.

LEED NC also includes credits that cover the recycling or salvaging of up to 75 per cent of demolition materials in general under the heading of 'Construction Waste Management'. This covers all non-hazardous materials derived from demolition that can either be recycled for use on-site or diverted for use in manufacture elsewhere (see Chapter 4.3 Construction Waste Management).

## Responsible sourcing

Sourcing of construction products occurs in at least two stages. The design team will specify products and materials to meet the cost and sustainability criteria to varying levels of detail depending on how critical they are to the design. This will be followed at the construction stage by value engineering and procurement based on exact specifications and measured quantities (see Chapter 4.2 Sustainable Procurement).

BREEAM and CSH recognizes this by rewarding 'Responsible Sourcing of Materials' for a list of eight main building elements that covers the structural frame, ground and upper floors, roof, external and internal floors, foundations, substructure and staircases.

Office fit-outs are assessed separately for materials used in stairs, windows, doors, skirting, panelling, furniture and fascias. There are no credits associated with the responsible sourcing of building services.

The BRE scheme is looking for materials that comply with measurable standards of environmental management in their manufacture from 'cradle to gate'. The scheme refers to four 'tiers' of quality, with Tier 1 meeting all of BRE's criteria. These require certification that certain standards of organizational and supply chain management and environmental and social responsibility have been met. There are a number of well-established national and international certification systems that do this, including the Forest Stewardship Council (FSC) and the Programme of Endorsement of Forest Certification (PEFC) schemes for timber. BRE have developed a framework standard that assesses all of these issues for specific materials or products and is being administered by the British Standards Institute (BRE, 2008). BSI awards certificates on a scale from Pass to Excellent, with Very Good and Excellent ratings qualifying for Tier 1 and Pass and Good for Tier 2. At the time of writing there is no central registration of certified companies or products, although a number of ready-mixed concrete and brick manufacturers are advertising their successful certification. In the absence of an officially recognized certification system, it is necessary to obtain evidence of a certified Environmental Management System (EMS) being in place for the whole of the supply chain (Tier 3) or if this only applies to the 'key process' then only a Tier 4 can be allocated. BRE has developed a calculator that enables credits to be awarded based on an area weighted scoring of the Tiers allocated to the applicable materials for the various building elements.

For information on specific products, including building services plant and equipment, there are numerous on-line product directories, some of which provide useful technical information and background, links to suppliers' and manufacturers' websites and key sustainability information, such as recycled content. A list of some of the leading UK sites is given below:

- www.constructionresources.com/;
- www.sigss.co.uk/?gclid=CKDu1MCosaACFZFo4wodzkuVTQ;
- www.greenbuildingstore.co.uk/;
- www.greenspec.co.uk/html/products/productscontent.html.

In the US the National Institute of Building Science has produced the on-line Whole Building Design Guide which includes the Federal Green Construction Guide for Specifiers.[9]

## Notes

1   www.bre.co.uk/greenguide/page.jsp?id=2069.
2   The draft methodology of the BRE profiling can be downloaded from www.bre.co.uk/filelibrary/greenguide/PDF/Environmental_Profiles_Methodology_2007_-_Draft.pdf.
3   www.bre.co.uk/filelibrary/greenguide/PDF/GreenGuideCalculatorGuidance.pdf.

4    ASTM E1765: Standard Practice for Applying Analytical Hierarchy Process (AHP) to Multiattribute Decision Analysis of Investments Related to Buildings and Building Systems.

5    www.hse.gov.uk/reach/.

6    The Avantec chemical data sheet can be accessed at: www.inventec.dehon.com/pdf_public?prod_unid=B7E1D5953A474F6CC125733D0065B9BC&file_name=n-pentane-s-english-version.

7    For an example see http://buildcarbonneutral.org/.

8    More information can be found at www.greenspec.co.uk/html/materials/embodied_energy.html.

9    The Whole Building Design Guide can be found at  www.wbdg.org/design/greenspec.php.

# References

BRE, 2008. *BES 600: 2008 Issue 1: Framework Standard for the Responsible Sourcing of Construction Products*. Watford: BRE Global

Crane Environmental for Sustainable Homes, 1999. *Embodied energy in residential property development – Guide to RSLs*. UK: Crane Environmental. Available at www.venablesconsultancy.co.uk/crane-environmental/Projectsheets/Sustainable-Homes.pdf, (accessed 22 March 2010)

Hammond, G., and Jones, C., 2008. *Inventory of Carbon and Energy*. UK: University of Bath. Available at www.bath.ac.uk/mech-eng/sert/embodied/

Harvey, L. D., 2007. Net climatic impact of solid foam insulation produced with halocarbon and non-halocarbon blowing agents. *Building & Environment*, vol 42, no 8, pp2860–2879

Lazarus, N., 2002a. *BedZED Toolkit Part 1: A guide to construction materials for carbon neutral developments*. UK: Bioregional

Lazarus, N., 2002b. *Beddington zero energy development: Construction materials report*. UK: Bioregional

Scheuer, C., and Keoleian, G., 2002. *NIST GCR 02-836: Evaluation of LEED™ Using Life Cycle Assessment Methods*. Ann Arbor: MI: University of Michigan. Available atwww.bfrl.nist.gov/oae/publications/gcrs/02836.pdf (accessed 1 March 2010)

# 3.13
## Pollution

## Introduction

As we have seen in previous chapters, buildings and their occupants are responsible for significant pollution to the atmosphere, both directly to the air around the building and indirectly to the upper atmosphere from power station chimneys. Pollution in effluent from buildings and oil or petrol leakage from motor vehicles used by a building's occupants can also impact on nearby land, watercourses and municipal sewers.

Of course global warming is primarily the result of pollution, and we have dealt with the factors that contribute to carbon dioxide emissions at length elsewhere. However, most refrigerants have significant global warming potential whilst the indirect impact of $NO_x$ and the chemical interactions that produce ozone also have a significant global warming potential ($NO_x$ has a GWP of around 300 over 100 years). $NO_x$ is a particularly damaging, highly reactive pollutant that, in the presence of sunlight and oxygen, forms ozone or else can combine with water to produce acid rain.

## Emissions to air

### Combustion products

Oxides of nitrogen are a product of combustion which is created at a concentration that depends on the type of fuel and combustion temperature. There are three mechanisms by which $NO_x$ is formed during combustion:

- oxidation of nitrogen at higher temperatures (> 1500°C);
- reaction of nitrogen, oxygen and hydrocarbon radicals, which is most important at lower temperatures;
- oxidation of nitrogen containing compounds in fuel, which is most important in solid fuels that have organically bound nitrogen, such as biomass.

All fuels produce $NO_x$ but the lowest concentrations arise from the efficient combustion of gas. Solid fuels tend to produce high concentrations, hence in the UK, because of the large number of coal-fired power stations and the inefficiencies of distribution, the $NO_x$ emissions associated with electricity consumption are on average around 1200mg/kWh. This compares with high efficiency gas boilers, which can produce $NO_x$ emissions of below 40mg/kWh.

**Table 3.6** *Conversion factors for NO$_x$ emissions*

| Unit | Conversion to dry NO$_x$ in mg/kWh |
|------|-----------------------------------|
| mg/m³ | ×0.857 |
| ppm | ×1.76 |
| mg/MJ, g/GJ | ×3.6 |
| mg/Nm³ | ×6.5 |
| wet NO$_x$ | ×1.75 |
| 3% excess O$_2$ | ×1.17 |
| 6% excess O$_2$ | ×1.4 |
| 15% excess O$_2$ | ×3.54 |

Boiler manufacturers usually quote NO$_x$ emissions in terms of 'dry NO$_x$ emissions at 0 per cent oxygen (O$_2$) in mg/kWh of heat output'. Alternative units are mg/m³, ppm, mg/MJ, g/GJ and mg/Nm³ whilst some manufacturers may refer to 'wet NO$_x$' or to NO$_x$ emissions at a given percentage of excess O$_2$, typically 3 per cent, 6 per cent or 15 per cent. Conversion factors are given in BREEAM and CSH guidance as set out in Table 3.6.

Both BREEAM and CSH reward the use of heat generation strategies that achieve an average dry emission below 100mg/kWh with maximum credits below 40mg/kWh. For boilers with rated heat outputs below 70kW there is a rating system as defined in BS EN 297: 1994 that includes emission classes which coincide with two of the ranges defined in CSH, that is Class 4 matches the 71–100mg/kWh range that qualifies for one credit, whilst Class 5 equates to the 41–70mg/kWh range that gives two credits. There is no Class under BS EN 297 that requires emissions below 40mg/kWh, which is awarded three credits under CSH. There is a large choice of condensing boilers across the full range of sizes that have rated NO$_x$ emissions below 40mg/kWh, with the lowest being around 10mg/kWh.

Combined heat and power plant that uses gas to generate heat and electricity thus offsetting electricity generation at the power station equivalent to 1200mg/kWh in the UK can be said to have zero net NO$_x$ emissions. Whereas heat pumps use electricity to generate heat, so that even at CoPs of 4.0 or higher the associated NO$_x$ emissions will not be below the threshold for a BREEAM credit. However, the picture will be very different in countries that have a high proportion of their electricity generated by zero NO$_x$ nuclear power or renewable energy sources.

Biomass firing on the other hand generates much higher emissions of NO$_x$ which vary depending on the type of biomass, but is typically between 300 and 600mg/kWh, tending to increase with plant size. A number of European countries and North American States have set limits for NO$_x$ emissions. Austria, for example, restricts the emissions from biomass plants smaller than 10MW to 662mg/kWh and larger plants to 533mg/kWh. The EMEP document referred to below and in Table 3.7 gives guidance on estimating emissions for a wide range of pollutants and includes a figure of 540mg/kWh for $NO_x$ emissions from biomass boilers. See also, the section on Biomass, bioliquids and energy from waste in Chapter 2.6 Energy Strategy and Infrastructure.

Fuel oils and solid fuels such as biomass also produce respirable particulate matter ($PM_{10}$ and $PM_{2.5}$), the subscripts for which indicate the aerodynamic diameter in microns. Inefficient combustion of these fuels can generate visible smoke, which is prohibited in British smoke control areas under the Clean Air Act. The finer the particles the deeper into the respiratory system they penetrate such that prolonged exposure can result in respiratory and cardiovascular illnesses.

Poor or partial combustion of coal, crude oil and waste timber impregnated with creosote can lead to the production of polycyclic aromatic hydrocarbons (PAH), exposure to which has been associated with cataracts, kidney and liver damage. Similarly partial combustion of any fuel will produce carbon monoxide (CO) which, although not identified as an environmental pollutant, can lead to poisoning if it escapes into an occupied space from a poorly vented appliance.

As referred to above, the European Monitoring and Evaluation Programme (EMEP) and European Environment Agency (EEA) have produced guidance on 'emission factors' for various fuels (EMEP/EEA, 2009). These represent current levels rather than a proposed standard; however they do give an indication of what emissions to expect from modern combustion plant (Table 3.7).

## Refrigerants

Most refrigerants become vapours at atmospheric pressure and hence diffuse into the surrounding air if they leak from a refrigeration system or storage cylinder. Table 3.1 in Chapter 3.2 Operational Energy and Carbon indicates the 100 year global warming potential of a range of refrigerants, many of which are orders of magnitude higher than $CO_2$. If a refrigerant leaks into a confined space, it will exclude the oxygen and asphyxiate anyone therein. Hence, apart from the high replacement cost of the product, there are very good reasons for preventing refrigerants from being released into the atmosphere. The UK Environmental Protection Act 1990 classifies refrigerants as controlled or hazardous substances.

For these reasons BREEAM includes credits that reward the prevention of leakage both through leak detection and through pump-down of the system contents into storage.

Where a refrigerant plant is located in a plant room, an automatic leakage detection system comprising multiple sensing heads set at 2000ppm or less can be used provided the room is relatively airtight and any ventilation does not dilute the leaking refrigerant. Leak detection should not create a fire risk and would normally be either of the infra-red, semi-conductor or electro-chemical type. Most detection systems can detect concentrations down to 10ppm or lower. Systems that are based on a fall in system pressure are not accepted under BREEAM criteria because of the 'natural fluctuations to the pressure of the refrigerant due to changes in volume and temperature of the system, and to the ambient temperature of the surroundings.'

**Table 3.7** *Emission factors for non-residential sources, automatic boilers burning wood*

| Tier 2 emission factors | | | | | |
|---|---|---|---|---|---|
| | Code | Name | | | |
| **NFR Source Category** | 1.A.4.a.i<br>1.A.4.c.i<br>1.A.5.a | Commercial / institutional: stationary<br>Stationary<br>Other, stationary (including military) | | | |
| **Fuel** | Wood | | | | |
| SNAP (if applicable) | | | | | |
| **Technologies/Practices** | Advanced wood combustion techniques <1MW - Automatic Boilers | | | | |
| **Region or regional conditions** | NA | | | | |
| **Abatement technologies** | NA | | | | |
| **Not applicable** | Aldrin, Chlordane, Chlordecone, Dieldrin, Endrin, Heptachlor, Heptabromo-biphenyl, Mirex, Toxaphene, HCH, DDT, PCP, SCCP | | | | |
| **Not estimated** | $NH_3$, Total 4 PAHs | | | | |
| Pollutant | Value | Unit | 95% confidence interval | | Reference |
| | | | Lower | Upper | |
| $NO_x$ | 150 | g/GJ | 90 | 200 | Guidebook (2006) chapter B216 |
| CO | 300 | g/GJ | 200 | 5000 | Guidebook (2006) chapter B216 |
| NMVOC | 20 | g/GJ | 10 | 500 | Guidebook (2006) chapter B216 |
| SOx | 20 | g/GJ | 15 | 50 | Guidebook (2006) chapter B216 |
| TSP | 70 | g/GJ | 60 | 250 | Guidebook (2006) chapter B216 |
| PM10 | 66 | g/GJ | 50 | 240 | Guidebook (2006) chapter B216 |
| PM2.5 | 66 | g/GJ | 50 | 240 | Guidebook (2006) chapter B216 |
| Pb | 20 | mg/GJ | 10 | 30 | Guidebook (2006) chapter B216 |
| Cd | 0.5 | mg/GJ | 0.3 | 2 | Guidebook (2006) chapter B216 |
| Hg | 0.6 | mg/GJ | 0.4 | 0.8 | Guidebook (2006) chapter B216 |

**Table 3.7** *Emission factors for non-residential sources, automatic boilers burning wood* (Cont'd)

| | | | | | |
|---|---|---|---|---|---|
| As | 0.5 | mg/GJ | 0.25 | 2 | Guidebook (2006) chapter B216 |
| Cr | 4 | mg/GJ | 2 | 10 | Guidebook (2006) chapter B216 |
| Cu | 2 | mg/GJ | 1 | 5 | Guidebook (2006) chapter B216 |
| Ni | 2 | mg/GJ | 0.1 | 200 | Guidebook (2006) chapter B216 |
| Se | 0.5 | mg/GJ | 0.1 | 2 | Guidebook (2006) chapter B216 |
| Zn | 80 | mg/GJ | 5 | 150 | Guidebook (2006) chapter B216 |
| PCB | 0.06 | mg/GJ | 0.012 | 0.3 | Kakareka et al (2004) |
| PCDD/F | 30 | ng I-TEQ/GJ | 20 | 500 | Guidebook (2006) chapter B216 |
| Benzo(a)pyrene | 12 | mg/GJ | 10 | 150 | Guidebook (2006) chapter B216 |
| Benzo(b)fluoranthene | 14 | mg/GJ | 10 | 120 | Guidebook (2006) chapter B216 |
| Benzo(k)fluoranthene | 8 | mg/GJ | 5 | 50 | Guidebook (2006) chapter B216 |
| Indeno(1,2,3-cd)pyrene | 6 | mg/GJ | 2 | 80 | Guidebook (2006) chapter B216 |
| HCB | 6 | µg/GJ | 3 | 9 | Guidebook (2006) chapter B216 |

Leak detection systems will normally activate a mechanical ventilation system in order to purge the refrigerant and safeguard anyone who happens to be in the plant room at the time of the leak.

Distributed refrigerant systems that typically comprise external condensing/compressor units and multiple evaporator coils scattered through a building will require many leak detection sensors in voids carrying refrigerant pipework throughout the building. External units must be encased for refrigerant leaks to be detected automatically.[1]

Automatic refrigerant pump down will help conserve the refrigerant should there be a leak. Leak detectors would either open solenoid valves to divert the entire contents of the system into a separate storage vessel or into an oversized heat exchanger vessel, which would then be isolated from the rest of the system.

## Emissions to watercourses and land

Surface water and soil may be contaminated by oil where there is a risk of leakage from vehicles or plant. The Environment Agency (EA) and its corresponding bodies in Scotland and Northern Ireland have produced Pollution Prevention Guidance (PPG3) that defines the type of site that needs to have measures in place to prevent this oil from polluting the environment (EA, 2006), that is:

- car parks typically larger than 800m$^2$ in area or for 50 or more car parking spaces;
- smaller car parks discharging to a sensitive environment;
- areas where goods vehicles are parked or manoeuvred;
- vehicle maintenance areas;
- roads;
- industrial sites where oil is stored or used;
- refuelling facilities;
- any other site with a risk of oil contamination.

Oil interceptors or separators provide a settlement chamber in which oil or petrol is allowed to float to the surface and separated for removal as part of an ongoing maintenance regime. There are a number of different types of separator depending on the application. Class 1 separators are designed to limit the concentration of oil in the discharge water to 5mg/litre and are suitable for discharging into most soakaways and storm water systems. Class 2 separators limit discharge oil concentration to 100mg/litre and will usually discharge into a foul sewer, provided permission is obtained from the sewer provider (Figure 3.56). Full retention separators treat the full flow from the drainage system whilst by-pass separators allow untreated water to bypass the separator once the rainfall exceeds 6.5mm/hour. They are suitable where the risk of spillage is low, such as for a short stay car park. Forecourt separators are used for petrol filling stations and are sized to take the potential spillage from a fully laden petrol tanker (typically 7,600 litres). All separators must be fitted with a visual and audible alarm to indicate when they have reached 90 per cent of capacity. In areas where there is a risk of silt accumulation, there should also be an alarm to indicate when silt levels are nearing their maximum.

The Environment Agency strongly recommends that sustainable urban drainage systems be used either in place of oil/petrol separators or to supplement them. In Scotland this is a legal requirement under the 2003 Water Environment and Water Services (Scotland) Act (WEWS) (see Chapter 3.14 Landscaping, Ecology and Flood Risk).

Fats, oil and grease (FOG) from cooking and food preparation and manufacture can accumulate in sewers and cause blockages and overflows. Grease interceptors or traps should be installed in the discharges from kitchens where there are likely to be high volumes of FOG in the wastewater. These devices use a very similar principle to an oil separator and similarly rely on

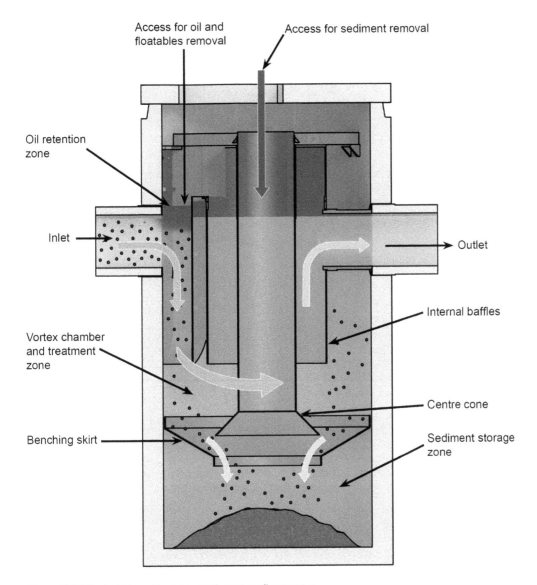

**Figure 3.56** *Typical Class 2 hydrodynamic vortex oil separator*

*Source:* Image courtesy of Hydro International (see also www.cpm-group.com/environmental/downstream-defenders.php)

good maintenance to ensure their efficient operation. If neglected the efficiency reduces to zero and they can become blocked with grease resulting in an overflow of wastewater. Grease recovery devices (GRD) (Figure 3.57) incorporate separate containers for the waste grease which can be removed for rendering into a fuel as a used cooking oil (UCO) bioliquid or transesterified into biodiesel (Chapter 2.6 Energy Strategy and Infrastructure).

**Figure 3.57** *Grease recovery device*

Source: Goslyn™ (www.greasetrap.ca)

## Note

1   Some manufacturer's refrigerant leak detection systems are subject to enhanced capital allowances, see www.eca.gov.uk/etl/find/_P_Refrigeration/38.htm.

## References

EA (Environment Agency), 2006. *Pollution Prevention Guidance. Use and design of oil separation in surface water drainage systems: PP3.* London: Environment Agency. Available at http://publications.environment.agency.gov.uk/pdf/PMHHO0406BIYL-e-e.pdf (accessed  23 March 2010)

EMEP/EEA, 2009. *Air pollutant emission inventory guidebook.* Available at www.eea.europa.eu/publications/emep-eea-emission-inventory-guidebook-2009/part-b-sectoral-guidance-chapters/1-energy/1-a-combustion/1-a-4-small-combustion-tfeip-endorsed-draft.pdf (accessed 20 February 2010)

Kakareka, S., Kukharchyk, T., Khomich, V., 2004. Research for HCB and PCB Emission Inventory Improvement in the CIS Countries (on an Example of Belarus)/Belarusion Contribution to EMEP. Minsk.

# 3.14

# Landscaping, Ecology and Flood Risk

## Introduction

Landscape does not only provide the setting for a building and amenity for the occupants, but also an opportunity to enhance the ecological value of the site whilst ensuring that rainwater is carried away in an environmentally sensitive way. It is important to consider the external surfaces of the building as part of the landscape, particularly for a constricted site. For example, roofs, terraces and even walls can be landscaped and provide amenity value, ecological niches and an opportunity to attenuate rainwater run-off.

## Ecology and biodiversity

As we saw in Chapter 1.2 Policy, Legislation and Planning, ecology and the protection of flora and fauna is at the heart of modern environmentalism. A recent publication from the UK Green Building Council (UK-GBC) reports that in the UK:

- as much as 39 per cent of habitats and 27 per cent of 'priority species' are in decline with some showing accelerated deterioration;
- bird numbers have been depleted by an average of 6 per cent in the last 30 years;
- butterfly populations have dropped an average of 55 per cent in the last 30 years; and
- major declines in bees, arable plants and amphibians have also been recorded. (UK-GBC, 2009)

The UK Government's 2009 Sustainable Construction Strategy's overarching biodiversity strategy is 'that the conservation and enhancement of biodiversity within and around construction sites (be) considered throughout all stages of a development' and that by 2012 'all construction projects over £1 million [in value are] to have biodiversity surveys carried out and necessary actions instigated'.

The UK-GBC report warns that 'our built environment has the potential to have major negative impacts on biodiversity', balanced by the assertion that, 'if done sensitively, the development and refurbishment of buildings can in fact increase the ecological value of the site.'

Since its launch in 1990 BREEAM has rewarded developments that incorporate measures that protect and enhance ecological value. The methodology has been adopted for use in CSH and the BREEAM suite of assessment schemes, whilst LEED uses a simpler approach which rewards designs that 'conserve existing natural areas and restore damaged areas to provide habitat and promote biodiversity.'

The assessment or prediction of ecological value requires an evaluation of the number of individual species of flora and fauna that are supported by a site. BRE provides guidance derived from DEFRA statistics from 1998 which indicate the average number of species that are typically supported for different landscape and planting types (DEFRA, 1998). For example, 'industrial derelict land' that has been unused for between 10 and 20 years and colonized by 'infertile grasses' including, typically tall herb vegetation, might have an ecological value of 15.8 species/hectare. This would make it very difficult to achieve credits for enhancing ecological value. Whereas sites which are either entirely covered by existing buildings and/or hard standings with up to a maximum of 20 per cent of the area having been used for single crop arable farming for the previous five years, or comprising regularly cut lawns or sports fields can be considered as having a low ecological value. In most cases it will be necessary to obtain an expert opinion on ecological value, both from a survey of the existing site and on what measures can be included in a design to maintain or enhance ecological value. Both BREEAM and CSH reward the use of a 'suitably qualified ecologist' (SQE) for this work. It is wise to use someone who not only has the required qualifications, professional memberships and ecological experience, but is also familiar with the BRE methodology and reporting pro forma.

There is much to be gained in developing a landscaping strategy with guidance from an SQE on species that are suitable for the location. The golden rule for maximizing ecological enhancement is to incorporate as far as possible species that are native to the area. The Natural History Museum has developed a database that gives a list of native plant species for a given postcode.[1] This lists plants under the headings of annual, biennial, perennial, shrub, tree, etc., giving information on family, provenance, protected status and whether suitable for garden use, as well as an image for most of those listed.

Greenfield and rural sites frequently pose the greatest challenges, particularly if there are mature trees and hedges, watercourses or wetland areas on the site. Trees or hedges taller than 1m or more than 100mm trunk diameter must be protected if credits are not to be lost. Similarly streams, rivers and wetland areas must not be interfered with or polluted during construction (see Chapter 4.4 Considerate Contracting and Construction Impacts).

Enhancing the ecological value of inner city sites can be achieved through the imaginative integration of native plants into courtyards, apron areas, roofs and walls (see below).


## Flood risk and sustainable drainage

Flood risk is to be assessed during the planning stage of a project and as part of the EIA where one is required (Chapter 2.2 Land Use and Density). The efficient and rapid removal of stormwater without overflow or back-up is also important in mitigating the risk of land pollution and soil erosion, with the related risks from undermining of foundations, damage to property and destruction of flora.

The UK Environment Agency, in the introduction to its guidance on sustainable drainage, states that

> surface water drainage from developed areas is increasingly affecting our river catchments. As development intensifies, so more water runs rapidly into rivers and less filters through the soil. This sealing of the ground can and does lead to localised flooding and water pollution, and will only get worse as our climate changes. We need a new approach to drainage that keeps water on site longer, prevents pollution and allows storage and use of the water. (EA, 2008).

As was mentioned in the previous chapter, sustainable urban drainage systems (SUDS) are becoming widely used in preference to the impermeable surfaces with gulley and pipe systems that have been used historically. Indeed all new surface water drainage systems in Scotland must now adopt the SUDS principles in their design and construction. In England and Wales many local authorities are incorporating the requirement for SUDS into their Local Development Frameworks (LDFs), whilst as a statutory consultee to planning applications the EA is committed to promoting SUDS.

The primary objective of SUDS is to attenuate flows through a combination of retention and infiltration using natural materials such as gravel and vegetation. There are a number of features that are combined to control flows and remove pollutants before allowing rainwater to soak away or discharge into watercourses.

According to the EA 'there are many SUDS design options to choose from and they can be tailored to fit all types of development, from hard surfaced areas to soft landscaped features. They can also be designed to improve amenity and biodiversity in developed areas. For instance, ponds can be designed as a local feature for recreational purposes and to provide valuable local wildlife habitat nodes and corridors.' The surface water management train approach should be used as shown in Figure 3.58.

The aim being to:

1   reduce the quantity of run-off from the site (source control techniques);
2   slow the velocity of run-off to allow settlement filtering and infiltration (permeable conveyance systems);
3   provide passive treatment to collected surface water before discharge into groundwater or to a watercourse (end of pipe systems).

Source control techniques include permeable paving and green roofs. Permeable conveyance uses swales, filter drains and infiltration trenches, which are shallow grass lined and gravel filled ditches respectively. Filter drains may have a perforated pipe in their base to assist drainage. Infiltration trenches and basins are normally dry and are used to hold surface water temporarily during high flow conditions. Balancing ponds and detention basins are terms used to describe ponds which are designed to provide temporary storage to attenuate stormwater. Lagoons on the other hand are settlement ponds, designed to

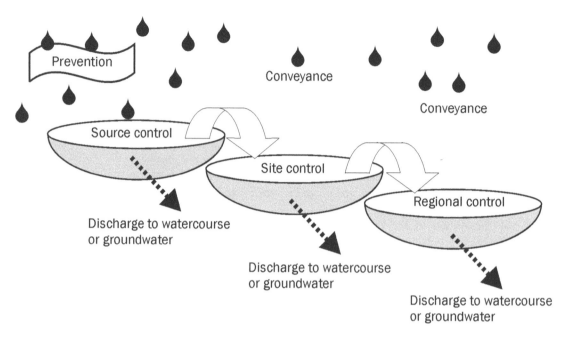

**Figure 3.58** *The surface water management train*

*Source:* CIRIA

remain full and allow sediment to settle. Wetlands are formed from shallow ponds that incorporate emergent vegetation such as reed beds that provide passive treatment over long retention times. See Figures 3.59 and 3.60 for examples of wetland and swale respectively.

All of these features are designed with shallow slopes so that, when full of water, the probability of a child accidentally getting out of their depth is reduced.

For detailed design guidance refer to the CIRIA SUDS Manual (CIRIA, 2007).

## Green and brown roofs

The first known historical references to man-made gardens on roofs were the ziggurats (stone pyramidal stepped towers) of ancient Mesopotamia, built from the 4th millennium until around 600 BC. The first modern examples of rooftop gardens were planted on a seventh floor terrace of the Rockefeller Centre in New York between 1933 and 1936.

However, the most recent incarnation of a 'green' roof is not intended primarily for amenity. In fact one of the primary aims of a green roof is to provide a niche for native species to attract insects and birds and hence enhance the ecological value of the site, which may not be compatible with providing continual access to humans. This concept was developed in Northern Europe in the 1960s and grew steadily, encouraged largely through the introduction of

**Figure 3.59** *Artificial wetland at Lanxmere, Culemborg, The Netherlands*

*Source*: Lamiot

**Figure 3.60** *Swale at University College Merced study housing, US*

*Source*: Lauren Jolly Roberts

municipal grants and incentives. In 2002 there were more than 70 local authorities across Europe either funding green roofs, or requiring them to be considered as part of a planning application. For example, the 2008 London Plan refers to green roofs as a means to recreate important elements of wasteland habitat within the design of new developments as well as an important sustainable drainage technique. The Corporation of London tracks the number of green roofs incorporated into new developments in the City each year as a 'Local Performance Indicator'.

Although there are numerous combinations of planting and landscaping that can be incorporated into a roof design, there are essentially two main types of green roof: intensive and extensive.

Intensive roofs are characterized by deeper soil, greater weight, greater variety of plants that can be supported, and consequent greater cost and maintenance. They normally require flat roofs and may be combined with accessible areas for amenity.

At the other end of the spectrum, extensive roofs are light and support a limited range of flora: predominantly plants such as sedum (Stonecrop). Costs and maintenance requirements are lower and roofs with a slope up to 30° can be used.

The key characteristics of green roofs can be summarized as follows:

- potential for enhancing ecological value by the incorporation of native species;
- the amount of stormwater retention depends on the depth of the growing medium, with an intensive roof having the greater retention capacity;
- absorption of solar radiation in summer and evaporation of retained water can significantly reduce heat gain to floor below;
- provide thermal insulation, although this may be compromised if the growing medium freezes;
- have the potential to reduce any heat island effect;
- plants absorb $CO_2$ and other gaseous pollutants, as well as trapping particulates;
- add to the sound insulation properties of roof;
- protect roof membranes from the damaging effects of solar radiation, increasing replacement intervals and potentially reducing life cycle cost to that of a conventional roof;
- aesthetic value;
- intensive roofs have amenity value, providing secure and pleasant space for building occupants.

All green roofs require long-term maintenance, especially the intensive type, whilst quality of construction is very important to minimize the risk of leakage of retained water into the spaces below. There may also be a risk of fire spread across the roof surface, although the potential for the fire to break through the roof, in either direction, is reduced by the density of the growing medium, particularly if it has retained moisture. For large roofs it may be necessary to incorporate fire breaks, typically 600mm wide and every 40m.

See Figures 3.61 to 3.63 for images of intensive and extensive green roofs.[2]

Brown roofs differ from green roofs in that they are created primarily to enhance biodiversity and generally do not incorporate bought-in plants. The substrate usually comprises local earth and materials, including stones and logs, whilst flora are allowed to colonize the roof opportunistically (Figure 3.64).

The composition of the roof might be designed to encourage specific local species, such as the black redstart, which is an endangered native of the Lea

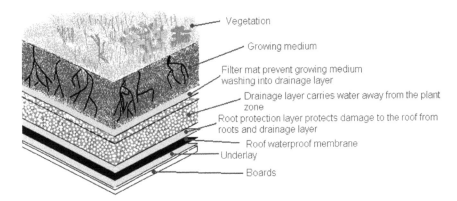

Vegetation

Growing medium

Filter mat prevent growing medium washing into drainage layer

Drainage layer carries water away from the plant zone

Root protection layer protects damage to the roof from roots and drainage layer

Roof waterproof membrane

Underlay

Boards

**Figure 3.61** *Composition of extensive green roof*

*Source*: www.delston.co.uk/greenroofs.htm

**Figure 3.62** *Intensive green roof at Edgedale Green, Singapore*

*Source*: Sengkang

**Figure 3.63** *Extensive green roof at Mountain Equipment Co-op, Toronto*

*Source*: sookie

**Figure 3.64** *Brown roof at Green Shop, Bisley, UK*

*Source*: thingermejig

**Figure 3.65** *Green wall in Paris, France*

*Source:* Kaldari

Valley and Docklands areas of London. Conservationists such as the London Wildlife Trust have been encouraging developers to incorporate its favoured rocky terrain into their roof designs.

Green walls have evolved from their more opportunistic ivy-clad forebears into some very exciting vertical green works of art, such as that shown in Figure 3.65. These require niches to be formed for planting to avoid potential damage to the facade.

## Notes

1   The Natural History Museum's postcode database can be accessed at www.nhm.ac.uk/nature-online/life/plants-fungi/postcode-plants/.
2   For more information on products and installers refer to www.greenroofs.com/.

## References

CIRIA (Construction Industry Research and Information Association), 2007. *The SUDS Manual.* London: Construction Industry Research and Information Association

DEFRA (Department for Environment, Food and Rural Affairs), 1998. *Digest of Environmental Statistics No 20.* London: HMSO. Available at www.defra.gov.uk/evidence/statistics/environment/pubatt/download/sim9697.pdf (accessed 24 March 2010)

EA (Environment Agency), 2008. *Sustainable Drainage Systems (SUDS).* Bristol, UK: Environment Agency

UK-GBC, 2009. *Biodiversity and the Built Environment.* London: Green Building Council

# 3.15
# Security and Flexibility

## Introduction

This chapter deals with two of the fundamental responsibilities of a building designer: to provide a secure environment for occupants and to future-proof against the changing needs of the occupants.

## Design of buildings for security

Under the heading of 'Secured by Design' in Chapter 2.4 Social Sustainability, we looked at the principles of secure design from the point of view of the masterplanner. These need to be factored into the design of individual buildings as recognized in both BREEAM and CSH.

In this section we are looking specifically at the design requirements for individual buildings. The Association of Chief Police Officers' (ACPO) Secured by Design service and award scheme provides a series of guides that cover new homes, sheltered accommodation, multi-storey dwellings and home security, as well as refurbished properties, railway stations, caravan parks, play areas, schools, hospitals, licensed premises, youth shelters and sports systems.[1] The ACPO have also developed the ParkMark[rtm] scheme, including guidance on the design of secure car parks.[2]

The Secured by Design (SBD) publications provide detailed specifications and reference standards for the key building components that impact on security. The SBD website also provides details of compliant products and access to manufacturer and supplier websites. Section 02 of the New Homes 2010 guidance (also downloadable from the SBD website) specifies doorset and locking system standards for all ground level doors for houses and apartment blocks, internal doors and loft hatches from communal halls and landings, windows, rooflights, viewing panels and party walls. It also covers garages, underground car parks and cycle storage, security lighting, access controls and intruder alarms.

It will be seen from the above that SBD does not have specific guidance for office buildings, although there is a credit within BREEAM Offices (Man 8) that rewards designs that adopt the principles of SBD, in liaison with a police Architectural Liaison Officer (ALO) or Crime Prevention Design Advisor (CPDA). The specific measures to be taken in the building design will depend on the location and the amount of public access expected. Some public sector office buildings might expect the same issues that are found in hospital buildings, hence it is worth cherry picking from the SBD – Hospitals guide, especially those sections dealing with access and building shell security.

In the UK special attention is paid to the buildings and facilities that make up the Critical National Infrastructure (CNI). The Centre for Protection of National Infrastructure (CPNI) has been established to 'provide expert advice to the critical national infrastructure on physical, personnel and information security, to protect against terrorism and other threats'.[3] CNI is defined as government, finance, communications, emergency services, health, transport, energy, water and food.

The CPNI's top ten guidelines, reproduced unedited in Box 3.6 from their website, should be factored into the design of all buildings that could be the subject of terrorist attack or the attention of those who wish to disrupt operations therein.

---

### Box 3.6 CPNI top ten guidelines

1 Carry out a risk assessment to decide on the threats you might be facing and their likelihood. Identify your vulnerabilities and the potential impact of exploitation.
2 If acquiring or extending premises, consider security at the planning stage. It will be cheaper and more effective than adding measures later.
3 Make security awareness part of your organisation's culture and ensure security is represented at a senior level.
4 Ensure good basic housekeeping throughout your premises. Keep public areas tidy and well-lit, remove unnecessary furniture and keep garden areas clear.
5 Keep access points to a minimum and issue staff and visitors with passes. Where possible, do not allow unauthorised vehicles close to your building.
6 Install appropriate physical measures such as locks, alarms, CCTV surveillance, complementary lighting and glazing protection.
7 Examine your mail-handling procedures.
8 When recruiting staff or hiring contractors, check identities and follow up references.
8 Consider how best to protect your information and take proper IT security precautions. Examine your methods for disposing of confidential waste.
10 Plan and test your business continuity plans, ensuring that you can continue to function without access to your main premises and IT systems.

---

Although some of the points in Box 3.6 relate to operational and housekeeping issues, it is the designer's responsibility to ensure that the building and its services assist the building operator in achieving these tasks.

Since the events in New York in September 2001, the security of high profile buildings in the US and elsewhere has been brought into sharp focus. A number of guidelines have been produced for application to federal buildings and are summarized in the National Institute of Building Sciences on-line Whole Building Design Guide,[4] including two recent publications from the Interagency Security Committee (ISC, 2008, 2009). The US General Services Administration has also produced a Site Security Design Guide for federal buildings (USGSA, 2007). This provides detailed guidance on the secure design of federal facilities based on 'protecting standoff perimeters, controlling site access and installing lighting for security and site surveillance' whilst integrating with other design objectives, such as sustainable drainage and community interaction. An example of this approach is shown in Figure 3.66, which is taken from the Vision and Hallmarks chapter of the Guide.

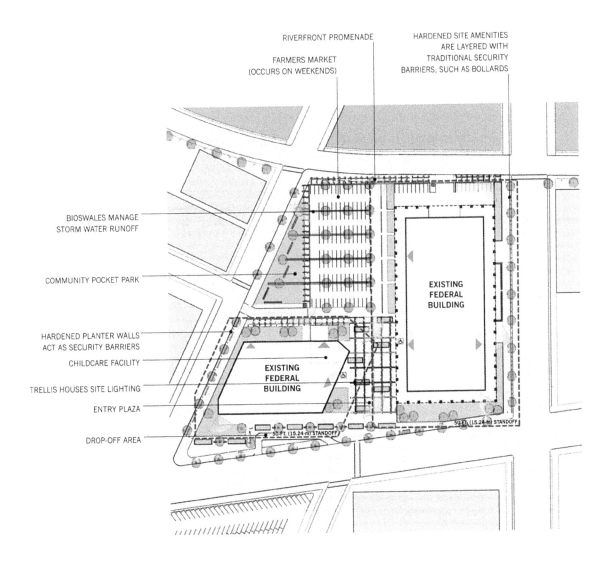

**Figure 3.66** *Secure design approach for federal buildings*

*Source*: Rios Clementi Hale Studios; Courtesy GSA (www.gsa.gov/graphics/pbs/GSA_Chapter_One_8-8-07.pdf)

## Lifetime homes

In the UK it is recognized that with an ageing population there is a need to future-proof homes against the needs of occupants with limited mobility. With this in mind the Habinteg Housing Association, the Helen Hamlyn Foundation and the Joseph Rowntree Foundation developed the Lifetime Homes criteria in the early 1990s.[5] The scheme requires the incorporation of 16 design features that together create a flexible blueprint for accessible and adaptable housing in any setting. It is essentially a set of criteria for making a dwelling wheelchair accessible and many of the requirements have since been incorporated into Part M of the Building Regulations (ODPM, 2004). A summary of the requirements and their relationship to Part M is given in Box 3.7.

---

### Box 3.7 Lifetime Homes and Approved Document M requirements

| Lifetime Homes | Approved Document M |
|---|---|
| **Criterion 1**<br>Where there is car parking adjacent to the home, it should be capable of enlargement to attain 3300mm width. | Requires 900mm access on top of 2400mm basic requirement for car parking, hence LH compliant |
| **Criterion 2**<br>The distance from the car parking space to the home should be kept to a minimum and should be level or gently sloping, with a slope of no more than 1 in 12 for <5m distance, 1 in 15 for 5–10m and 1 in 20 for >10m, with minimum path width of 900mm, 1200mm landings plus any gate or door swing. | Also requires that all surfaces be firm and even. |
| **Criterion 3**<br>The approach to all entrances should be level or gently sloping, with slopes treated as above. | Also requires that all surfaces be firm and even. |
| **Criterion 4**<br>All entrances should:<br>(a) be illuminated<br>(b) have level access over the threshold and<br>(c) have a covered main entrance. | No additional requirements |
| **Criterion 5**<br>(a) Communal stairs should provide easy access and<br>(b) where homes are reached by a lift, it should be fully accessible.<br>Minimum dimensions for communal stairs:<br>• Uniform rise not more than 170mm<br>• Uniform going not less than 250mm | No additional requirements |

- Handrails extend 300mm beyond top and bottom step
- Handrail height 900mm from each nosing

Minimum dimensions for lifts:

- Clear landing entrances 1500mm × 1500mm
- Min. internal dimensions 1100mm × 1400mm
- Lift controls between 900 and 1200mm from the floor and 400mm from the lift's internal front wall.

Criterion 6

The width of the doorways and hallways should conform to the following specifications:

No additional requirements

| Doorway clear opening width (mm) | Corridor/passageway width (mm) |
|---|---|
| 750 or wider | 900 (when approach is head-on) |
| 750 | 1200 (when approach is not head-on) |
| 800 | 1050 (when approach is not head-on) |
| 900 | 900 (when approach is not head-on) |

The clear opening width of the front door should be 800mm. There should be 300mm to the side of the leading edge of doors at entrance level

Criterion 7

There should be space for turning a wheelchair in dining areas and living rooms and adequate circulation space for wheelchairs elsewhere.

A turning circle of 1500mm diameter or 1700mm × 1400mm ellipse is required in dining areas and living rooms.

No additional requirements

Criterion 8

The living room should be at entrance level.

No additional requirements

Criterion 9

In houses of two or more storeys, there should be space on the entrance level that could be used as a convenient bed-space.

No additional requirements

Criterion 10

There should be:

(a) a wheelchair accessible entrance-level WC, with
(b) drainage provision enabling a shower to be fitted in the future.

Dwellings of three or more bedrooms or on one level:

- The WC must be fully accessible. A wheelchair user should be able to close the door from within the closet and achieve side transfer from a wheelchair to at least one side of the WC.
- There must be at least 1100mm clear space from the front of the WC bowl.
- The shower provision must be within the closet or adjacent to the closet (the WC could be an integral part of the bathroom in a flat or bungalow).

For other dwellings of two or fewer bedrooms on more than one level, where the design has failed to achieve the above fully accessible standard WC, the Part M standard WC will meet this standard.

No additional requirements

Criterion 11

Walls in bathrooms and toilets should be capable of taking adaptations such as handrails.

Wall reinforcements should be located between 300 and 1500mm from the floor.

No additional requirements

| | |
|---|---|
| Criterion 12<br>The design should incorporate:<br>(a) provision of a stair lift<br>(b) a suitably identified space for a through-the-floor lift from the ground to the first floor, for example to a bedroom next to a bathroom.<br>There must be a minimum of 900mm clear distance between the stair wall (on which the lift would normally be located) and the edge of the opposite handrail/balustrade. Unobstructed 'landings' are needed at top and bottom of the stairs. | No additional requirements |
| Criterion 13<br>The design should provide a reasonable route for a potential hoist from a main bedroom to the bathroom. | No additional requirements |
| Criterion 14<br>The bathroom should be designed to incorporate ease of access to the bath, WC and wash basin.<br>Although there is not a requirement for a turning circle in bathrooms, sufficient space should be provided so that a wheelchair user can use the bathroom. | No additional requirements |
| Criterion 15<br>Living room window glazing should begin at 800mm or lower and windows should be easy to open/operate.<br>People should be able to see out of the window whilst seated. Wheelchair users should be able to operate at least one window in each room. | No additional requirements |
| Criterion 16<br>Switches, sockets, ventilation and service controls should be at a height usable by all (i.e. between 450 and 1200mm from the floor). | No additional requirements |

## Future-proofing office space

It is likely that at least 75 per cent of the buildings that currently exist in the UK will still be in use in 2050 (Ravetz, 2008). Hence it is important that new buildings are built to be robust and flexible, enabling them to adjust to changes in user requirements, technologies and working practices. Although this may not stretch to shopping centres having to be suitable for conversion into dwellings, for example, clearly at any one time there is over-supply in some sectors that could help meet demand in others. In coming years new methods of working could result in more home working and 'hot desking', that is workstations that are designed for visitors, most likely using laptop computers. Open-plan office space, although unpopular with some, enables greater flexibility, particularly when combined with bookable meeting space.

The key to future-proofing office space is to ensure that the space and access provided for building services allows for all anticipated scenarios for change of use. A successful speculative office building designed for multi-tenancy will normally achieve this anyway.

Access floors provide the ultimate in flexible connectivity for servicing workstations and allow for exposed soffits and mixed-mode natural ventilation/air conditioning. Unfortunately, future-proofing is likely to entail a degree of summer cooling to cater for increased summer time temperatures and heat gain from IT equipment.

## Notes

1   All of the ACPO Secured by Design guides are downloadable from www.securedbydesign. com/professionals/guides.aspx).
2   Details of the ParkMark[rtm] scheme can be found at www.britishparking.co.uk/files/ saferparkingscheme/guidelines_manual_web_ready.pdf. The ParkMark[rtm] scheme has its own website at www.parkmark.co.uk, which includes an addressable database of registered car parks.
3   www.cpni.gov.uk/aboutcpni188.aspx.
4   www.wbdg.org/design/provide_security.php.
5   www.lifetimehomes.org.uk/.

## References

ISC (Interagency Security Committee), 2008. *Facility Security Level Determinations*. Washington, DC: Interagency Security Committee
ISC, 2009. *Physical Security Criteria for Federal Facilities*. Washington, DC: Interagency Security Committee
ODPM (Office of the Deputy Prime Minister), 2004. *Approved Document M: Access to and Use of Buildings, 2004 Edition*. London: TSO
Ravetz, J., 2008. State of the stock – What do we know about existing buildings and their future prospects? *Energy Policy*, vol 36, pp4462–4470: Elsevier. Available at www.foresight.gov.uk/Energy/EnergyFinal/Ravetz%20paper-section%205.pdf (accessed 20 February 2010)
USGSA (US General Services Administration), 2007. *The Site Security Design Guide*. US General Services Administration. Available at www.gsa.gov/Portal/gsa/ep/contentView. do?contentType=GSA_BASIC&contentId=23429 (accessed 20 February 2010)

# Part 4

# Sustainable Construction

# 4.1
# Tendering Process

## Introduction

The best laid plans of the design team will come to nothing if they are lost in the construction process. Hence the appointment of contractors that understand the challenges of sustainable construction and have experience with the principles and technologies involved is vital to the success of the project.

This is driven most strongly by the requirement to maintain the environmental assessment rating achieved at the design stage through to the handover of the building. LEED has always incorporated this, but it has only recently become an intrinsic requirement of BREEAM and CSH. Hence it has become an essential performance requirement to include within a construction contract.

There is no single model for a form of construction contract that is used internationally. Indeed, most countries have multiple choices that have evolved primarily to achieve specific commercial and legal objectives. For example, a guide to construction contracts available in the UK produced by the Local Government Taskforce highlights 15 different forms produced by Joint Contracts Tribunal Ltd (JCT), the Institution of Civil Engineers (ICE), the Association of Consultant Architects (ACA), the International Federation of Consulting Engineers (FIDIC), the New Engineering Contract (NEC) and Government Contracts (GC) (Local Government Taskforce, undated). These include contracts designed for civil engineering projects, Government projects, minor works, construction management, design and build, prime cost, measured term and partnering, as well as the traditional standard forms. In addition many of the larger public sector construction projects in the UK are procured through Public/Private Partnerships (PPP) such as the Private Finance Initiative and Prime Contract which incorporate a period of facility operation and management once the buildings are completed, introducing operating cost as an ingredient in the procurement equation.

The US has a similar wide choice of contract types, with the American Institute of Architects (AIA) alone offering 80 different forms, whilst the Associated General Contractors of America (AGC) and National Society of Professional Engineers (NSPE) also offer standard forms of contract.

# Tendering for sustainable construction

The composition of the construction contract has a major impact on the success of translating the design team's sustainability objectives into practice. Some key decision points have been proposed by JCT as follows:

- The long- and short-term objectives you wish to achieve in terms of sustainability.
- What you must achieve in terms of sustainability i.e. legal and regulatory requirements, both project specific and general.
- Who is to take the lead in delivering the sustainability requirements. This will have a major influence on the procurement path to be followed, e.g. management team, design team or construction team led.
- How to involve the supply chain in the design and construction process as early as practicable.
- Whether the providers of services and works are to be incentivized. If so, what mechanisms are to be used, e.g. bonus, gain share, damages.
- What documentation is required, e.g. specification, contract schedule, for the various providers and how the requirements are to be spelled out – are they to be generic provisions or specific measurable provisions.
- What items are to be provided for – consider the practicality of delivering sustainability requirements.
- Whether performance indicators are to be used, if so, which indicators and what target levels.
- The contract(s) to be used and which contract clause options related to sustainability are to be included. Insofar as it can, ensure the contract achieves certainty so that it is clear whether the requirements are fulfilled (JCT, 2009).

Some of the key issues that govern the success of this process that should be taken into account when developing forms of contract and programmes for making appointments are set out below:

- A demonstrable commitment to sustainability from the client is vital. For example the Office of Government Commerce is the client for public sector projects in the UK. They have set very high standards in sustainable procurement through their 'Achieving Excellence Construction Procurement Guide 11' which sets out a framework for sustainable construction procurement (OGC, 2007).
- Managing sustainability requires translating design specifications into purchase and installation, whole life cost control, control of construction impacts, management of considerate contracting issues (Chapter 4.4), waste management (Chapter 4.3), value engineering (Chapter 4.5), document control, commissioning, handover (Chapter 4.6) and final assessment. Most of these issues are mirrored in the management of health and safety on site and hence many of the control measures and responsibilities set out in such legislation as the UK's Construction Design and Management (CDM) Regulations can easily be adapted to sustainability and management of construction impacts.

- As with CDM, all of these processes have to be trickled down through the supply chain through a combination of early involvement and incentivization, established in the forms of contract. For example, 'Design and Build' contracts require the participation of contractors and their supply chains during design development.
- A contract based on central construction management providing site control to a single management contractor, having responsibility for oversight of construction impacts and considerate contracting, as well as performance in the post-construction BREEAM, CSH or LEED assessment and trickle down of all performance requirements to the supply chain.

Examples of forms of contract that have addressed sustainability are set out in the JCT Guidance note referred to above. For example clause 16 of their Framework Agreement states that:

> The Provider will assist the Employer and the other Project Participants in exploring ways in which the environmental performance and sustainability of the Tasks might be improved and environmental impact reduced. For instance, the selection of products and materials and/or the adoption of construction/ engineering techniques and processes which result in or involve:

- reductions in waste;
- reductions in energy consumption;
- reductions in mains water consumption;
- reductions in $CO_2$ emissions;
- reductions in materials from non-renewable sources;
- reductions in commercial vehicle movements;
- maintenance or optimization of biodiversity;
- maintenance or optimization of ecologically valuable habitat; and
- improvements in whole life performance.

Clause 17 of the same agreement also tackles value engineering which, as we will see in Chapter 4.5, can be a process by which sustainable performance can easily be eroded. The following text may go some way to mitigating this:

> The Provider is encouraged to suggest changes to Tasks which, if implemented, would result in financial benefits to the Employer. Such benefits may arise in the form of:

- a reduction in the capital cost of the project of which the Tasks form part;
- a reduction in the life cycle and/or operating costs associated with the project;
- completion of the project at an earlier date or in a manner which will result in savings; and/or
- any other financial benefit to the Employer.

> The Parties will discuss the details of any changes and any cost, time, quality and performance implications of them and will negotiate with a view to agreeing the implementation of any changes and the financial effects of them provided that such changes remain compliant with the pricing procedures set out in the Pricing Documents.

In this context 'performance' should include rating under the relevant environmental assessment scheme as well as any other relevant performance indicators set out in the contract documentation. JCT has approved the following clauses to cover this requirement:

> The Employer shall monitor and assess the Contractor's performance by reference to any performance indicators stated or identified in the contract documents.

> The Contractor shall provide to the Employer all information that he may reasonably require to monitor and assess the Contractor's performance against the targets for those performance indicators.
>
> Where the Employer considers that a target for any of those performance indicators may not be met, he may inform the Contractor who shall submit his proposals for improving his performance against that target to the Employer.

It is usual for performance indicators to be incorporated into the contract document either as part of the specification or as a schedule to the contract conditions.

In this context the BREEAM, or equivalent, assessment spreadsheet makes a very useful tracking tool for the employer to monitor the performance of the construction team in achieving sustainability criteria. A good method of ensuring the consistency and independence of this process is to use the same assessor to undertake periodic checks of the rating, reporting directly to the employer, through to the final post-construction assessment.

Other performance indicators that are worth considering for inclusion in the schedule referred to above include Building Emission Rating (BER) as required for the Energy Performance Certificate and Building Regulations compliance in the UK, including any commitment for renewable energy generation as may have been required by the local planning authority, Site Waste Management Plans (see Chapter 4.3) and relevant Constructing Excellence Key Performance Indicators (KPIs).[1]

There are several hundred KPIs, many of which address aspects of building and contractor performance that are not related to sustainability, hence it is important to relate relevant KPIs to the commitments made for the BREEAM assessment and development design, including socio-economic aspects, most of which will be location specific.

It should be noted that the summer performance of a naturally ventilated building cannot be defined within tightly controlled temperatures and hence has to be defined by the parameters used in the dynamic computer simulation model.

## Case history

### Lion House, Alnwick

This new office building in Alnwick, Northumberland, built for DEFRA was opened in 2008 with the aim of setting a new benchmark for the procurement of public buildings. It has since won the CIBSE low carbon building award and achieved a score of over 80 per cent under BREEAM Offices 2006. The following is an extract from the OGC case study for the project:

**Figure 4.1** *DEFRA offices, Lion House, Alnwick*

*Source:* OGC

DEFRA adopted the NEC form of contract to procure the new Lion House. NEC is a major attempt to draft a simple and direct standard form from first principles. Its aim is to:

- achieve a higher degree of clarity when compared to other existing contracts;
- use simple language and avoid legal jargon to reduce the instance of disputes;
- produce core conditions and exclude contracts-specific data;
- precisely and clearly set out key duties and responsibilities.

NEC's reported strength is in encouraging a partnering approach between contracting parties whilst placing emphasis upon proactive project management and the early identification of project risks. This partnering element was key to delivering a sustainable new Lion House. DEFRA used the NEC3 contract to create a framework that would allow appointment of one contractor to deliver two projects that had sustainability at their heart: the refurbishment of the DEFRA HQ in York and the construction of a new building in Alnwick. The framework rationale provides economy of scale while cutting down on time spent tendering. The new Lion House is one of a number of projects in the North East collectively known as NEPS (North East Premises Solution). The NEPS Framework Agreement falls within EU Threshold Limits and is bound by Public Procurement Rules. In response to an OJEU (Official Journal of the European Union) advertisement, Kier Northern was appointed as the Framework Contractor in April 2006. The agreement provided DEFRA with a Framework Stage 1 Service on a fee basis leading up to a Framework Stage 2 Offer to deliver a new building using design-and-build procurement.

Choosing a design-and-build route permitted early involvement of the construction supply chain wile capitalizing on Kier Northern's construction knowledge of 'buildability'. Given the complexity of achieving BREEAM Excellent and the need to meet other ambitious sustainability targets, there were significant risks to be managed early in the programme. DEFRA realized, as client, that certain elements of the project brief were unlikely to be delivered with certainty if they were specified merely in performance terms to a design-and-build contractor. Accordingly, Appleyards as the client's project manager and lead consultant was asked to progress the design to a point that would address this concern. This required the advice of sub-consultants within the client's team to progress the project from concept stage to scheme design and to deliver two key objectives:

1. Develop the design solution sufficiently to safeguard the achievement of BREEAM Excellent and deliver on the sustainability objectives for the building.
2. Enable planning approval to be achieved by the end of summer 2006 and allow Kier Northern to take responsibility for detail design and construction.

After progressing these design issues the client's design team compiled a performance specification for the contractor in the form of a tender enquiry inviting a Framework Stage 2 Offer to design and build the new facility. This was then evaluated for its compliance with DEFRA's requirements before proceeding under the NEC contract. The dual structure of contractor team and client team while ensuring a sustainable design solution is a key element within the NEC3 form of contract. It also ensured the collective management of risk.[2]

## Notes

1    Constructing Excellence KPIs can be found at www.kpizone.com/.
2    Case History taken from www.ogc.gov.uk/documents/LionHouse_Case_Study.pdf.

## References

JCT (Joint Contracts Tribunal Ltd.), 2009. *Guidance Note: Building a sustainable future together*. London: Thomson Reuters (Legal) Ltd. Available at www.jctltd. co.uk/assets/Building%20a%20sustainable%20future%20together%202009%20 Web.pdf (accessed 24 March 2010)

Local Government Taskforce, undated. *A Guide to Standard Forms of Construction Contract*. Available at www.constructingexcellence.org.uk/download.jsp?url=%2Fp df%2Fdocument%2Fstandard_forms.pdf (accessed 24 March 2010)

OGC (Office of Government Commerce), 2007. *Achieving Excellence in Construction Procurement Guide 11: Sustainability*. London: Office of Government Commerce. Available at www.ogc.gov.uk/documents/CP0016AEGuide11.pdf (accessed 24 March 2010)

# 4.2
# Sustainable Procurement

## Introduction

This chapter builds on the guidance to the design team provided in Chapter 3.12 Materials Specification. It is primarily intended for those in the construction team responsible for interpretation of the design specifications, tendering of subcontract packages and purchase of materials, plant and equipment.

The UK Government in their 2008 Strategy for Sustainable Construction makes the following statement as part of its commitment to sustainable procurement:

> A successful procurement policy requires ethical sourcing, enables best value to be achieved and encourages the early involvement of the supply chain. An integrated project team works together to achieve the best possible solution in terms of design, buildability, environmental performance and sustainable development. (BERR, 2008)

In the same document the overarching target for construction materials is 'that the materials used in construction have the least environmental and social impact as is feasible both socially and economically'. With this in mind the UK Government is committed to promoting framework standards to facilitate the development of sector 'responsible sourcing schemes' on the lines of the FSC scheme for timber. The Government is already committed to purchasing timber that meets the criteria set out in the EC FLEGT initiative. DEFRA has set up a Sustainable Products and Materials division which is developing construction product roadmaps, or life cycle inventories, for selected products, including windows, plasterboard, domestic lighting, electric motors and WCs.[1] DEFRA is developing a voluntary action plan to reduce the life cycle impacts of each of these products.

The UK Treasury's procurement arm, the Office of Government Commerce (OGC) has established a Centre of Expertise on Sustainable Procurement and, together with the NHS, the MoD and Local Government, has developed a Sustainable Procurement Action Plan (OGC, 2007).

## Tendering of subcontract packages

Most construction projects can be divided into a number of specialist subcontract packages, such as foundations, building services, access flooring and landscaping. The number and type of packages will depend on the size and type of development and the skills available to the main contractor as direct labour. In the UK there are essentially two categories of subcontractor: domestic subcontractor and nominated subcontractor. The first is generally employed by the main or managing contractor, whilst the second would normally be employed by the client (Employer).

In either case the schedule or specification clauses relating to sustainability in the main form of contract must be passed on to the subcontractors, and hence trickle down to any sub-subcontractors that they appoint.

Most projects will benefit from the early involvement of contractors and specialist subcontractors in the design process, especially where there are complex and innovative technologies required to meet sustainability objectives. A difficult balancing act exists when there are few subcontractors in the marketplace that have experience with a specialist field, whilst still ensuring best value for money through competitive tendering. It may be necessary to enter into a separate agreement for design services during design development, whilst ensuring that the eventual specification can be tendered by the appropriate number of competitors. In many instances it is very useful to use a contractor to undertake a 'buildability' check on a design, whilst preparing a construction statement to submit with a planning application (refer to Chapter 4.1 Tendering Process).

## Purchasing strategy

One of the primary objectives of a construction company is to make a profit, which means that in a highly competitive marketplace, when very tight margins are required to win projects, every opportunity to reduce expenditure has to be explored. Historically this has usually required shopping around for savings when purchasing materials and value engineering down to the last nut and bolt. Unfortunately, the number of suppliers that meet the BREEAM and CSH Materials credits is currently quite small and this reduces the latitude the contractor has for shopping around. It is very important that contractors recognize this when tendering for projects, and ensure that the specifications are precise in their requirements and that the tender responses are thoroughly checked to ensure that all tenderers have met the requirements.

The contractor does have other opportunities to increase profit however, mostly from reducing waste and running an efficient site. BRE set out the following useful pointers for sustainable purchasing:

- Think before you buy and plan your building supplies to keep wastage of materials to a minimum (just-in-time delivery).
- Calculate the quantities of materials you will need for the job and do not order more than you need for the sake of convenience.
- Timely ordering of materials ensures that supplies arrive when required, and reduces storage where they could be damaged or stolen.
- If you require a small amount of a material which is only available in bulk, use a materials exchange scheme or plan to use the material for future jobs.
- Think about specifying materials that have a good recyclate content as this will help to reduce your carbon footprint. (BRE, undated)

Just-in-time delivery reduces the amount of storage space required on-site and also the amount of packaging required to protect materials during long-term storage. Packaging represents some 60 per cent of skipped materials on a typical building site (Chapter 4.3 Construction Waste Management). Stored materials are much more likely to be damaged than when newly delivered – some 13 per cent of skipped waste is unused materials. It is also wise to ensure that suppliers will accept returns. It is worth noting that some regions of the UK have on-line materials exchange programmes.[2]

Apart from achieving the specified BRE Green Guide rating and Tier level for responsible sourcing, materials and products should, where possible, be purchased from local suppliers and manufacturers. This will also help in achieving targets for minimizing $CO_2$ emissions associated with deliveries established from the Construction Site Impacts credits in BREEAM and CSH (see Chapter 4.4 Considerate Contracting and Construction Impacts). 'Local' can be defined as within 30km, however, a realistic aim should be to procure from as close as possible to the site and avoid products that are manufactured overseas as far as possible. Of course for specialist products this may be impossible, for example the architect for a recent project in Wales wished to use cedar as a cladding material and a locally grown source was identified. However, when a sample of the product was obtained it was realized that there was a high probability it would split once in situ, hence cedar was imported from a sustainable FSC certified source in Canada, despite the higher carbon footprint.

Where the designer has not specified a specialist product but has provided a performance specification that includes sustainability criteria, it is recommended that the purchaser visits the websites given at the end of Chapter 3.12 Materials Specification. These provide links to manufacturers' and suppliers' websites as well as a lot of background information on a wide range of construction products and materials.

## Pre-fabrication and modern methods of construction (MMC)

One method of reducing waste on-site and placing less reliance on-site labour is to pre-fabricate major building components off-site. This can include anything from entire houses to timber or steel frames, as well as bathroom pods pre-cast concrete systems and structural insulated panel systems (SIPS).

Research from 2006 carried out by the UK National Audit Office (NAO) 'showed that, compared to more traditional techniques, modern methods of construction can reduce on-site labour requirements to less than a quarter and on-site time to less than a half. These reductions are not magic: they result from some work taking place off-site. They bring real benefits because off-site work involves different labour that is not as stretched as the on-site workforce and off-site work can be taken off the overall critical path providing it is planned properly. Modern methods of construction can therefore make better use of scarce labour and reduce total development time.'[3]

The NAO found that MMC tended to be more expensive than on-site assembly of the same components, although the gap was closed significantly due to faster completion and consequent earlier rent receipts and/or shorter

**Figure 4.2** *Pre-fabricated house during construction*

*Source:* H Raab

**Figure 4.3** *Modular flats in Amersfoort, The Netherlands*

*Source:* Image Copyright Spacebox.nl (www.spacebox.nl/index.cfm?lng=en&mi=3&pmi=12)

borrowing period. However, the economics and convenience of MMC were more evident where there was restricted space. The project risks for MMC are heavily front-loaded and late changes to the design cannot be accommodated. Clearly the accuracy of measurements taken on-site are also very important since tolerances are small and significant inaccuracies could lead to problems with leakage of water, air, noise or pollutants.[4]

The USA has a long history of pre-fabrication, dating from the second half of the 19th century when the newly built railroad allowed settlers to transport entire houses across North America. As early as the 1860s modular timber houses were being built in Boston and shipped across the continent.

The pre-fabrication sector in the US has grown to a multi-billion dollar industry covering not just housing but hotels, prisons, agricultural buildings, institutional buildings, schools and hospitals.[5]

## Notes

1    www.defra.gov.uk/environment/business/products/roadmaps/.
2    For example, www.eastex.org.uk.
3    www.nao.org.uk/our_work_by_sector/housing,_property/modern_methods_of_construction.aspx.
4    For more information on MMC products and contractors in the UK visit www.mmccentre.com/.
5    For an extensive review of pre-fabricated projects, products, suppliers and contractors worldwide visit www.scrapbookscrapbook.com/DAC-ART/modular-kit-houses.html.

## References

BERR (Department for Business, Enterprise and Regulatory Reform), 2008. *Strategy for Sustainable Construction*. London: Department for Business, Enterprise and Regulatory Reform

BRE, undated. *Sustainable Construction: Simple ways to make it happen*. Watford, UK: Building Research Establishment

OGC (Office of Government Commerce), 2007. *Sustainable Procurement Action Plan*. London: Office of Government Commerce

# 4.3

# Construction Waste Management

## Introduction

At more than 100 million tonnes per annum, construction and demolition waste represents approximately one-third of the total waste produced in the UK[1] and about 25 per cent of the total mass of materials used in construction. Of this, around 30 million tonnes goes to landfill. The Government target for 2012 is to reduce this by 50 per cent.

No recent equivalent statistics on the generation of construction and demolition (C&D) waste could be found for the US, however, the US EPA refers to figures from 1996 for 'building debris' of 136 million tons, of which between 20 and 30 per cent was recycled. An estimate of >200 million tons for 'quantities from infrastructure and land clearing' is given, although no data for recycling were given.[2]

The EPA has identified C&D waste as a priority area for reduction and has set out the following goals:

- characterize, measure, and increase knowledge and understanding of the C&D materials stream;
- promote research and development on best practices for C&D materials reduction and recovery;
- foster markets for construction materials and other recycled materials that can be incorporated into building products;
- work with key players in the construction, remodelling, and demolition industries to implement more resource-efficient practices; and
- incorporate C&D materials issues and projects into broader 'green building' programmes.[3]

The EPA WasteWise Building Challenge is a voluntary scheme that has been operating since 2002 to which 'partners' sign up, although its uptake appears to have been relatively limited at the time of writing. The principles adopted are those of the 'three R's' that underpin all waste reduction strategies – Reduce, Re-use, Recycle – and are outlined in the following steps promoted by the EPA and already touched on in Chapter 4.2 Sustainable Procurement:

- Incorporate environmental specifications into your building contracts and guidelines.
- Develop standard operating procedures for C&D reuse and recycling at your construction site.
- Rehabilitate an existing structure in place of planned demolition.
- Use deconstruction techniques rather than demolition if a building must be torn down.
- Employ efficient framing to reduce the amount of lumber used without sacrificing structural integrity.
- Invest in durable products to ensure that materials last as long as possible.
- Return unused construction material to vendors.
- Consider the end-of-life management, or recyclability, of building products at the start of a project.
- Salvage C&D waste for sale and reuse.
- Purchase recycled-content building materials including insulation, carpet, cement, paint, floor tiles, shower and restroom dividers, laminated paperboard, and structural fiberboard.

## Site waste management plan (SWMP)

In England and Wales, since 2008, there has been a requirement to produce a 'site waste management plan' for all construction projects with a projected cost of more than £300,000 on one site at the start of the project, other than those covered by the Environmental Permitting (England and Wales) Regulations 2007, which includes a requirement for waste management.

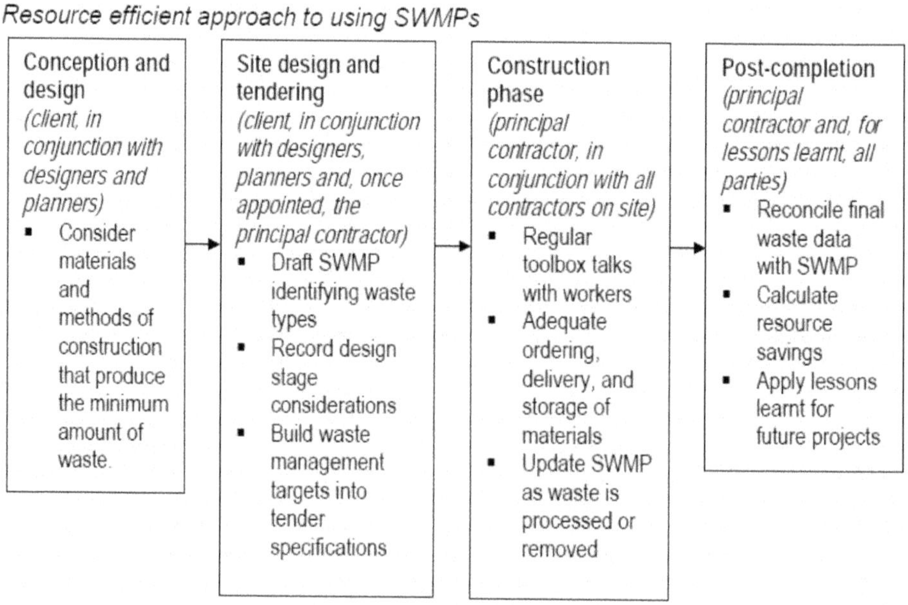

*Resource efficient approach to using SWMPs*

| Conception and design (client, in conjunction with designers and planners) | Site design and tendering (client, in conjunction with designers, planners and, once appointed, the principal contractor) | Construction phase (principal contractor, in conjunction with all contractors on site) | Post-completion (principal contractor and, for lessons learnt, all parties) |
|---|---|---|---|
| ■ Consider materials and methods of construction that produce the minimum amount of waste. | ■ Draft SWMP identifying waste types ■ Record design stage considerations ■ Build waste management targets into tender specifications | ■ Regular toolbox talks with workers ■ Adequate ordering, delivery, and storage of materials ■ Update SWMP as waste is processed or removed | ■ Reconcile final waste data with SWMP ■ Calculate resource savings ■ Apply lessons learnt for future projects |

**Figure 4.4** *Waste minimization strategy from guidance to SWMP regulations 2008*

A greater level of detail is required for all projects with a start price of £500,000 or more.

The primary objectives of the Regulations are to improve resource efficiency in construction projects and reduce fly tipping.

Figure 4.4 shows the strategy recommended for a 'resource efficient' approach to waste management set out in DEFRA's guidance to the SWMP Regulations.[4] In summary, the SWMP Regulations require the following as a minimum:

- The 'client' is responsible for preparation of the SWMP before construction starts and ensuring that its requirements are incorporated into contract documents for appointment of the 'principal contractor'.
- A 'principal contractor' must be identified who takes ownership of the SWMP and is responsible for keeping it up to date, trickling its requirements down to subcontractors, ensuring direct labour is trained in its requirement and complying with the waste management licensing, waste duty of care and waste carrier registration regimes. This is likely to be the same individual identified as 'principal contractor' in the Construction (Design and Management) Regulations 2007.
- The plan will need to describe a set of 'waste management estimates' against which performance will be compared as the construction project progresses. First, the quantity of each type of waste likely to be produced on-site will have to be forecast, along with the proportion that will be re-used or recycled on-site, or removed from the construction site for re-use, recycling, recovery or disposal elsewhere. As a minimum the waste will have to be classified as inert, non-hazardous or hazardous (as defined in the Hazardous Waste Regulations 2005).
- During construction the principal contractor must update the plan as waste is disposed of, re-used or recycled. In this way the SWMP becomes a 'living' document that describes progress against the waste management forecasts contained in the Plan.
- The duty of care requires the principal contractor to take care of waste while it is in their control, which includes:
  - checking that the person to whom the waste is given is authorized to receive it, making sure they hold the appropriate licence or permit;
  - completing, exchanging and keeping waste transfer notes when the waste is handed over and, for projects of £500,000 and more, details of the destiny of waste once off-site; and:
  - taking all reasonable steps to prevent unauthorised handling or disposal by others.
- Performance against the SWMP should be reported following completion of the project. A more detailed analysis is required for projects of £500,000 or more, including an estimate of cost savings from site waste management.
- The SWMP must be kept for at least two years following completion of the project.

In summary the SWMP must contain as a minimum:

- Responsibilities:
  1   The client
  2   The principal contractor
  3   The person who drafted the plan

- Description of the construction works:
  1   The location of the construction site
  2   The estimated cost of the project

- Materials resource efficiency:
  1   Any decision taken before the SWMP was drafted to minimize the quantity of waste produced on-site.

- Waste management:
  1   Description of each waste type expected to be produced during the project
  2   For each waste type estimate of the quantity of waste that will be produced
  3   For each waste type the waste management action proposed (including re-use, recycling, other types of recovery and disposal)

- Waste controls and handling:
  1   A declaration that all waste produced on the site will be dealt with in accordance with the waste duty of care
  2   A declaration that materials will be handled efficiently and waste managed appropriately

The Building Research Establishment (BRE) has developed a scheme called SmartWaste that includes a tool (SmartWaste Plan) that supports the development of a SWMP.[5] It uses benchmarks from a database of past projects to enable prediction of waste arisings for new build projects in terms of volume per 100m² floor area or £100,000 project value. Refer to Table 4.1 for BRE's averaged figures for various different building types.

BRE has also developed for DEFRA the 'true cost of waste calculator' that 'calculates the carbon impact of wasted construction materials using embodied carbon data from Life Cycle Analysis (LCA) for the materials listed. LCA is an objective process to evaluate the environmental burdens associated with a product, process or activity by identifying energy and materials used, emissions and wastes released to the environment.[6] From the total weight of material to be used it estimates the total weight of wasted material from default values, along with the lost embodied carbon and costs of wasted material, labour and materials disposed of.

BRE has also developed a directory of UK waste management facilities that can be located on-line within a specified distance from a given postcode.[7]

BREEAM and CSH include credits for Construction Site Waste Management that go beyond the minimum requirements of legislation. For an office building the targets set by BRE are between 13 and 16.6m³ per 100m² floor area for 1 credit (compared to 15m³ for an average office building in Table 4.1) and 9.2m³ per 100m² for 3 credits. CSH, on the other hand, awards a credit for setting targets using the Construction Excellence Key Performance Indicators.[8] This

**Table 4.1** *Waste benchmarking data for new build projects*

| Project type | Number of projects data relate to | Average m³/100m² | Number of projects data relate to | Average m³/£100K |
|---|---|---|---|---|
| Residential | 116 | 15.2 | 112 | 18.3 |
| Public buildings | 6 | 26.1 | 8 | 22.2 |
| Leisure | 3 | 12.3 | 5 | 20.6 |
| Industrial buildings | 5 | 20 | 5 | 11.3 |
| Healthcare | 14 | 15 | 12 | 13.4 |
| Education | 20 | 13.4 | 21 | 17.3 |
| Commercial retail | 24 | 20.1 | 22 | 14.9 |
| Commercial offices | 27 | 15 | 24 | 10.4 |
| Civil engineering | 9 | 24.3 | 6 | 20.3 |
| Total number of projects | 224 | | 215 | |

*Source:* www.smartwaste.co.uk/filelibrary/benchmarks%20data/Waste_Benchmarking_Data_for_new_build_projects_only__updated_31_Aug_2008_by_project_type.pdf

can be used to benchmark and target energy/carbon emissions, mains water use and commercial vehicle movements as well as waste volumes removed from the site and to landfill.

Credits are available for using best practice to reduce waste volumes and diverting 'key waste groups' from landfill. These are defined in the 2002 European Waste Catalogue (EWC) which categorizes different types of waste under various industry groups, assigning each group with a unique reference number.[9] Relevant groups are covered by section 17: Construction and Demolition Wastes and include bricks, concrete, gypsum, packaging, etc. BRE add a number of items not included such as oils, liquids and hazardous waste; the latter being covered under separate legislation of course.

Where the project involves demolition or refurbishment BREEAM/CSH rewards an audit of available materials with the aim of identifying those that can be recovered and either re-used or recycled on-site or sold on for reclamation.

LEED similarly rewards the diversion of construction and demolition debris from disposal in landfills and incinerators, the redirection of recyclable recovered resources back to the manufacturing process and reusable materials to appropriate sites.[10]

## Waste reduction and recovery

A strategy for reducing construction waste has already been discussed in the sections dealing with purchasing strategy and pre-fabrication in Chapter 4.2 Sustainable Procurement. However, the full list of waste groups given in BREEAM and CSH needs to be reviewed in order to ensure that all opportunities for reducing waste have been explored.

Data from BRE indicates that for residential projects the dominant waste groups by volume per 100m² floor area, in order of magnitude, are packaging, concrete, site buildings, plaster/cement, timber, ceramics/bricks, plastics and insulation:

## Packaging

Packaging comprises some 17.4 per cent of the total waste by volume and comprises a mix of cardboard, polythene, polystyrene, plastics and timber, much of which can be recycled. However, a major limitation to the recycling of packaging waste is the space required to allow appropriate segregation and storage. This is compounded by the variety of packaging wastes arising. Materials such as pallets can be segregated and recovered through a well-developed recovery infrastructure. However, packaging comprising plastic, hessian bags and cardboard can be difficult to store, although they can be compacted or baled using on-site facilities to reduce space requirements. Where specific waste packaging is generated in significant quantities, they should be segregated for collection by recycling companies. Envirowise has developed a free resource to assist in the development of a strategy to reduce packaging waste.[11] It includes posters, guidelines for toolbox talks, a suite of checklists and an Excel spreadsheet to predict volumes of packaging waste that will be generated.

Waste can be reduced by choosing suppliers that do not use excessive packaging and/or take back packaging material for reuse. Off-site construction can reduce the packaging required for transportation of the individual components to site.

## Concrete

Concrete waste represents some 14.3 per cent of residential construction waste according to the BRE figures referred to above.

The main cause of concrete wastage is inaccurate ordering of quantities. The practice is a nuisance to ready-mixed concrete suppliers, who have to find a use for unwanted loads of concrete. Few plants have the resources to deal with this waste in their yard, and it is difficult to match highly specific mix types with suitable customers at short notice. Most ready-mixed concrete companies charge their customers a premium for returned concrete. This can be as much as £100 per cubic metre. Avoiding this waste should result in savings through reduced water consumption, savings in time required to wash out and handle waste, lower waste disposal and wastewater discharge costs, reduce build-up in mixer drums and lower exposure to substantial environmental fines and liabilities.

## Site buildings and fencing

In most instances there should be no reason why site facilities and fencing are thrown away at the end of a project. These should be robust and easily transportable modules manufactured from sustainable timber and moved from site to site.

## Plasterboard

In the UK there has been particular concern about the amount of plasterboard wasted, which has been some 300,000 tonnes of the 2.5 million tonnes of new plasterboard installed per annum, with up to a further 1 million tonnes of material sent to landfill from demolition and refurbishment. Clearly there has been a massive amount of over-ordering from contractors, whilst an increase in pre-fabrication would significantly reduce wastage. Under the Ashdown Agreement, many manufacturers have signed up to recycling plans that require on-site separation of plasterboard waste for distribution to designated recycling facilities.[12]

## Timber

According to the WRAP, 'approximately 50 million cubic metres of timber is used in the UK annually'. Our per capita consumption is one of the highest in the world. The construction industry is by far the largest consumer, using up to 70 per cent of the softwood consumed in the UK. A further 16 per cent is used in packaging such as pallets and packing crates.

'Of the wood that enters UK construction sites, 39 per cent leaves as waste. It is estimated that approximately four million tonnes of waste wood are generated from construction and demolition sites, together with over one million tonnes from packaging. Virtually all timber has the potential for some kind of recovery, however, of the waste currently produced, just 15 per cent is re-used and 31 per cent recycled, while 37 per cent is landfilled and 17 per cent incinerated.'[13]

Actions which could minimize wood waste include:

- Pallets should be reused or returned to suppliers. Broken pallets should be mended if possible rather than disposing of them as waste (see Packaging above).
- Wood can be easily damaged during handling or storage. Designated protected storage areas and training in the correct handling of materials can minimize accidental damage.
- Wood waste can be reduced by up to 40 per cent by using pre-fabricated timber frames, walls and flooring for houses, flats and small commercial buildings (see Chapter 4.2 Sustainable Procurement). Waste can be more easily controlled in a factory environment and panels can be tailored, removing the generation of off-cuts on-site and reducing the amount of over-ordering.
- Better quality control and more accurate measurement on-site should result in optimized cutting patterns for boards and panels. Careful positioning of windows and doors should also reduce wastage from off-cuts.

## Ceramics, bricks and blocks

The ceramics waste stream includes damaged sanitary ware and tiles, for which the most common cause of waste is damage during handling, storage and after installation. This can be reduced significantly through a combination of just-in-time delivery, better protected storage and improved site procedures.

Brick and block waste can also be reduced through careful storage, reducing the amount of damaged material. Bricks are classified as non-hazardous inert waste and have a lifetime of greater than 200 years, so undamaged material has commercial value and can often be reused. Bricks and blocks that are damaged can be recycled as aggregate for use as general fill, highway sub-base, etc. New bricks should be specified that have a high content of recycled or secondary materials, such as fly ash and recycled glass.

## Plastics

Plastics are used in pipework, insulation, wiring, windows, flooring, wall coverings and packaging. The construction industry consumes 23 per cent of the 5 million tonnes of plastics used in the UK, of which only 19 per cent is recovered. Of the total volume of packaging used for construction products, 25 per cent is manufactured from plastic.

Not all plastics can be recycled and separate processes are required for different types. However, the most common plastics, including PVC, polyethylenes, polypropylene, polystyrene and ABS copolymer can be recycled.[14]

The most common sources of waste are over-ordering and damage after delivery. As with other materials, the waste reduction strategy should comprise reducing margins when ordering, just-in-time delivery, pre-fabrication and separating waste for re-use or recycling.

## Insulation

Insulation is used for reducing heat flow, absorbing sound and fire protection and the materials used vary considerably in their characteristics and the amount of waste generated. The range of materials used include fibre glass, mineral wool, polystyrene, foams such as polyurethane, sprayed foams and granules, sheep's wool and recycled paper. They are available in flexible sheets, rigid boards and in-situ blown material. The type of insulation used determines how much waste is generated during installation. Rigid insulation is estimated to produce 10–15 per cent waste, flexible 8 per cent and blown 5 per cent.

The following methods are recommended to reduce waste:

- Use of pre-formed materials cut to the required dimensions and shape. Waste can be more easily managed in a factory environment.
- Just-in-time ordering can reduce over ordering and minimize handling. Unused surplus material is often disposed of rather than re-used on the next project as it is not seen as economic to transport and store it.
- Careful storage and handling. Rigid boards in particular can be damaged by poor storage and handling. Designating a secure storage area can protect materials from weather and accidental damage. Off-cuts can also be stored for reuse.
- Re-use of off-cuts. This is easier with flexible materials such as mineral wool than with rigid boards. Partially used rolls of insulation can be used in another project.

## Glass

Approximately 500,000 tonnes of flat glass waste are produced each year from the UK construction industry, of which 20–30 per cent is recycled. The most common causes of wastage are, not surprisingly, breakage and over-ordering. With such a fragile material the benefits of just-in-time ordering and protective storage could not be more obvious. The processes for separation and collection for recycling are well-established.

## Recycling demolition material

We have already discussed the use of recycled aggregates from the perspective of the designer under the heading of Re-use and recycling in Chapter 3.12 Materials Specification. As we saw, it is recognized in both BREEAM and LEED, provided certain criteria are met. The biggest challenge for the constructor is ensuring that demolition material is of the correct size and quality for use in what BREEAM defines as 'high grade' aggregates.

Approximately 275 million tonnes of aggregates are used each year in the UK as raw construction materials, with around 70 million tonnes derived from recycled or secondary sources.

Recycled aggregates are derived from material arising from construction and demolition (concrete, bricks, etc.), highway maintenance (asphalt planings), excavation and utility operations. If not produced on-site, they can be obtained directly from demolition sites or from suitably equipped processing centres. The quality of the recycled aggregate is dependent upon the quality of the materials that are processed, the selection and separation processing used, and the degree of final processing that these materials undergo.

Significant cost savings are possible with on-site sorting and crushing, including reduced transport costs, along with the environmental benefits from reduced lorry movements. However, if it is necessary to purchase or hire crushing plant and/or lease a nearby site to locate it, the cost savings may be eroded considerably. Crushing demolition material is a noisy process and the plant takes up considerable space, so it may be difficult to arrange for on-site aggregate recycling where there are major space restrictions. Figure 4.5 shows a typical crushing plant.

Secondary aggregates (post-industrial or post-consumer) are derived from a very wide range of materials. Many arisings of secondary materials have a strong regional character. For example, china clay sand in Southwest England, slate waste in North Wales, and metallurgical slag in South Wales, Yorkshire and Humberside.[15]

WRAP launched its Aggregate Programme funded from the Aggregates Levy Sustainability Fund in 2002.[16] It provides capital funding to companies seeking to set up or expand aggregate recycling facilities and includes a residual value guarantee scheme (eQuip) on the aggregate recycling equipment purchased for leasing.

Most of the components of an existing building have the potential for being recovered, either for re-use or processing and recycling. For example, one recent project for the Glasgow Housing Association recovered more than 95 per cent of demolition waste transported to the contractor's off-site sorting and recycling facility.[17]

**Figure 4.5** *Typical crusher for recycling demolition material*

*Source:* Author

## Notes

1   Office for National Statistics at http://www.statistics.gov.uk/cci/nugget.asp?id=1304.
2   www.epa.gov/osw/rcc/resources/action-plan/act-p2.htm#beneuse1c.
3   www.epa.gov/epawaste/conserve/rrr/imr/cdm/programs.htm.
4   www.defra.gov.uk/environment/waste/topics/construction/pdf/swmp-guidance.pdf.
5   www.smartwaste.co.uk/page.jsp?id=1.
6   www.wastecalculator.co.uk/login.jsp.
7   www.bremap.co.uk/bremap/advancedSearch.jsp.
8   The Construction Excellence Key Performance Indicators can be purchased as a CSH KPI Toolkit from www.constructingexcellence.org.uk/tools/sustainablehomes/.
9   http://environmentagency.info/static/documents/GEHO1105BJVS-e-e.pdf.
10   Scotland has its own construction waste management strategy which is promoted through the Waste Aware Construction website at www.wasteawareconstruction.org.uk/.
11   See www.envirowise.gov.uk/uk/Our-Services/Publications/GG606-Managing-packaging-waste-on-your-construction-site.html.
12   www.wrap.org.uk/construction/construction_materials/plasterboard_gypsum/index.html.
13   www.wrap.org.uk/construction/construction_materials/wood/index.html.
14   The European Recovinyl initiative has resulted in sites being established for recycling PVC across Europe, including 30 sites in the UK (www.recovinyl.com/).
15   Refer to WRAP's aggregate recycling website at http://aggregain.wrap.org.uk/.
16   www.wrap.org.uk/downloads/WRAP_Aggregares_Programme_2_.a20aea8f.4078.pdf.
17   www.wrap.org.uk/downloads/Demolition_Glasgow_Housing_Association.9ab98776.4874.pdf.

## References

*Construction (Design and Management) Regulations 2007*, No 320. Available at www.opsi.gov.uk/si/si2007/uksi_20070320_en_1 (accessed 30 March 2010)

*Environmental Permitting (England and Wales) Regulations 2007*, No 3538. Available at www.opsi.gov.uk/si/si2007/uksi_20073538_en_1 (accessed 30 March 2010)

*Hazardous Waste Regulations 2005*, No 894. Available at www.opsi.gov.uk/si/si2005/20050894.htm (aAccessed 30 March 2010)

*Site Waste Management Plans Regulations 2008*, No 314. Available at www.opsi.gov.uk/si/si2008/uksi_20080314_en_1 (accessed 30 March 2010)

# 4.4
# Considerate Contracting and Construction Impacts

## Introduction

A building site can be a hostile place and its impact on the environment and its neighbours in particular is likely to be orders of magnitude greater than the completed building. It is the responsibility of the 'principal contractor', usually the site manager, to minimize these impacts to an acceptable level. There are numerous activities carried out on any given site, from making tea to piling, which vary enormously in their potential impact. Many of these activities are potentially disruptive and some can involve generating noise or pollution to land, watercourses or air. Some of these are governed by statute and, in England and Wales for example, will fall under the ambit of the Environmental Protection Act and require permission under the Pollution Prevention and Control Regulations with licences issued by the Environment Agency. Others will fall under the Health and Safety at Work, etc. Act and the Construction (Design and Management) Regulations.

A significant proportion of the embodied carbon of the development is generated from site activities and deliveries, whilst large quantities of water can be consumed by the 'wet' trades and in controlling dust.

In this chapter we will consider two interrelated methods of assessing construction impacts and the measures required by contractors to control them. The first deals with consideration: to the environment, to neighbours and to the public at large. The second looks at measurable environmental impacts associated with site operations and deliveries to the site. These are both assessed under BREEAM and the CSH, whilst LEED includes pollution control of site activities as a prerequisite.

## Considerate contracting

The City of London launched its pioneering Considerate Contractor Scheme in 1987, which has since become mandatory for all construction projects within the City.[1] At about the same time the City of Westminster set up its own Considerate Builder Scheme, which is still going strong.[2]

However, the reputation of the construction industry in Britain remained at a very low ebb, leading to the 1994 Latham Review 'Constructing the Team', from which was established an Implementation Forum in 1996. This Forum, learning from the established schemes run by the Cities of London and Westminster, developed a Code for Considerate Practice which became the Considerate Constructors Scheme (CCS) in 1997.[3]

In the same year the Office of the Deputy Prime Minister (ODPM) set up the Construction Task Force under the chairmanship of Sir John Egan against the backdrop of a chronically under-performing construction industry. His influential report 'Rethinking Construction', published a year later, led to a number of important initiatives, including the quasi-governmental Constructing Excellence organization which was 'charged with driving the change agenda in construction'.[4]

Both BREEAM and CSH reward compliance with any of the above schemes, the default being the CCS, which is applicable for any site in the UK. CCS is reviewed annually and it is important to use the most recent checklist when establishing what actions and resources are required to achieve the targeted score under the scheme. The most recent Site Registration Monitors' Checklist can be downloaded from the CCS's dedicated website.[5]

CCS is a voluntary scheme that encourages the considerate management of construction sites. The scheme is operated by the Construction Confederation and points are awarded in increments of 0.5 over the following eight sections:

- considerate;
- environment;
- cleanliness;
- good neighbour;
- respectful;
- safe;
- responsible; and
- accountable.

To achieve certification under this scheme a score of at least 24 is required with a minimum of 3 points under each heading. Both BREEAM and CSH award one credit if a score of 24 to 31.5 is achieved and two credits for a score of 32 or more. In addition BREEAM requires a score of at least 36 (out of 40) as a prerequisite for an overall BREEAM Exemplary rating. The Construction Federation rewards the top 7.5 per cent of schemes every year with bronze, silver and gold awards.

The scheme uses trained 'monitors' to review the performance of a site against the checklist referred to above and based typically on two visits. Outlined below are the issues that they would consider and what they might be looking for. Each issue needs to be validated by reference to clear and verifiable records. Although there are five points available under each heading, they would expect quite exceptional performance under a heading to consider awarding the full five points. The first points bulleted under each heading are the minimum required to achieve 3 points. All references below are from the 2010 CCS Site Registration Monitors' Checklist:

## Considerate

For points to be awarded under this heading, evidence is required that those working on the site are 'fully aware of all those who may be affected by the work and ... what efforts are made to minimize any nuisance or inconvenience'. As a minimum the evidence must be available that:

- all those affected by site activities have been informed of the project and provided with information on the type of activities and programme;
- if roads and footpaths have to be obstructed, how this is to be monitored and what arrangements provided for minimizing inconvenience to pedestrians etc.;
- road names and other signs are still clearly visible;
- diversions are clear to pedestrians, cyclists and motorists;
- unloading and parking is managed around the site to minimize disruption;
- impact of site traffic on vehicle routes, including access for deliveries, is minimized through an on- and off-site Traffic Management Plan;
- measures are in place to reduce or remove the negative impact of smoking on the image of the site.

Further points can be gained for:

- clear identification of site access, with a system in place to ensure that all visitors are provided with directions to the site;
- information provided in other languages where relevant to those affected by the site;
- the provision of an access route to the site office from the site entrance suitable for use by people of all abilities, or reasonable alternative arrangements can be made through contact with the Site Manager;
- site neighbours are provided with questionnaires during or on completion of the project, with feedback analysed to improve future performance.

## Environment

The points under this heading are primarily to reward environmental awareness and ensure that measures are in place to minimize impacts and make a positive contribution (where possible). Measures should address energy, waste, pollution, resources, ecology, etc. As a minimum, evidence is required that:

- the main contractor has an environmental policy which is clearly displayed;
- site specific environmental issues have been identified, with evidence of the control measures used and how their effectiveness is being monitored;
- measures are taken for reducing energy and water consumption, with records to demonstrate how effective they have been;
- measures are in place for avoiding, reducing, re-using and recycling waste, including a Site Waste Management Plan (SWMP) and required monitoring and records (a legal requirement – see Chapter 4.3 Construction Waste Management);
- re-used, recycled and sustainable materials and products have been sourced (Chapter 4.2 Sustainable Procurement);
- waste is diverted from landfill;

- plants, trees, watercourses and wildlife have been identified and protected, and that this is monitored (see Impact on site ecology below);
- that procedures are in place to reduce or eliminate noise, light, water and air pollution (see Construction impacts below).

Most of the above are covered in more detail in both BREEAM and CSH under the heading of Construction Site Impacts.

Further points can be gained if it can be demonstrated that:

- efforts have been made to use local labour, suppliers, subcontractors and materials;
- environmental issues are covered in site induction and toolbox talks; hazardous substances, such as oils, paints and chemicals are stored in secure and designated areas, with suitable equipment (and procedures) available for dealing with spills (Control of Substances Hazardous to Health (COSHH) Regulations may apply);
- alternative energy sources have been considered for the construction process;
- targets are in place for improving and monitoring environmental performance;
- the site's carbon footprint is known and measures are in place to reduce it;
- activities have been undertaken that give environmental and ecological benefits to the area, with suitable publicity and promotion;
- efforts have been made to source materials and products with low embodied energy and to monitor and record this (Chapters 3.12 Materials Specification and 4.2 Sustainable Procurement).

## Cleanliness

All of these points relate to site housekeeping, which also impacts on safety and waste management. As a minimum it must be ensured that:

- the first impression is of a clean, tidy and well presented site;
- procedures are demonstrably in place to maintain the cleanliness of:
  - the site perimeter, including, where necessary, collection of non site-related rubbish;
  - roads and access points are kept clean and mud-free;
  - site office, welfare facilities and around them;
  - work area of the site;
- skips are tidy, not overfilled and covered when not in use;
- dust prevention measures are in place and can be demonstrated to be effective;
- storage of plant and materials is tidy and gives a good impression of the site;
- graffiti and vandalism is monitored and addressed on the site and at the boundary.

In addition:

- operatives are discouraged from dropping litter, with suitable procedures in place for monitoring and addressing any problems identified;
- arrangements are in place for on-site waste collection and management and are demonstrably effective. This should be in the SWMP (Chapter 4.3 Construction Waste Management).
- procedures are in place for encouraging operatives to keep site facilities clean;
- appearance and cleanliness of site plant and vehicles is considered.

## Good neighbour

This heading deals with the quality of communication between the site and those interested or affected by the activities thereon and what impression will be left after the project is completed. The following issues must be addressed:

- the measures taken to reduce noise and inconvenience affecting neighbours, including limiting working hours if necessary (noise impact is discussed further under Construction impacts below);
- hoarding/fencing is sensitive to the site surroundings and well maintained;
- neighbours are notified prior to noisy or disruptive work, or changes in site activity that impact on them.
- arrangements are in place to enable a telephone response during working hours;
- procedures are in place to respond to and record compliments, comments or complaints, with suitable contact details;
- complaints are properly dealt with by the site manager;
- operatives are informed of relevant compliments received from the public.

In addition points can be gained if:

- measures have been taken to enhance the image of the site and to respect neighbours' privacy, particularly where scaffolding is used;
- active involvement with the community has been facilitated, such as public relations events;
- regular progress updates are made available to interested and affected parties;
- viewing points are provided in the site hoarding with regular checks to ensure that the view gives the intended impression;
- site lighting is not directed into neighbours' properties (or create hazardous glare for road users – refer to section on Light pollution in Chapter 3.6 Light and Lighting);
- a 24-hour 'hotline' is displayed on the site boundary or otherwise easily available to the public, directly advised to neighbours and monitored to ensure suitable responses;
- goodwill gestures have been made (such as funding community projects, improvements to local facilities, a gift of public art, etc.);

- a positive influence has been exerted in the area (for example projects with a benefit to local community).

## Respectful

This topic is looking for a culture on-site that results in all involved with site activities demonstrating a positive image of their company and the construction industry as a whole. It is expected that this will be shown in the condition of the site and accommodation thereon. The following must be achieved for three points to be awarded:

- adequately sized and equipped welfare facilities provided within a reasonable distance of the work area;
- welfare facilities and especially toilets are screened from public view;
- operative appearance is appropriate and monitored, giving a professional impression, including a requirement to 'cover up' in summer;
- induction procedure provides rules dealing with shouting, abusive and offensive language and requires courteous behaviour towards the public;
- measures are in place to prevent the display of potentially offensive material;
- appropriate restrictions are implemented and monitored on the use of music players, mobile phones and, where necessary, cameras;
- suitable changing, drying and canteen facilities are provided;
- site personnel are required to be in reasonably clean dress for use in off-site facilities such as shops and public transport;
- operatives are discouraged from taking their breaks in public view.

Further points can be gained by introducing the following measures:

- free provision of branded company work-wear to all workers;
- action to protect operatives from excessive exposure to the sun;
- provision of clean personal protective equipment (PPE) for use by site visitors (in a range of sizes);
- cater for the reasonable requirements of site visitors of all abilities and gender;
- provide lockers as appropriate and introduce a system to encourage and manage their use;
- provide showers as appropriate and introduce a system to encourage and manage their use.

## Safe

Safety is a fundamental requirement of site management covered by the Health and Safety at Work, etc. Act and CDM Regulations. These points are awarded primarily for the quality of safety culture on-site and the way this is communicated to the outside world The HSE website provides a guide to UK health and safety law.[6] Key actions that must be demonstrated are as follows:

- implementation of the Construction Phase Health and Safety plan, including its periodic review and audit;
- access to site accommodation is protected and segregated from site vehicle access, well-lit and signposted and kept usable at all times;
- measures in place to ensure the safety and security of pedestrians and road users at the site boundary;
- temporary works outside the site boundary have been assessed for all security and safety risks and all preventive measures instigated;
- plant movements outside the site boundary are monitored and managed to ensure public safety;
- measures are in place to ensure the public is protected from falling debris;
- a system is in place for recording accidents and near misses and managing outcomes (this is a legal requirement under the Reporting of Injuries, Diseases and Dangerous Occurrences Regulations, 1995 (RIDDOR), and serious accidents have to be reported to the HSE);
- types and causes of accidents and near misses are analysed and action taken to prevent similar occurrences in the future.

Further points can be obtained if evidence can be provided that:

- temporary road crossings are provided where necessary, along with ramps (suitable for wheelchair use) where works create a break in the pavement;
- health and safety information is provided to operatives and visitors (as required under the The Health and Safety Information for Employees Regulations 1989);
- scaffolding is boxed in, protected or taped and well maintained where likely to affect pedestrians;
- the improvement of safety on-site is promoted and incentivized;
- emergency procedures are up-to-date and practised at suitable intervals;
- it is established whether all operatives can read and understand English (if not arrangements will need to be made to cater for non-English speakers).

## Responsible

This heading covers the responsibility of site management as an employer to the operatives and to the public in general. It also looks at the role contractors can play in the recruitment and training of the construction industry's future workforce. As a minimum evidence should be provided that the following arrangements are in place:

- site personnel have been made aware of the location of the nearest accident and emergency hospital and Minor Injuries Unit, with this information clearly displayed for use during an emergency;
- trained first aiders have been identified, their details clearly displayed and approved first aid equipment kept in accessible locations (as required by the Health and Safety (First Aid) Regulations, 1981);
- sub-contractor first aiders are acknowledged and identified;
- all visitors are inducted and escorted when on-site;

- inductions are site specific;
- all operatives' skills are recorded, including enrolment in the Construction Skills Certification Scheme (CSCS) and skills card type;
- operatives' medical conditions and contact details are recorded and available at short notice in the event of an emergency;
- procedures require all visitors to sign in and details from their CSCS card recorded, if applicable.

In addition points can be gained if evidence is provided that:

- drivers are required to sign in and details from their CSCS cards recorded, if applicable;
- appropriate measures are in place for out of hours security, including access arrangements;
- contacts have been made with local schools, colleges or universities to arrange activities, events or visits relating to safety and/or careers in the construction industry;
- opportunities for apprenticeships, work experience or placements on the site have been identified, where possible;
- operatives are provided with appropriate personal identification where compatible with the nature of their work;
- equal opportunities and diversity policies are in place and evidence is available on how they are implemented at site level;
- opportunities are provided for operatives to improve their literacy and numeracy, where required;
- occupational health advice is provided to operatives;
- all operatives are covered by drugs and alcohol policies;
- information is available about the site on-line.

## Accountable

This heading deals with promotion of the project to the wider public, the accountability and accessibility of the contractor and the creation of a sense of pride in working in construction amongst the workforce. In order to demonstrate these achievements it is expected that:

- the site manager is aware of the expectations from being registered under the CCS and the efforts being made to ensure the core requirements are met;
- CCS posters and banners are displayed and correct information provided;
- a system is in place for making site personnel, including subcontractors and consultants familiar with CCS, its requirements and their responsibilities under the scheme;
- training needs have been identified for everyone on-site;

Further points can be gained if evidence is provided that:

- the main contractor's contact information is prominently and publicly displayed, including on CCS posters;

- records of CCS activities are identifiable and easily accessed;
- it is clear how the main contractor's head office personnel are involved in supporting the scheme, including directors and management;
- the benefits of scheme registration have been promoted to the site's neighbours;
- site signage is clearly visible at night;
- it is clear how the client is involved in supporting the scheme;
- particular attention is being given to working with utility companies involved in the project to promote its image.

In order to do exceptionally well under the scheme, the CCS Monitor will be looking for initiatives and innovations that make the project stand out, as well as very few complaints from the public, a low turnover of staff, a low accident rate and a good uptake of the CSCS scheme.

## Construction impacts

The emissions associated with construction activities are addressed in a number of ways. They are assessed under various headings in the EIA submitted with the planning application, whilst the carbon emissions associated with delivery and site activities form part of the life cycle assessment and carbon footprint of the development. In this section we will deal with activities that can be controlled, monitored and recorded by the contractor and that are covered by the Construction Site Impacts credits in BREEAM and CSH. Some of these impacts are regulated and in some cases require licensing or permits and these will be discussed in context.

The issues covered by BREEAM and CSH fall under the headings of:

- Monitoring, reporting and setting targets for:
  - $CO_2$ emissions associated with site activities;
  - $CO_2$ emissions associated with transport to and from site;
  - water consumption associated with site activities.
- Implementation of best practice policies to minimize:
  - air (dust) pollution from site;
  - water (ground and surface) pollution from site activities;
  - use of sustainable site timber.

BREEAM also requires the contractor to have a policy for the sourcing and procurement of 'environmental' materials and an EMS in place.

### $CO_2$ emissions associated with site activities

The UK Government's Strategy for Sustainable Construction calls for a '15% reduction in carbon emissions from construction processes and associated transport compared to 2008 levels' (BERR, 2008).

BREEAM and CSH are looking for evidence that targets have been set for $CO_2$ emissions associated with energy use arising from site activities using the key performance indicators established by Constructing Excellence and available by subscription from the KPI website www.kpizone.com/ (also known as the DTI Environmental Key Performance Indicators). These targets only cover electricity and gas consumption and do not include fuel consumed by site vehicles. BRE require regularly updated meter readings displayed in the site office as a graphical representation of $CO_2$ emissions against time. However, since the $CO_2$ emissions associated with site vehicles are required to calculate the site carbon footprint (see the section on CCS above), it is strongly recommended that fuel use is monitored and targets set based on a continual reduction month on month. BRE provide the following conversion factors for $CO_2$ emissions associated with fuel use:

| | |
|---|---|
| Petrol | 2.30kg$CO_2$/litre |
| Diesel | 2.65kg$CO_2$/litre |
| Compressed natural gas | 2.65kg$CO_2$/kg |
| Liquid petroleum gas | 1.49kg$CO_2$/litre |

The conversion factors for mains electricity are given in Chapter 3.2 Operational Energy and Carbon as:

| | |
|---|---|
| Electricity | 0.591kg$CO_2$/kWh |
| Gas | 0.206kg$CO_2$/kWh |

These figures are correct for the UK in 2010, but will vary from country to country and as the contributions from different fuels in generation of electricity changes with time.

The carbon footprint calculation will also require an estimate of the fuel used by operatives commuting to site by car or public transport that can be estimated using various carbon footprinting and reporting protocols, and conversion factors produced by AEA Environment for DEFRA (DEFRA, 2008).

BRE does not require targets to be met to achieve the post-construction credits, however, the aim of these credits is to establish a site culture that encourages improvement in environmental performance.

## $CO_2$ emissions associated with transport to and from site

The site manager will need to establish a procedure for monitoring and recording deliveries that includes the mode of transport and the distance travelled for each delivery. In order to estimate the $CO_2$ emissions associated with deliveries, it is necessary to use the conversion factors in the DEFRA document referred to above and included in the guidance to the CSH, which includes factors for various types of commercial vehicle, fuel type and percentage of maximum laden weight, reproduced in Table 4.2 and 4.3.

The distance used to estimate fuel consumption will depend on whether the delivery is specifically for the site, in which case the round trip distance is used, or whether multiple drop-offs have been made, in which case the distance from the last delivery and to the next must be recorded.

Targets for $CO_2$ emissions associated with deliveries can be developed from the Constructing Excellence/DTI Environmental KPIs and tailored for the project.

Performance in this category is highly dependent on the success of the design team and contractor in identifying suppliers for construction materials

**Table 4.2** *Conversion factors for vans*

Van/light commercial vehicle road freight mileage conversion factors: vehicle km basis

| Type of van | Gross vehicle weight (tonnes) | Total vehicle km travelled | Multiply by a factor of | kg $CO_2$ per vehicle km | Total kg $CO_2$ |
|---|---|---|---|---|---|
| Petrol | up to 1.25 | | × | 0.224 | |
| Diesel | up to 3.5 | | × | 0.272 | |
| LPG or CNG | up to 3.5 | | × | 0.272 | |
| Average | up to 3.5 | | × | 0.266 | |
| Total | | | | | |

*Source:* DEFRA (2008).

and products that are located close to the site. For example, the BedZED project in Beddington had a policy of purchasing locally where possible, whilst reclaimed structural steel was obtained from demolition sites within 35 miles of the development (Lazarus, 2002).

## Water consumption associated with site activities

The UK Government's Sustainable Construction Strategy includes a target for 'water usage in the manufacturing and construction phase to be reduced by 20% (by 2012) compared to 2008 usage'. Targets for site water consumption taken from meter readings can also be developed from the Constructing Excellence/DTI Environmental KPIs.

Water wastage on construction sites can be very high, particularly during the summer months when large volumes can be used for reducing dust generation or for washing wheels when site conditions are muddy. Procedures and training need to be in place to ensure that taps and stopcocks are not left open unnecessarily. Where possible, wastewater from canteen, shower and hand washing should be recovered for use on-site. Water fittings and appliances for canteens, showers and toilet fittings should be designed to minimize demand (Chapter 3.9 Water Conservation).[7]

## Dust pollution from site

BRE have developed best practice measures for the control of dust from construction and demolition activities (Kukadia et al, 2003). Dust generation is highly weather dependent and if possible the major dust generating activities such as demolition, crushing, excavation, concrete grinding, earth moving and sand blasting should not be carried out either when it has been dry for prolonged periods or during high winds. Water should be used to suppress dust during these operations and on dusty surfaces. Any stored or stockpiled materials that are prone to dustiness should be covered if possible, or stabilized through chemical binders, mulches or seeding. Structures being demolished should be sheeted and materials disassembled as far as possible before

**Table 4.3** *Conversion factors for HGVs*

**Diesel HGV road freight mileage conversion factors: vehicle km basis**

|  | Gross vehicle weight (tonnes) | % weight laden | | Total vehicle km travelled | × | kg $CO_2$ per vehicle km | Total kg $CO_2$ |
|---|---|---|---|---|---|---|---|
| Rigid | >3.5–7.5 | 0 | | | × | 0.525 | |
| | | 50 | | | × | 0.571 | |
| | | 100 | | | × | 0.617 | |
| | | 41 | *(UK average load)* | | × | *0.563* | |
| Rigid | >3.5–7.5 | 0 | | | × | 0.525 | |
| | | 50 | | | × | 0.571 | |
| | | 100 | | | × | 0.617 | |
| | | 41 | *(UK average load)* | | × | *0.563* | |
| Rigid | >7.5–17 | 0 | | | × | 0.672 | |
| | | 50 | | | × | 0.768 | |
| | | 100 | | | × | 0.864 | |
| | | 39 | *(UK average load)* | | × | *0.747* | |
| Rigid | >17 | 0 | | | × | 0.778 | |
| | | 50 | | | × | 0.949 | |
| | | 100 | | | × | 1.119 | |
| | | 56 | *(UK average load)* | | × | *0.969* | |
| All rigids | UK average | | | | × | 0.895 | |
| Articulated | >3.5–33 | 0 | | | × | 0.672 | |
| | | 50 | | | × | 0.840 | |
| | | 100 | | | × | 1.008 | |
| | | 43 | *(UK average load)* | | × | *0.817* | |
| Articulated | >33 | 0 | | | × | 0.667 | |
| | | 50 | | | × | 0.889 | |
| | | 100 | | | × | 1.111 | |
| | | 41 | *(UK average load)* | | × | *0.929* | |
| All artics | UK average | | | | × | 0.917 | |
| All HGVs | UK average | | | | × | 0.906 | |

*Source:* DEFRA (2008).

demolition. Walls and roofs should be guided so as to collapse inwards, with all surfaces being damped down. All skips and vehicles being used to hold dusty loose material should be securely sheeted and damped down during loading or unloading. All vehicles that drive through the construction site should have their wheels and body undersides washed before exit.

## Water pollution from site activities

Guidance has been developed by the Environment Agency, and its Scottish and Northern Irish equivalents, on the prevention of pollution to watercourses (EA, PPG1, PPG5, PPG6, undated). PPG6 specifically deals with 'working at construction and demolition sites' and the following key points should be addressed in developing a strategy for avoiding water pollution:

- Ensure that all contaminated land has been remediated, including underground storage tanks containing oil or hazardous substances.
- Ensure underground service zones have been identified and clearly marked prior to demolition or excavation; e.g. colour code manhole covers.
- If it is intended to discharge dirty water to a foul sewer, ensure approval has been obtained from the undertaker; ensure this cannot be discharged into dedicated surface water drains under any circumstances.
- Fresh concrete and cement are very alkaline and corrosive and can cause serious pollution in watercourses. It is essential to ensure that the use of wet concrete and cement in or close to any watercourse is carefully controlled so as to minimize the risk of any material entering the water, particularly from shuttered structures or the washing of equipment (EA, PPG5, undated).
- Wash waters from mobile pressure washers should not be discharged to surface water drains, watercourses or soakaways. Even if described as bio-degradable, detergents are not suitable for discharge to surface drains, so such activities should be carried out in designated areas draining to the foul sewer (EA, PPG1, undated).
- Any storage tank or drums used for the storage of oil, fuel or chemicals should be sited on an impervious base within a suitable bund having no drainage outlet. All fill pipes, draw pipes and sight gauges should be enclosed within the bund, and the tank vent pipe should be directed downwards into it.
- Refuelling of vehicles on-site should be carried out on a designated impermeable surface remote from any surface drains or watercourses.
- Wastes should be stored in designated areas isolated from surface drains. Used chemical containers may need special treatment and manufacturer's instructions should be followed. Used oil and fuel filters should be stored in a bunded area for separate collection.
- Water containing silt from wheel washing, etc. cannot be discharged directly into surface water drains, rivers or streams. EA prefers silty water to be discharged into a foul sewer, although approval may be obtained if the silt is allowed to settle in a lagoon, which may also be used for recycling purposes.

- Any spillages should be contained with an absorbent material such as sand or soil and the relevant agency notified immediately.

## Use of sustainable site timber

BREEAM and CSH both reward the use of timber from sustainably managed and legal sources (Tiers 1 and 2 under the Responsible Sourcing criteria – see Chapter 3.12 Materials Specification) for 80 per cent of uses during construction, including formwork, site hoardings, site huts and other temporary site timber used for the purpose of facilitating construction. This is an ideal opportunity to re-use timber from other sites.

## Environmental materials

BREEAM is looking for evidence that the main contractor has a policy to procure materials from environmentally sustainable sources and that this has been trickled down to subcontractors through contract obligations (Chapter 4.2 Sustainable Procurement).

## Environmental management

BREEAM rewards evidence that the main contractor operates an EMS covering its main operations. The EMS must be either third party certified, to ISO14001/ EMAS or equivalent standard, or have a structure in compliance with BS 8555: 2003. The EMS must have reached Phase 4 of the implementation stage – 'implementation and operation of the environmental management system' – and have completed phase audits 1–4, as defined in the British Standard.

# Noise from site activities

The noise generated from construction activities can be a major source of conflict with neighbours and has to be considered under the 'Good neighbour' heading in order to achieve CCS certification. There are no specific requirements set out in BREEAM or CSH, although it will have to be assessed to satisfy most local authorities. Many, such as the City of Westminster, have specific requirements.[8] These generally refer to BS 5228: 2009 which comprises codes of practice on assessing both noise and vibration (BSI, 2009a and b).

Under the Control of Pollution Act, 1974, local authorities can restrict hours of working and will require the contractor to demonstrate that appropriate controls are in place. Below are some typical requirements summarized from the City of Westminster information sheet:

- All equipment should be maintained in good mechanical order and fitted with the appropriate silencers, mufflers or acoustic covers.
- Stationary noise sources should be sited as far as possible from noise sensitive development, and where necessary, acoustic barriers used to shield

them. Such barriers may be proprietary types, or may consist of site materials such as bricks or earth mounds.

- Piling should be driven using the method that generates the least possible noise and vibration; sheet steel piling should be driven by vibratory jacking or box-silenced percussion systems.
- The movement of vehicles to and from the site must be controlled and should not take place outside the permitted hours unless with prior approval.
- Employees should be supervised to ensure compliance with the noise control measures adopted.

## Impact on site ecology

The impact of construction activities are included under the Environment heading of the CCS, but are primarily governed by the strategy adopted for the Ecology credits of BREEAM and CSH (see the Ecology and biodiversity section of Chapter 3.14 Landscaping, Ecology and Flood Risks). The contractor must put into place procedures for protecting those ecological features that are to be retained for the proposed development and monitor the activities of site operatives to ensure that no damage is done. This not only requires fencing of trees, hedges and waterways, but ensuring that they are not poisoned or polluted by site activities.

## Notes

1   www.cityoflondon.gov.uk/corporation/LGNL_Services/Business/Business_support_and_advice/considerate_contractor_scheme.htm.
2   www.westminster.gov.uk/services/business/businessandstreettradinglicences/highways/considerate_builders/.
3   www.ccscheme.org.uk/.
4   www.constructingexcellence.org.uk/aboutus/.
5   See   www.ccscheme.org.uk/images/stories/news_articles/news-2009/downloads/2010 checklist.pdf.
6   See www.hse.gov.uk/pubns/hsc13.pdf for a guide to UK health and safety law.
7   The Water Regulations Advisory Scheme provides advice on mains connection to construction sites and methods for avoiding backflow (www.wras.co.uk/PDF_Files/Construction%20 site%20advice.pdf).
8   www.westminster.gov.uk/services/business/noise_info_sheet2/.

## References

BERR (Department for Business, Enterprise and Regulatory Reform), 2008. *Strategy for Sustainable Construction*. London: BERR

BSI, 2009a. *BS 5228: 2009. Code of Practice for noise and vibration control on construction and open sites. Part 1: Noise*. London: British Standards Institution

BSI, 2009b. *BS 5228: 2009. Code of Practice for noise and vibration control on construction and open sites. Part 2: Vibration*. London: British Standards Institution

BSI, 2003. *BS 8555: 2003. Environmental Management System. Guide to the phased implementation of an environmental management system including the use of environmental performance evaluation.* London: British Standards Institution

DEFRA (Department for Environment, Food and Rural Affairs), 2008. *Guidelines to Defra's GHG Conversion Factors.* London: DEFRA. Available at www.defra.gov.uk/environment/business/reporting/pdf/ghg-cf-guidelines-annexes2008.pdf (accessed 26 March 2010)

EA (Environment Agency), undated. *Pollution Prevention Guideline, PPG1: General guide to the prevention of pollution.* Rotherham, UK: EA/SEPA/Environment and Heritage Service

EA, undated. *Pollution Prevention Guideline PPG5: Works in, near or liable to affect watercourses.* Rotherham,UK: EA/SEPA/Environment and Heritage Service

EA, undated. *Pollution Prevention Guideline PPG6: Working at Construction and Demolition Sites (under review).* Rotherham, UK: EA/SEPA/Environment and Heritage Service. Available at www.environment-agency.gov.uk/static/documents/Business/PPG_6.pdf (accessed 26 March 2010)

*Health and Safety (First Aid) Regulations 1981.* SI 1981 No 917. Available at www.uki.net/php/files/edp.uki.net/health%20and%20safety%20(first%20aid%20regulations).pdf (accessed 30 March 2010)

*Health and Safety Information for Employees Regulations 1989* SI 1989 No 682. Available at www.opsi.gov.uk/si/si1989/Uksi_19890682_en_1.htm (accessed 30 March 2010)

Kukadia, V., Upton, S. and Hall, D., 2003. *Control of Dust from Construction and Demolition Activities.* Watford, UK: Building Research Establishment

Lazarus, N., 2002. *BedZED: Toolkit Part I: A guide to construction materials for carbon neutral developments.* UK: Bioregional Developments

*Pollution Prevention and Control (England and Wales) Regulations 2001.* SI 2000/1973. London: HMSO

*Reporting of Injuries, Diseases and Dangerous Occurrences Regulations 1995.* Available at www.opsi.gov.uk/SI/si1995/Uksi_19953163_en_1.htm (accessed 30 March 2010)

# 4.5
# Value Engineering and Management

## Introduction

The UK Sustainable Procurement Task Group agreed a generalized definition of sustainable procurement which stated that 'Sustainable procurement is a process whereby organisations meet their needs for goods, services, works and utilities in a way that achieves value for money on a whole life basis in terms of generating benefits not only to the organisation, but also to society and the economy, whilst minimising damage to the environment' (DEFRA, 2007). For development of new federal buildings in the US the principle of the early integration of value engineering into the design process is well established. For example the introduction to a key guide to designers of federal facilities states that:

> Sustainable development as an integrated concept for buildings seeks to reverse the trends in the architectural and engineering communities that focus on first costs and treat each discipline's contribution to the whole building as separate, independent efforts. (FFC, 2001).

The assessment of value for money on a whole life basis requires ongoing management and scrutiny, of which value engineering and management are key components.

Value engineering (VE) is a structured process used to assess a project design to ensure that project goals are met efficiently and effectively. It is included here as part of the construction process because historically it has been driven by the contractor. This has resulted in the process being condemned as a vehicle for stripping cost out of the project so that contractors can increase profit. It is not unusual for VE to be used as a panic measure once it is realized that a project is going over budget. The principles should be integrated throughout the design process as life cycle or whole life costing. Although VE can be facilitated by the project cost consultant, there are merits in using an independent facilitator at key decision-making stages during the design and procurement processes. However, the later in the process the VE exercises are carried out the less the latitude for making fundamental changes to the design. This is another very powerful argument for the contractor to be involved from a very early stage of the design process (refer to Chapter 4.1 Tendering Process).

Because of the association of VE with cost minimization, the concept of 'value management' has been developed (Green, c1996). Green defines value management in the following terms:

> Value management is concerned with defining what 'value' means to a client within a particular context. This is achieved by bringing the project stakeholders together and producing a clear statement of the project's objectives. Value for money can then be achieved by ensuring that design solutions evolve in accordance with the agreed objectives. In essence, value management is concerned with the 'what', rather than the 'how'.

This is compatible with the philosophy of integrated sustainable design that runs through this book and that must not be lost through the construction process.

## Value management and sustainability

The template for value management should be shaped during the conceptual stages of the design and tested as the design develops through to production of tender packages and procurement of products and materials. This reflects the process for sustainability and environmental assessment so that all key decisions can be tested against a menu of key objectives, such as carbon impact, running cost, environmental impact, functionality, buildability, maintainability, reliability, life expectancy, programme implications, aesthetic value and BREEAM (or equivalent) rating. Other issues can be added to the list for specific items, such as sourcing from within a specified distance from the site, responsible sourcing Tier level, Green Guide Rating, etc. (see Chapters 3.12 Materials Specification and 4.2 Sustainable Procurement). Capital cost, life cycle assessment and hence value for money are then used to evaluate and compare options.

Value management integrates value engineering into design methodologies. Green has developed this approach into what he calls the 'simple multi-attribute rating technique' (SMART) which structures the agreed objectives into a value hierarchy. An evaluation matrix is developed by the integrated design and construction team that comprises the above objectives. Each design decision can be examined using the matrix that allows individual weighting of each objective and design options rated using an agreed scoring system. The sensitivity of the decision to changes in score can then be tested. Capital costs can be incorporated into the matrix or, more usually, examined once the design objectives have been discussed. Similarly, the impact on certain key issues, such as BREEAM rating and carbon emissions (SAP or SBEM for example) may need to be looked at separately since a single design change may cause the target BREEAM or emission rating to be lost, or this may be the cumulative impact of a number of design changes. Similarly there will be some issues that must be ring-fenced to meet planning obligations, such as carbon emissions offset by on-site renewable technologies, use of sustainable drainage (SUDS) and number of Lifetime Homes compliant dwellings.

For a detailed description and examples of using 'value analysis' tools, refer to BSRIA's 1996 guide on 'Value Engineering of Building Services' (Hayden and Parsloe, 1996). This incorporates a suite of pro forma matrices, reports and analysis forms.

## References

DEFRA (Department for Environment, Food and Rural Affairs), 2007. *UK Government Sustainable Procurement Action Plan*. London: Department for Environment, Food and Rural Affairs

FFC (Federal Facilities Council), 2001. *Federal Facilities Council Technical Report no.142: Sustainable Federal Facilities: A Guide to Integrating Value Engineering, Life-Cycle Costing, and Sustainable Development*. Washington, DC: National Academy Press. Available at www.wbdg.org/ccb/SUSFFC/fedsus.pdf (accessed 26 March 2010)

Green, S., *c*1996. *A Smart Methodology for Value Management*. Reading, UK: Department of Construction Management and Engineering, University of Reading. Available at www.personal.rdg.ac.uk/~kcsgrest/hkivm2.html (accessed 26 March 2010)

Hayden, G. W. and Parsloe, C. J., 1996. *BSRIA Application Guide 15/96: Value Engineering of Building Services*. Bracknell, UK: Building Services Research and Information Association

# 4.6
# Commissioning and Handover

## Introduction

A building that has been designed and built to best practice sustainable design standards can still fail if it is not handed over having been properly commissioned and with all of the documentation required to understand the operation of the building.

This is recognized in the British Building Regulations and Standards which require evidence that the building services have been commissioned properly and the as-built performance has been assessed. These also require that a log book be provided giving information on the building services, together with an Energy Performance Certificate. The Construction (Design and Management) Regulations (2007) require a health and safety file to be produced that includes operating and maintenance instructions for all relevant plant, equipment and services.

On top of these BREEAM incorporates credits that reward best practice in allocating resources and undertaking commissioning, as well as providing information on its sustainable operation, covering a wide range of information in a Building User Guide (or Home User Guide in the case of CSH).

LEED Homes, on the other hand, includes credits for providing guidance and training both to home owners (or tenants) and building managers.

## Commissioning

Building services must be designed so that they can be easily adjusted to perform as intended. In the UK, Regulation 20c in Part L of the Building Regulations (CLG, 2009) requires that commissioning of heating and hot water systems be confirmed with the local authority on completion. The associated guidance refers to the procedures in CIBSE Commissioning Code M having been approved by the Secretary of State, including the production of a 'commissioning plan' along with the HVAC duct leakage test. Part L requires that the reported carbon emissions (building or dwelling emission rating – BER or DER) reflect the actual performance of the building and services, taking into account measured air leakage through the fabric, fan performance and duct leakage, for example. The requirement for a commissioning plan is a recent addition and requires details of the systems to be tested and the tests to be carried out to be submitted to building control along with the design stage $CO_2$ emission calculations.

BREEAM encourages a rigorous approach to commissioning that starts once the design team is in place through the appointment of 'appropriate project team member(s) ... to monitor and programme pre-commissioning, commissioning and, where necessary, re-commissioning on behalf of the client'.

Evidence will be required that commissioning has been carried out 'in line with current Building Regulations and BSRIA and CIBSE guidelines, where applicable'. BSRIA has produced a series of guides that provide guidance on design and installation considerations as well as commissioning and testing procedures covering the following:

- TM1/88.1: Commissioning HVAC Systems;
- AG2/89.2: Commissioning of water systems in buildings;
- AG3/89.3: Commissioning of air systems in buildings;
- AG1/91: Commissioning of VAV systems in buildings;
- AG20/95: Commissioning of pipework systems;
- AG1/2001.1: Pre-commission cleaning of pipework systems;
- AG 5/2002: Commissioning Management;
- BG8/2009: Model Commissioning Plan.

The most recent of these publications provides guidance on preparing a commissioning plan in compliance with the latest requirements of the Building Regulations.

As well as providing authoritative procedures for management of the commissioning process the CIBSE Commissioning Codes cover:

- CC A: Air Distribution – including guidance on setting up fans and balancing of ductwork systems to achieve design airflow rates;
- CC B: Boilers – including guidance on design and specification and procedures for programming, pre-commissioning, testing, dry and live runs, training, log books and handover;
- CC C: Automatic Controls – including design for commissionability, specification, project management, pre-commissioning, commissioning and handover;
- CC L: Lighting – including commissioning management, pre-commissioning checks, commissioning, documentation, training and handover;
- CC R: Refrigeration – including guidance on good practice for designers, manufacturers, contractors and clients, specification for commissioning and standards for installers;
- CC W: Water Distribution – including guidance on setting up pumps and balancing of water distribution systems to achieve design flow rates.

CIBSE Commissioning Code M provides general guidance on good practice for commissioning including statutory requirements, safety, selection and appointment of a commissioning management organization (CMO), design for commissionability, witnessing, logbooks and handover.

BREEAM also specifies the following be in place in order for the Commissioning credits to be awarded:

- The main contractor's main programme (and budget) must take full cognizance of the commissioning programme.
- A specialist commissioning manager must be appointed during the design stage (by either client or contractor) for complex systems such as air conditioning, mechanical ventilation, displacement ventilation, complex passive ventilation, building management systems (BMS), renewable energy technologies, microbiological safety cabinets, fume cupboards, cold storage enclosures and refrigeration plant. The manager's responsibilities should include:
  - advice on the commissionability of designs;
  - commissioning management input to construction programming;
  - commissioning management input during installation stages;
  - management of commissioning, performance testing, handover and post handover stages.
- Where a BMS is installed, the following commissioning procedures must be carried out:
  - commissioning of air and water systems to be undertaken after all control devices are installed, wired and functional;
  - commissioning records to include as a minimum final air and water flow rates, room air temperatures, off coil temperatures, and other key parameters as appropriate.
- The BMS and/or controls installation should be running in automatic mode with satisfactory internal conditions prior to handover.
- All BMS screen schematics and graphics must be fully installed and functional to user interface before handover.
- The occupier must be fully trained in the operation of the system.
- The specialist commissioning manager's appointment must include the following responsibilities for seasonal commissioning over a minimum 12 month period, once the building becomes occupied:
  - Testing of all building services under full load conditions, i.e. heating equipment in midwinter, cooling/ventilation equipment in mid-summer, and under part-load conditions in spring or autumn;
  - Where applicable, testing should also be carried out during periods of extreme (high or low) occupancy;
  - Interviews should be carried out with building occupants (where they are affected by the complex services) to identify problems or concerns regarding the effectiveness of the systems;
  - Re-commissioning of systems should be undertaken following any work needed to serve revised loads (and/or respond to the interviews), incorporating any revisions in operating procedures into the operation and maintenance (O&M) manuals.
- For buildings that have no complex systems and are fully naturally ventilated the following seasonal commissioning can be carried out by a specialist consultant or facilities manager:
  - Review thermal comfort, ventilation and lighting, at three-, six- and nine-month intervals after initial occupation, either by measurement or occupant feedback.

- Take all reasonable steps to re-commission systems following the review and incorporate any relevant revisions in operating procedures into the O&M manuals.

Seasonal commissioning is usually the most difficult aspect of the BREEAM commissioning requirements to define and plan, particularly for naturally ventilated buildings that can only be tested for thermal performance from space temperature measurements during periods when there is a demand for heating. However, the alternative strategy of post occupancy evaluation is only possible for spaces that are fully occupied, which may not occur in speculative buildings for an indeterminate time after completion. Testing during periods of high occupancy will also be problematic under these circumstances, although it may be possible to simulate the sensible heat gain from occupants, latent gains are more difficult to replicate.

# Handover

As can be discerned from the previous section the exact time when the building can be defined as fully commissioned and hence ready for final handover may vary from project to project. Clearly there will be a warranty period for each project, during which seasonal commissioning can be carried out and residual defects remedied.

For contractual purposes 'handover' involves the handing over of the keys to the building and all the documentation required for its operation and maintenance.

Through its requirement for a comprehensive Building or Home User Guide, BRE is looking for the supply of information that goes well beyond that specified by Building and CDM Regulations for energy performance certificate (EPC), log books, health and safety file and associated O&M instructions.

The list in Box 4.1 indicates the type of information that BRE expects to be provided in a Building User Guide to meet the needs of the facilities management (FM) team or building manager and the general users.

---

### Box 4.1 Contents of Building User Manual (from BREEAM Offices 2008, Man 4)

1   Building services information
   (a) General user – Information on heating, cooling and ventilation in the building and how these can be adjusted, e.g. thermostat location and use, implications of covering heating outlets with files, bags, etc., and use of lifts and security systems.
   (b) FM – As above, plus a non-technical summary of the operation and maintenance of the building systems (including BMS if installed) and an overview of controls.

2   Emergency information
   (a) General user – Include information on the location of fire exits, muster points, alarm systems and firefighting systems.
   (b) FM – As above, plus details of location and nature of emergency and firefighting systems, nearest emergency services, location of first aid equipment.

3   Energy and environmental strategy
   This should give owners and occupiers information on energy efficient features and strategies relating to the building, and also provide an overview of the reasons for their use, e.g. economic and environmental savings. Information could include:
   (a) General user – Information on the operation of innovative features such as automatic blinds, lighting systems, etc., and guidance on the impacts of strategies covering window opening and the use of blinds, lighting and heating controls.
   (b) FM – As above, plus information on airtightness and solar gain (e.g. the impact of leaving windows/doors open in an air-conditioned office, or the use of blinds in winter with respect to solar gain); energy targets and benchmarks for the building type, information on monitoring such as the metering and sub-metering strategy, and how to read, record and present meter readings.

4   Water use
   (a) General user – Details of water saving features and their use and benefits, e.g. aerating taps, low flush toilets, leak detection, metering, etc.
   (b) FM – As above, plus details of main components (including controls) and operation. Recommendations for system maintenance and its importance, e.g. risk of legionella.

5   Transport facilities
   (a) General user – Details of car-parking and cycling provision; local public transport information, maps and timetables; information on alternative methods of transport to the workplace, e.g. car sharing schemes, local 'green' transport facilities.
   (b) FM – As above, plus information on conditions of access, maintenance and appropriate use of car parking and cycling facilities, e.g. number of spaces provided.

6   Materials and waste policy
   (a) General user – Information on the location of recyclable materials storage areas and how to use them appropriately.
   (b) FM – As above, plus information on recycling, including recyclable building/office/fit out components, waste storage and disposal criteria, examples of waste management strategies and any cleaning/maintenance criteria for particular materials and finishes.

7 Re-fit/re-arrangement considerations
   (a) General user – An explanation of the impact of re-positioning of furniture, i.e. may cover grilles/outlets, implications of layout change, e.g. installation of screens, higher density occupation etc.
   (b) FM – As above, plus environmental recommendations for consideration in any refit. Relevant issues covered in BREEAM should be highlighted, e.g. the use of natural ventilation, use of Green Guide 'A' rated materials, re-use of other materials, etc., the potential impact of increasing occupancy and any provision made in the original design to accommodate future changes.

8 Reporting provision
   (a) General user – Contact details of FM/manager, maintenance team and/or help desk facility; and details of any building user group if relevant.
   (b) FM – As above, plus contact details of suppliers/installers of equipment and services and their areas of responsibility for reporting any subsequent problems.

9 Training
   Details of the proposed content and suggested suppliers of any training and/or demonstrations in the use of the building's services, features and facilities that will be needed. This could include:
   (a) General user – Training in the use of any innovative/energy saving features.
   (b) FM – As above, plus training in emergency procedures and setting up, adjusting and fine tuning the systems in the building.

10 Links and references
   This should include links to other information including websites, publications and organizations. In particular, the Carbon Trust programme should be referenced and links provided to its website and good practice guidance.

11 General
   Where further technical detail may be required by the FM team or manager, there should be references to the appropriate sections in the Operation and Maintenance Manual.

The information required to meet the Man 1 credit in CSH for a Home User Guide is very similar, although it is written with the occupants in mind, and goes beyond the requirements for a Home Information Pack. The requirement falls into two parts under the headings of:

1 Part 1 – Operational issues
   • Environmental strategy, design and features – including the operation and implications of passive design features, renewable technologies etc.

- Energy – including information required by Building Regulations on how to efficiently operate and maintain energy systems, information on high efficacy lamps and ecolabelled white goods.
- Water use – including details and advice on water saving measures.
- Recycling and waste – including information on waste collection and advice on reducing waste.
- Sustainable DIY – advice on reducing the environmental impact of home improvements.
- Emergency information – including information and advice on smoke detectors.

2   Part 2 – Site and surroundings
- Recycling and waste – including advice on what to do with waste not covered by local collection schemes, such as redundant fridges, batteries, etc.
- Sustainable drainage – including details of SUDS within site boundary and maintenance requirements.
- Public transport – details of public transport, cycle storage, car parking and access to local amenities.
- Local amenities – location of local amenities.
- Responsible purchasing – advice on purchasing sustainable white goods, electrical equipment, timber products and groceries.
- Emergency information – contact details for local minor injuries clinics, A&E departments, police and fire station.

In addition CSH suggests including links, details and information for relevant sources and organizations that can help with running the home more efficiently, choices for living sustainably and understanding the occupants' relationship with the local community.

Up until May 2010 the provision of a Home Information Pack (HIP) was a legal obligation for vendors of homes in the UK including, as a minimum, EPC and CSH certificates, as well as a property information questionnaire (PIQ), sales statement, evidence of title, standard search details and copy of lease or commonhold, if relevant. HIPs have been abandoned but the requirement for EPC and CSH certificates retained.[1]

Detailed guidance on the handover of commercial buildings is provided by BSRIA in their 2007 publication 'Handover, O&M Manuals and Project Feedback'. This includes chapters detailing the handover information required for building services, a model specification for operating and maintenance manuals, guidance on the preparation of logbooks and energy certificates, suggested content of a condition survey report and tools for obtaining feedback, such as 'occupant satisfaction surveys' for post occupancy evaluation.

# Note

1   For more information see www.direct.gov.uk/en/HomeAndCommunity/BuyingAndSelling YourHome/Homeinformationpacks/index.htm

# References

BSRIA (Building Services Research & Information Association), 2007. *BG 1/2007: Handover, O&M Manuals and Project Feedback*. Bracknell, UK: Building Services Research and Information Association

CLG (Department for Communities and Local Government), 2009. *Proposals for amending Part L and Part F of the Building Regulations – Consultation Volume 2: Proposed technical guidance for Part L*. London: CLG. Available at www.communities.gov.uk/documents/planningandbuilding/pdf/partlf2010consultation (accessed 21 February 2010)

*Construction (Design and Management) Regulations 2007*. Available at www.opsi.gov.uk/si/si2007/uksi_20070320_en_1 (accessed on 30 March 2010)

*Home Information Pack (Amendment) Regulations 2009*. SI 2009 No 341 Available at www.opsi.gov.uk/si/si2009/em/uksiem_20090034_en.pdf (accessed on 30 March 2010)

# Index

9 781138 972841